# CRISPR-/Cas9 Based Genome Editing for Treating Genetic Disorders and Diseases

*Editor*

**Luis María Vaschetto**

Oncativo, Córdoba
Argentina

CRC Press
Taylor & Francis Group
Boca Raton  London  New York

CRC Press is an imprint of the
Taylor & Francis Group, an **informa** business

A SCIENCE PUBLISHERS BOOK

First edition published 2022
by CRC Press
6000 Broken Sound Parkway NW, Suite 300, Boca Raton, FL 33487-2742

and by CRC Press
2 Park Square, Milton Park, Abingdon, Oxon, OX14 4RN

© 2022 Taylor & Francis Group, LLC

*CRC Press is an imprint of Taylor & Francis Group, LLC*

| Library of Congress Cataloging-in-Publication Data |
|---|
| Names: Vaschetto, Luis M. editor. |
| Title: CRISPR-/Cas9 based genome editing for treating genetic disorders and diseases / editor, Luis María Vaschetto. |
| Description: First edition. | Boca Raton, FL : CRC Press, 2022. | Includes bibliographical references and index. |
| Identifiers: LCCN 2021033275 | ISBN 9780367542863 (hardcover) |
| Subjects: MESH: CRISPR-Cas Systems | CRISPR-Associated Protein 9 | Gene Editing--methods | Genetic Therapy |
| Classification: LCC QP623 | NLM QU 475 | DDC 572.8/8078--dc23 |
| LC record available at https://lccn.loc.gov/2021033275 |

ISBN: 978-0-367-54286-3 (hbk)
ISBN: 978-0-367-54287-0 (pbk)
ISBN: 978-1-003-08851-6 (ebk)

DOI: 10.1201/9781003088516

Typeset in Times New Roman
by Innovative Processors

# Preface

The Nobel Prize in Chemistry 2020 was awarded jointly to Emmanuelle Charpentier and Jennifer A. Doudna who have discovered the CRISPR/Cas9 genetic scissors. The CRISPR/Cas 9 genome editing system is a highly versatile and easy-to-use technology that has revolutionized both basic and applied research in many disciplines ranging from medicine to agriculture, democratizing genome editing in laboratories around the world. In medicine, CRISPR/Cas9 has provided a powerful tool for research and holds an enormous potential for the treatment of genetic disorders and non-inherited diseases. Accordingly, this book offers an updated in-depth review on the implementation status of CRISPR/Cas9 technology in research and clinical settings.

Moreover, this book brings to readers a selection of CRISPR-based gene editing platforms, as well as perspectives from professionals in non-medical fields pertinent to the use of CRISPR-Cas9. Each of the chapters in ***CRISPR-/Cas9 Based Genome Editing for Treating Genetic Disorders and Diseases*** has been selected to provide a full understanding of the various (bioethical, legal, technical) challenges related to the applicability of CRISPR-Cas9 in human cells. The book sets a benchmark for the involvement of CRISPR-Cas9 system in bioethics and regulatory issues, neuroscience research, gene therapy, epigenome editing, cancer research, telomere length, genome mapping and virus infection control.

I sincerely wish to thank the reviewers, Prof. Hallam Stevens, Prof. Zdenka Ellederova, Prof. Bernhard Schmierer, Prof. You Lu, Dr. Manh T. Tran, Dr. Marta Martinez-Lage, Prof. Jonathan Ploski, and Prof. Hiroshi Yamaguchi, whose work and insightful comments have enriched the quality of this book.

**Luis María Vaschetto**

# Contents

# The CRISPR/Cas9 Genome-editing System: Principles and Applications

Cecilia Pop-Bica[1], Andreea Nutu[1], Roxana Cojocneanu[1], Sergiu Chira[1] and Ioana Berindan-Neagoe[1,2]*

[1] Research Center for Functional Genomics, Biomedicine and Translational Medicine, "Iuliu-Hatieganu" University of Medicine and Pharmacy, Cluj-Napoca, Romania
[2] Department of Functional Genomics and Experimental Pathology, The Oncology Institute "Prof. Dr. Ion Chiricuţă", Cluj-Napoca, Romania

## Introduction

CRISPR/Cas9 represents an adjustable and widespread genome-editing tool adapted from the bacterial immune system, which is presently used in basic and applied biomedical research. Genetic alterations are frequently the cause of genetic diseases. Even though recent technological advances eased the discovery of disease-associated gene alterations, general therapeutic options are usually designed to treat the symptoms, not to restore the altered genetic sequences responsible for the disease phenotype. In this context, gene therapies appeared to support the concept of restoring the genetic mutations in order to prevent or treat the disorders caused by these alterations (Mirgayazova et al. 2020). In what concerns the CRISPR/Cas system (clustered regularly interspaced short palindromic repeats-CRISPR associated), the fundamental principle includes genome editing and the regulation of physiological phenomena of various organisms and cells. This system was first discovered in bacteria, where it functions as an adaptive "immune" strategy used to destroy foreign genetic material or develop resistance to phage infections (Wright et al. 2016, Yadav et al. 2021). The CRISPR/Cas systems is characterized by a sequence of about 20-50 bp organized as direct repeats, isolated by spacers of comparable length, and tailed by an AT-rich "leader" region (Jansen et al. 2002, Kunin et al. 2007). This system has two main components – a guide RNA (gRNA), comprised of a CRISPR RNA (crRNA) and trans-activating crRNA (tracrRNA) (Figure 1 and Figure 2), which acts as a guide RNA for the complementary DNA sequence, and the Cas proteins (encoded by *cas* genes), which are involved in the degradation of the target DNA/RNA sequence (Strich and Chertow 2019). The principle of the complementary

---

* Corresponding author: ioana.neagoe@umfcluj.ro

binding of the crRNA prompted the researchers to use this system to design sequences that would bind to and cut the sequences of interest in different pathologies (Jinek et al. 2012). Furthermore, this technology allowed researchers to identify the function of a gene through the introduction of genetic alterations (Shalem et al. 2015). The reduced time-frame needed to perform CRISPR/Cas9 gene editing, together with the reported efficiency of this system in terms of reduced off-target effects, makes this a superior technique compared to others based on ZFNs (Zinc Finger Nucleases) and TALENs (Transcriptor Activator-Like Effector Nucleases), in which the efficiency is exclusively based on the nucleases affinity and specificity (Qu et al. 2013).

# The bacterial origin of CRISPR/Cas

## The discovery of CRISPR/Cas nucleases system

The coexistence of prokaryotes and viruses generated various defense mechanisms such as restriction modifications, toxin-antitoxin systems, and, in the later years, CRISPR/Cas systems were discovered (Labrie et al. 2010). The discovery of CRISPR as a component of the prokaryotic "immune" system, and the repurposing of this system as a genome editing tool, determined a broad use of this technology in molecular biology applications, making it one of the most used technologies in active research in biology (van Soolingen et al. 1993, Bolotin et al. 2005, van der Oost et al. 2009, Mali et al. 2013a, Cho et al. 2013) (Figure 1). Even though the first CRISPRs were observed decades ago in bacteria (Ishino et al. 1987), and subsequent studies revealed the presence of CRISPRs in archaea (Mojica et al. 1993), it was only in the early 2000s that researchers discovered the sequence similarities between the viruses, bacteriophages, and plasmids and the spacer regions in CRISPR, managing to uncover the defense function of CRISPR (Mojica et al. 2005, Bolotin et al. 2005). Independent parallel studies revealed a set of genes associated with CRISPR, consequently named *cas* (CRISPR-associated), and, in 2008, Marakova et al. suggested the existence of a CRISPR/Cas complex that acts as an acquired immune system to protect the bacterial cell against invading phages or other exogenous genetic material (Jansen et al. 2002, Makarova et al. 2006).

## Components of the CRISPR/Cas system

CRISPR/Cas system ensures the immunity of the bacteria in three steps that require target recognition and cleavage (Barrangou and Marraffini 2014, Sorek et al. 2013). The first step is the adaptation, which allows the copy and paste of the foreign nucleic acids into the 'spacers' of the CRISPR arrays, thus providing acquired resistance against the invading phage (Barrangou et al. 2007). The next step includes the biogenesis of the crRNA (expression stage), in which the small interfering RNAs are generated through transcription and further processing. In the interference stage, the gRNA heads the Cas enzymes to cleave the DNA (Marraffini and Sontheimer 2008). The CRISPR-Cas system can be classified into two main categories according to the effectors: first, where all functionalities in the effector complexes are carried out using a protein, and second, the multi-unit effector complexes (Shmakov et al. 2017). These two classes are further divided into six types of CRISPR-Cas systems

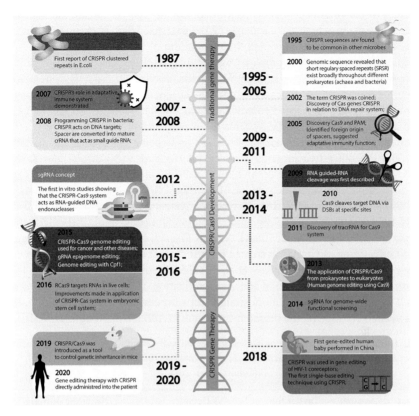

**Figure 1.** Timeline progression and development of CRISPR and a brief description of fundamental discoveries. In 1987, during a study focused on alkaline phosphatase *loci* in *E. coli*, the first observation of invariably repeated sequences flanking specific sequences, later described as CRISPR arrays (in 2002 (Jansen et al. 2002)), was made by Ishino et al. (Ishino et al. 1987). Later, CRISPR sequences were discovered in other microorganisms (Mojica et al. 2000), and in 2005 foreign sequences (spacers) from phages and plasmids were identified in CRISPR (Bolotin et al. 2005, Haft et al. 2005). The bacterial "immune" function (Marraffini and Sontheimer 2008) of CRISPR, and the crRNAs that serve as gRNAs for DNA targeting (Brouns et al. 2008) are discoveries made in the following years. Between 2009-2011, CRISPR/Cas system to cleave RNA molecules (Hale et al. 2009), the specific DSB DNA cleavage (Garneau et al. 2010), and tracrRNA-crRNA structure associated with Cas9 were first described (Deltcheva et al. 2011). In 2012, another major breakthrough was made, as two separate studies indicated that CRISPR system operates as RNA-guided endonucleases (Gasiunas et al. 2012), respectively, that the Cas9 are reprogrammable and this system could potentially serve as a tool for genome-editing (Jinek et al. 2012). Another breakthrough is represented by the transfer of this technology from prokaryotes to eukaryotic cells in 2013, and in 2014 the first sgRNA libraries were used for genome-wide screening using Cas9 (Cong et al. 2013, Mali et al. 2013a, Wang et al. 2014, Shalem et al. 2015). Recent studies proved the efficiency of CRISPR/Cas systems in cancer biology (Sanchez-Rivera and Jacks 2015), and embryonic stem cells systems (Jin and Li 2016). In 2018, the first gene editing of HIV-1 co-receptors using CRISPR was made (Allen et al. 2018). In 2019, CRISPR/Cas9 was used to control genetic inheritance in mice (You et al. 2019), and in 2020, it was the first time when CRISPR was used as a genetic scissor directly administered into a patient (Ledford 2020).

according to the presence of the signature genes. In type I systems, the signature protein is Cas3, in which the cleavage of the external DNA is carried out by the nuclease and helicase domains, and the recognition of the target sequence is made by the multi-protein-crRNA complex Cascade. In type II systems, the unique protein required for the interference is Cas9 (signature protein). In type III systems, the signature gene is *cas10*, a gene encoding a multidomain protein (Cas10) that is assembled into an interference complex needed for the identification and cleavage of the target sequence. Type IV systems are usually not linked to a CRISPR array. The effector complex is represented by Csf1 (*csf1* being considered the signature gene) and the other two proteins encoded by *cas* genes (Makarova and Koonin 2015). Type V systems comprise a unique Cas9-like nuclease, which can be Cpf1, C2c1, or C2c3 according to the subtype of the CRISPR/Cas type V system (Shmakov et al. 2017, Zetsche et al. 2015). In type VI systems, the protein C2c2 with two HEPN RNase domains is the signature protein (Shmakov et al. 2017). All these systems are classified as either Class1 systems (Type I, III, IV) as they have a multi-subunit effector or as Class 2 systems (Type II, V, VI) as they are characterized by a single-subunit effector (Shmakov et al. 2017, Makarova and Koonin 2015). Table 1 illustrates a brief classification and characterization of the CSISPR/Cas systems. Class 2 systems are seen as a more straight-forward device for genetic editing, thus being intensively used by the research community (Adli 2018, Wang et al. 2016).

**Table 1.** Overview of classification and characteristics of CRISPR/Cas systems (Shmakov et al. 2017, Pyzocha and Chen 2018, Makarova et al. 2020, Perez Rojo et al. 2018, Tamulaitis et al. 2017)

| Class | System type | Target for cleavage | Effector | Nuclease domain |
|-------|-------------|---------------------|----------|-----------------|
| I | I | DNA | Cas proteins – Cas3, Cas5-Cas8, Cas10, and Cas11 | HD fused to Cas3 |
| | III | DNA/RNA | | HD fused to Cas10 |
| | IV | DNA | | Unknown |
| II | II | DNA | Cas9 | RuvC and HNH |
| | V | DNA | Cas12a/b/c | RuvC and Nuc |
| | VI | RNA | Cas13a/b/c | HEPN domains |

HD – Histidine-aspartate domain; HEPN – Higher eukaryotes and prokaryotes nucleotide binding domain

CRISPR technology originated as an adaptive prokaryotic "immune" system that ensures protection against exogenic nucleic acids, and is now utilized as a precise genetic/genomic tool in molecular biology (Barrangou et al. 2007, Haapaniemi et al. 2018). The defense response generated by the CRISPR-Cas system consists of three points, which are as follows: adaptation, expression and maturation, and interference. In the adaptation step, a complex of Cas proteins is responsible for the identification of the target sequence in the DNA (protospacer-adjacent motif – PAM), binding, and the generation of double-strand breaks in this sequence. This target DNA releases a short sequence that will be acquired by the CRISPR array as a 'spacer'. The expression step consists of the transcription of the CRISPR array into a precursor RNA that is further processed by the Cas proteins and other accessory factors into smaller entities of RNA – CRISPR RNAs (crRNAs) that, together with the Cas proteins, form the Cas-crRNA complex. The last step – the interference – includes the action of the Cas-crRNA complex to recognize and cleave the invading nucleic acid (Amitai and Sorek 2016, Makarova et al. 2006).

Depending on the type of CRISPR-Cas system, there are particular stages of expression and interference. In the type I system, the Cas complex acts by cleaving at the junction of the ssRNA (single-stranded RNA) and the dsRNA (double-stranded RNA) formed by the hairpin loops (Makarova and Koonin 2015). In type II CRISPR-Cas systems, the double break in the target DNA is induced by the Cas9 protein (guided by the tracrRNA:crRNA complex). Specifically, each DNA strand is cleaved by two different Cas9 endonuclease domains (Jinek et al. 2012). In type III systems, there are two main nucleases – one sequence-specific ribonuclease to degrade the ssRNA, and one deoxyribonuclease to target and degrade the ssDNA (Tamulaitis et al. 2017). The mechanism is based on the DNA sequence recognition, and does not require hairpin loops (Janik et al. 2020, Garneau et al. 2010, Gasiunas et al. 2012).

## Features of the CRISPR/Cas9 for genome engineering

Cas9 is specific for the Type II CRISPR *locus*. Given the variety of the *cas* genes, this type II CRISPR *locus* comprises three subtypes (IIA, IIB, IIC). *Cas9, cas1,* and *cas2* together with the CRISPR array and the tracrRNA are the main component of type II CRISPR *loci* (Makarova et al. 2011). The three genes are shared by all three subtypes of type II CRISPR *loci*, whereas type IIA has an additional *csn2* gene and type IIB an additional *cas4* gene (Chylinski et al. 2013, Makarova et al. 2011). Nonetheless, this classification is not relevant for the structural diversity of the Cas9 proteins. These nucleases are characterized by sequence homology and variable lengths (Hsu et al. 2014). Despite of this variation in length (usually associated with the REC domain), Cas9 proteins are characterized by a structure consisting of two nuclease domains – RuvC (responsible for the cleavage of the non-target strand) and HNH (responsible for the cleavage of the target strand), REC domains, and the bridge helix (Makarova et al. 2011, Fonfara et al. 2014). The preference for this type II CRISPR/Cas system in applications for genome editing relies on the fact that there are few specifications that few specifications need to be fulfilled to ensure a programmable DNA targeting, such as the presence of Cas9 endonuclease and the existence of a guide RNA (gRNA) that can be customized (Jinek et al. 2012). The DNA cleavage depends on two factors: the crRNA that targets the protospacer, and the PAM, a sequence located downstream of the protospacer in the non-target strand (tracrRNA) (Garneau et al. 2010). CRISPR/Cas9 systems are now used as efficient gene/genome-editing tools in different organisms, including human cells, facilitating the multiplex genome engineering by editing multiple sites at the same time by using multiple gRNAs (Mali et al. 2013b, Wang et al. 2013, Cong et al. 2013). In addition, in mammalian genomes, it was shown that the adaptation of Cas9 into a nickase facilitates homology-directed repair (HDR) (Mali et al. 2013a, Ran et al. 2013). Nowadays, there are a few RNA-guided Cas9 isolated from *Streptococcus pyogenes, Streptococcus thermophilus*, and *Neisseria meningitides* converted and used as gene-editing tools (Charpentier and Doudna 2013, Horvath and Barrangou 2013, Hou et al. 2013, van der Oost 2013).

# Molecular mechanism of CRISPR/Cas9 for genome editing

The use of CRISPR-Cas9 system in gene editing is based on the development of guide RNA (gRNA) sequences, which, in turn, compel for a unit fusion molecule – such as single-guide RNA (sgRNA) – or a specific crRNA with the trans-activating CRISPR RNA (tracrRNA) required (Yadav et al. 2021). Specifically, to target and cleave a double-stranded DNA (dsDNA) sequence, this system is based on the formation of the complex of sgRNA and the Cas protein (Figure 2). CRISPR/Cas9 employs Cas9 endonuclease to recognize and cleave the DNA strands using one of the two domains (RuvC and HNH) responsible for the cleavage of the target and non-target DNA strands (Garneau et al. 2010, Gasiunas et al. 2012, Jinek et al. 2012). During this stage, the tracrRNA – a small noncoding RNA – directs the maturation of crRNAs by pairing with the repeat sequence and forming a dual-RNA hybrid structure meant to guide the Cas9 at the target sequence (usually a 20 nucleotide sequence) adjacent to the PAM (Deltcheva et al. 2011, Gasiunas et al. 2012, Jinek et al. 2012). Jinek et al. (2012) showed that by combining the tracrRNA and the crRNA into a chimeric sgRNA would simplify the system and maintain Cas9 function for DNA cleavage at a specific target. Therefore, the modification of the spacer sequence within the crRNA would result in a simplified CRISPR-Cas system that can be engineered to target the desired DNA sequence and induce double-strand breaks (Jinek et al. 2012). The generated double-strand breaks (DSB) are then restored either by error-prone nonhomologous end joining (NHEJ) – this repair generates small random indels at the DSB site – or by high-fidelity homology directed repair (HDR) – this type of repair is based on the use of a homologous template that produces precise genome modification at the cleavage site (Lieber 2010, San Filippo et al. 2008) (Figure 2). The insertion of frameshift mutations leading to the impairment of the protein function makes NHEJ less reliable and, therefore, makes its use restricted. On the other hand, HDR is the mechanism preferred in the CRISPR technology as the long stretches of homologous sequences used to repair the DNA lesions do not cause mutations (Deng et al. 2013).

## Mechanisms of targeted modifications

The adaptation of the CRISPR/Cas9 for the editing of several eukaryotic models implies the exploration of this system to generate DSB that would facilitate the insertion of mutations at specific locations in the genome. Therefore, to achieve loss-of-function mutations, a DSB should be targeted in a constitutively spliced coding exon (Shalem et al. 2015). Repair of the DSB through NHEJ can result in the introduction of indels and the generation of frameshift mutations and premature stop codons, producing the nonsense-mediated decay of the transcript. The insertion of indels that introduce frameshift mutations can also result in non-

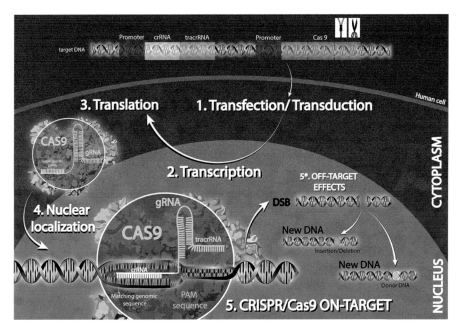

**Figure 2.** Mechanism of action in CRISPR/Cas9 genome-editing system (after Chira et al. (Chira et al. 2017)). The discovery of the original CRISPR/Cas9 system in bacteria facilitated its transformation into a genome-editing tool. The tracr-RNA-crRNA forming a dual-RNA hybrid structure serving as gRNA can be inserted into the target cell, directed towards the nucleus. Then the cas9 gene is transcribed and exported outside the nucleus and translated into protein (Cas9). The interaction between the gRNA and the functional Cas9 protein generates a ribonucleic-protein effector complex able to target the gDNA (genomic DNA) at a specific *locus*, nearby to the PAM sequence. DSBs are then introduced, and the restoration of these brakes can be performed either through NHEJ (which increases the possibility of introducing indel mutations), or via HDR (which requires the presence of a donor DNA).

functional proteins, not necessarily nonsense-mediated decay. In this respect, the preferred exons for sequence targeting using CRISPR/Cas9 are the early exons due to the high probability of introducing premature stop codons if indels or frameshift mutations are inserted (Doench et al. 2014). Various studies focused on the use of RNA-mediated programmed Cas9 on genome-scale knockout in mammalian cell cultures highlighted the fact that this approach offers promising phenotypic effects, high reagent consistency, and significant validation rate of the knockout efficiency (Shalem et al. 2015, Wang et al. 2014, Koike-Yusa et al. 2014, Zhou et al. 2014).

In addition to the knockout function for genes, deactivated Cas9 (dCas9) or other complexes using Cas9 or sgRNAs fused with transcriptional repressors, activators, or recruitment domains are used to regulate gene expression at a specific site. This approach provides the possibility of gene-editing without introducing irreversible changes in the DNA sequence. For this purpose, CRISPRi and CRISPRa are dCas9 systems employed for the activation and inhibition of transcription at a specific target site (Larson et al. 2013, Bikard et al. 2013). Oi et al. (2013) developed a CRISPRi system to induce the repression of gene expression in HEK293 human cells. Their system uses dCas9 and can be employed for effective gene modulation (repression/activation) (Qi et al. 2021). Another study demonstrated that the binding of dCas9 at the promoter region inhibits transcription, most likely by preventing RNA polymerase from binding. Moreover, by engineering this system to direct dCas9 binding at open reading frames (ORF), transcription elongation is obstructed. Another highlight of this study was the fusion of one RNA polymerase subunit with dCas9, thus transforming it into a transcription activator. This strategy is efficient in enhancing the transcription of underexpressed genes (Bikard et al. 2013, Dove and Hochschild 1998). On the other hand, gain-of-function screens are restricted to cDNA overexpression libraries (Yang et al. 2011). Given the partial coverage of these libraries due to the impediments of cloning large cDNA constructs, the deficiency in covering the complexity of transcript isoforms, these gain-of-function screens are limited compared to the loss-of-function screens. Various transcriptional activators (VP64, p65) are fused with dCas9 to ease gain-of-function screens (Konermann et al. 2013, Maeder et al. 2013, Perez-Pinera et al. 2013).

## CRISPR/Cas9 applications

Recent years showed an increased interest in using CRISPR/Cas9 systems in basic research studies, biotechnological applications, and for the development of new therapeutic strategies for complex diseases (Komor et al. 2017). Current developments include functional gene screening (Shalem et al. 2015), transcriptional studies (Konermann et al. 2015), generation of cellular and animal models (Wang et al. 2013, 2016), genome fluorescence imaging (Chen et al. 2013), and the development of sequence-specific anti-microbials/anti-virals (Bikard et al. 2014) (Figure 3).

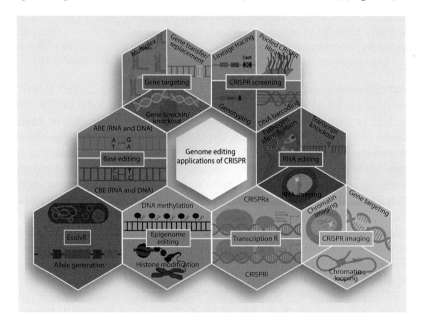

**Figure 3.** General overview of CRISPR applications.

## CRISPR/Cas9 as a tool for biomarkers discovery

Molecular biomarkers offer valuable information regarding the biological/molecular characteristics of a patient that can be reflected in the progression of the disease, in the outcome of the patients, or can predict the efficiency of a targeted therapy. CRISPR/Cas systems can be used for the identification of these biomarkers by evaluating the genetic/epigenetic alterations that occur in the tumor tissue (Matsuoka and Yashiro 2018, Zamborsky et al. 2019).

In a study by Xiao et al. (2018), CD44 (Cluster of Differentiation 44) was identified as a prognostic biomarker, its overexpression predicting a poor overall survival. The researchers used CRISPR/Cas9 system to knockout CD44 in drug-resistant osteosarcoma cell lines, followed by a decrease in the levels of P-glycoproteins. This resulted in an increased intracellular absorption of doxorubicin in the cells. These results reveal that using CRISPR/Cas9, CD44 was highlighted as being involved in the response to chemotherapy in osteosarcoma cell lines (Xiao et al. 2018). Mutations in the WNT signaling pathway induce its constitutive activation and are usually associated with drug resistance (Duchartre et al. 2016). Using CRISPR/Cas9 technology to create xenograft mouse models (tumors derived from HCT116 cells in nude mice) with *TIAM1*(T-cell lymphoma invasion and metastasis 1) knocked-down in colorectal tumors, Izumi et al. (2019) observed an increased sensitivity of these cells in the treatment with 5′ FU (5′ fluorouracil) and reduced tumor size and mass measured in comparison with the controls. These results indicated a direct correlation between upregulation of *TIAM1* and drug resistance, making this gene a potential therapeutic target that can be used in colorectal patients in the attempt to overcome drug resistance (Malliri et al. 2006, Izumi et al. 2019). In lung cancer studies, the knockdown of *YES1* gene using CRISPR/Cas9 technology was correlated with reduced metastasis rate and diminished tumor burden, indicating *YES1* as a prognostic biomarker for lung cancer patients (Garmendia et al. 2019).

CRISPR/Cas9 is also effective in editing non-coding molecules, such as microRNAs (miRNAs). These molecules modulate a vast number of molecular pathways by targeting a variable number of genes, acting either as tumor suppressors or as oncogenes, based on their downregulation or upregulation encountered in cancer progression (Berindan-Neagoe et al. 2014). Therefore, in a study published by Zhou et al. (2017), the authors demonstrated that upregulation of miR-3188 in hepatocellular carcinoma is associated with an unfavorable outcome for the patients. Using CRISPR/Cas9 to knockout the expression of this microRNA in hepatocellular carcinoma cell lines by aiming the sequences next to the Drosha processing sites harbored in the miR-3188 secondary loop, the authors observed that underexpression of miR-3188 suppressed malignant characteristics. These results indicated miR-3188 as a potential biomarker for early diagnosis and therapeutic target for the treatment of hepatocellular carcinoma (Zhou et al. 2017).

In epigenetic studies, CRISPR/Cas9 was used to mediate the DNA methylation of *DERARE* (Distal Element Multiple Retinoic Acid Response Element), which inhibited the expression of *HOXB* and diminished leukemogenesis. This was possible by using the epigenetic editing tool developed by Vojta et al. (2016), in which the group constructed a dCas9-DNMT3A (deactivated Cas9-DNA methyltransferase

3A) fusion protein by merging the 3' DNMT3A catalytic domain with the dCas9 C terminus. This tool allows DNA methylation next to the sgRNA binding *locus* in *DERARE*. Therefore, CRISPR/Cas9 was used as a screening technique that would allow the use of DNA methylation patterns to identify the suitable drugs or therapeutic approaches for cancer therapy (Vojta et al. 2016, Qian et al. 2018).

## Applications of CRISPR/Cas9 system in gene therapy

The introduction of foreign gene copies into a target cell harboring a damaged or mutated gene to treat various diseases is called gene therapy (Naldini 2015). The use of CRISPR/Cas9 systems for multiplexed targeted genome engineering represents an easier approach, as the design of the gRNAs to specifically target a genome *locus* is more accessible than developing a protein-based sequence identification motif for every specific target (Doudna and Charpentier 2014).

In monogenic diseases, CRISPR/Cas9 system for genome editing is preferred because it can achieve permanent removal of the genetic defect, while other gene therapy methods work by adding a functional copy of the gene to the cells, but do not eliminate the dysfunctional copy of the gene. There are multiple studies in which CRISPR/Cas9 system was used to repair or to adjust target sequences in the DNA or to remediate monogenic diseases in different types of tissue culture cells (Schwank et al. 2013) or other postnatal organs including the brain (Swiech et al. 2015), liver (Yin et al. 2016), and heart (Long et al. 2016, Nelson et al. 2016). Therefore, in a mouse model for Duchenne Muscular Dystrophy (DMD) - disease caused by mutations in the *dystrophin (DMD)* gene, which encodes dystrophin protein - an essential component of the muscle cell (Fairclough et al. 2013), CRISPR/Cas9 system was used for gene-editing in the germline. This resulted in genetically mosaic offsprings with variable percentages (2-100%) of somatic cells harboring the functional copy of the impaired gene (Long et al. 2014). Cataract mouse models were used with CRISPR/Cas9 systems directed to correct the deletion in *Crygc* (crystallin gamma C) gene that results in a frameshift mutation causing the cataract phenotype. In this study, germline editing of the gene corrected the cataract phenotype by repairing the dysfunctional *Crygc* gene. Furthermore, this study revealed that the repaired *Crygc* allele could be transmitted to the next generation through germline (Wu et al. 2013). CRISPR/Cas9 can be used as therapeutic strategy in human stem cells (including induced pluripotent stem cells). Restoration of the defective *CFTR* (Cystic fibrosis transmembrane conductance regulator) gene in intestinal stem cells from cystic fibrosis patients through CRISPR/Cas9 and homologous recombination kept the pluripotency of these cells, a fact proven by the presence of organ-like expansions observed in the cell culture (Schwank et al. 2013).

The treatment for infectious diseases evolved in recent years and CRISPR/Cas9 can now be used for the treatment of viral infectious diseases. In this respect, gene therapy using this technology is used to transform the host cells to either avoid the infection or to impede the viral multiplication and transmission (Lee 2019). The human immunodeficiency virus (HIV) infection still represents a major problem worldwide especially in terms of treatment costs, side effects, and drug resistance. The infection with HIV is characterized by the targeting of CD4+ T lymphocytes by the retrovirus resulting in the integration of the virus into the host cell genome and inducing acquired immunodeficiency syndrome (AIDS). There are multiple examples of research studies using CRISPR/Cas9 technology to target either HIV-1 genome or host factors (CCR5, CXCR4, and restriction factors) (Xiao et al. 2019). The first documentation regarding the possible use of CRISPR/Cas9 systems for the HIV-1 treatment via elimination of the viral genes was in 2013, when a group of researchers used CRISPR/Cas9 to inhibit the expression of the viral genes by specifically targeting HIV-1 LTR (long terminal repeat) in Jurkat cell lines. They successfully inhibited the transcription and replication of the HIV-1 provirus by using as target sites the NF-kB binding cassettes from the U3 region and the TAR sequences in R region (Xiao et al. 2019). In recent years, progress has been made regarding the applications of CRISPR/Cas9 systems to overcome HIV treatment issue including its use in *in vitro, ex vivo* and also animal models (Hu et al. 2014, Kaminski et al. 2016a, b, Dampier et al. 2014).

CRISPR potential in cancer therapy makes this technique preferred as it brings efficient results in terms of genetic manipulation. This technique offers researchers the possibility to identify new therapeutic targets allowing the investigation of the functional particularities of the genes or different signaling pathways (Azangou-Khyavy et al. 2020). The most thoroughly used CRISPR system for gene editing is CRISPR/Cas9. This system uses Cas9 as the RNA-guided nuclease and the crRNA as the guiding RNA. Besides, it also employs tracrRNA, an essential component for the Cas9 binding, cleavage of the DNA target sequence, and crRNA processing (Deltcheva et al. 2011, Jinek et al. 2012).

The malignant phenotype is characterized by several hallmarks that appear as a result of genetic alterations in key regions of the genome (Hanahan and Weinberg 2011). CRISPR/Cas9 is a technology that facilitates gene editing in such a manner that one can hinder the activity of an oncogene or can restore the activity of the tumor suppressor gene. The most common gene suffering mutation in cancer is *TP53* (tumor protein 53), the so called "guardian of the genome", with about 50% of the cancer cases having an altered form of *TP53* (Lane 1992, Chira et al. 2018). In a publication by Chira et al (2018), the authors introduced a hypothetical concept for a *TP53* delivery strategy, in which a fully functional *TP53* could substitute the altered gene by HDR mechanism of CRISPR/Cas9 system. This concept might be reasonable, as CRISPR/Cas9 systems are known to empower the insertion of large nucleic acid sequences (Zhang et al. 2015). The effect would be the restoration of the normal expression of tp53 (Chira et al. 2018, Zhang et al. 2015).

In cancers with EGFR mutations (Epidermal Growth Factor Receptor), even though therapeutic compounds targeting this receptor (TKIs – Tyrosine Kinase Inhibitors) are used in the clinical practice in patients with *EGFR* mutations, patients often develop resistance to therapy within two years (Politi et al. 2015). Tang et al. (2016) proposed the use of CRISPR/Cas9 technique as a molecular surgery tool to restore or to impair the function of mutated *EGFR* in NSCLC patients. The proposed strategy implies sampling tumor-tissue biopsies from NSCLC patients for the identification of the mutational status in *EGFR* to develop sgRNA to target the altered sequences (exons) – such as L858R found in exon 21, E19del found in exon 19, or the T790M found in exon 20, associated with resistance to therapy (Tang and

Shrager 2016). Another strategy to overcome drug resistance is to target *NFE2L2* gene that encodes the NRF2 protein, known to be involved in resistance to chemotherapy. In drug-resistant lung cancer cell lines, a group of researchers employed CRISPR/Cas9 system to impair the function of *NFE2L2*. The reported results indicated that loss of function for this gene improved the sensibility of the cells to cisplatin, carboplatin, and vinorelbine (Bialk et al. 2018). In a clinical study (NCT02793856), the safety and practicability of using CRISPR/Cas9 system was investigated for T-cell therapy by targeting *PD*-1 in patients diagnosed with NSCLC, using a nucleofection method to deliver the CRISPR/Cas9 elements into primary T-cells. The experiment revealed limited off-target effects (0.05%) at 18 candidate sites using NGS technology (Next Generation Sequencing). The overall result indicated CRISPR/Cas9 as a feasible and safe technique for *ex vivo* T-cell therapy in advanced NSCLC patients (Lu et al. 2020).

Advanced research highlighted the importance of epigenetic changes in tumor development and progression (Jones and Baylin 2007). The epigenetic targeting of different genes to restore/destroy their function can be achieved through CRISPR/Cas9. Thus, in a study published by Wang et al. (2018), the group fused the C-terminus in the dCas9 (deactivated Cas9) to the DNMT3a, EZH2 (histone 3 K27 methyltransferase), and KRAB (Krüppel associated box) to develop three CRISPR/Cas9 systems that would epigenetically target *GRN* (granulin) in hepatoma cells. Their research indicated that all three epigenetic suppressors (dCas9-DNMT3a/dCas9-EZH2/dCas9-KRAB) inhibited the expression of *GRN*, thus suppressing invasion and sphere formation in hepatoma cell lines (Wang et al. 2018).

Genetically engineered mouse models are the most refined *in vivo* models to study breast cancer, offering a reliable simulation of the progression of a normal cell to a transformed one, providing the biological context in terms of the stromal compartment and immune response. After developing a knock-in mouse model with Cre-conditional BE3 allele, the CRISPR/Cas9 system was used to induce loss of function in *PTEN* and restoration of *AKT1* and *PIK3CA* function, genes are known to be drivers of BRCA-1 associated breast cancers, therefore assessing the roles of these molecules in tumor formation (Annunziato et al. 2020).

## Other applications of CRISPR/Cas – diagnosis of infectious diseases

Infectious diseases include various types of pathologies caused by viruses, bacteria, fungi and different parasites, which easily spread between individuals and within communities, having the potential to cause serious public health issues such as epidemics of different amplitudes, which usually take a high toll on human lives and impose serious and destabilizing burdens on society and economics. Recent times have shown the severe impact of these outbreaks caused by pathogens such as SARS, Ebola, the Zika virus, or the current SARS-CoV-2 pandemic.

According to recent World Health Organization reports, communicable diseases are still present in the top 10 leading causes of death worldwide. Thus, WHO's 2019 Global Health Estimates show that three out of the top 10 most lethal killers were infectious pathologies, with even higher impact in lower income countries. Most of these diseases are of zoonotic origin, and are disseminated into the susceptible human population either through symptomatic exposure or by occult carriers. At present time, the gold standard for the identification of the pathogen in the case of communicable diseases is the quantitative Real-Time Polymerase Chain Reaction assay (qRT-PCR), which offers high sensitivity and specificity, and reduced instances of false positives. On the downside, performing qRT-PCR based diagnosis relies on the completion of several steps performed in specialized molecular biology laboratories, by trained personnel, on dedicated machines, during a timespan which, when it comes to deadly outbreaks, can translate into more casualties and a higher degree of spreading across larger populations. Therefore, there is a serious need for a better understanding of the pathogenic mechanisms that constitute the foundations of these diseases, and for the development of novel tools for diagnosis and treatment, which, in turn, will generate improved outcomes of these pathologies (Chiu 2018).

One such approach employs the CRISPR-Cas system, and the ability of some Cas enzymes to recognize and specifically cut either ssRNA or ssDNA which, when connected to fluorescent probes, can generate a signal that can be detected and interpreted for diagnosis. The high applicability of this technology to portable, easy to use lateral flow tests has led to the development of exact, fast and adjustable "point of care" diagnostic prototypes based on CRISPR. Chen et al. from University of California, Berkeley developed such an instrument based on the CRISPR-Cas technology, which uses isothermal amplification and the ssDNase capabilities of Cas12 in order to detect exogenous DNA found at attomolar concentrations. Thus, they used DETECTR (DNA endonuclease-targeted CRISPR trans reporter) for the rapid and precise identification in premalignant lesions of the two most important types of human papillomavirus, HPV16 and HPV18, which are responsible for malignant transformation (Woodman et al. 2007). The DETECTR capabilities were validated by testing patient samples that were previously characterized via PCR, and the accuracy for HPV16 was 100%, and 92% in the case of HPV18. Taking into account the fact that the entire assay was performed within an hour, in a "one pot" type reaction, the authors emphasized the potential of this tool to detect DNA sequences with increased sensitivity and specificity (Chen et al. 2018).

Similarly, Gootenberg and colleagues from the Broad Institute have upgraded their portable CRISPR-Cas13 based platform named SHERLOCK (specific high-sensitivity enzymatic reporter unlocking), which was capable of discriminating between small quantities of nucleic acid sequences that differ by only one nucleotide. By addressing limitations concerning deficient quantitation and the need for specialized equipment for fluorescence detection, the team has developed the second version of the platform, which was capable of detecting ssRNA from two of the pathogens that caused devastating outbreaks in recent times, the Zika (ZIKV) and the Dengue (DENV) viruses. They resolved several issues, such as multiplexing, precise quantitation capabilities at attomolar levels, robustness and sensitivities, ability to detect mutations in liquid biopsies, all in a portable, easy to use lateral flow format (Gootenberg et al. 2018). The same research team further improved the SHERLOCK platform by developing HUDSON (heating unextracted diagnostic samples to obliterate nucleases), a field deployable tool that managed to rapidly and precisely identify several serotypes of DENV, and specific ZIKV strains directly from bodily fluids (Myhrvold et al. 2018).

The challenges of achieving a point of care diagnosis tools have driven the development of various approaches that present several advantages, such as the ability to perform the assay in a single reaction, over a short period of time (less than two hours), adaptability and flexibility, direct use of clinical samples without the need of processing or nucleic acid extraction, portability, fluorescence based or lateral flow readout. Although the process of generating a foolproof diagnosis instrument for infectious diseases is still underway, the efficiency of such attempts as the ones previously described has proved its potential in the current SARS-CoV-2 pandemic, when a group of researchers from San Francisco, California, have used the DETECTR platform to develop an instrument capable of identifying the presence of the virus in patient samples obtained from nasopharyngeal or oropharyngeal swabs in UTM (universal transport medium). Thus, by combining CRISPR-Cas12 technology with simultaneous reverse transcription and isothermal amplification (RT-LAMP), the assay was able to detect in under one hour the presence of the E and N genes of SARS-CoV-2, in the form of a lateral flow strip (Broughton et al. 2020).

Although these diagnosis tools need further optimization, their potential for the rapid diagnosis of various infectious diseases, or for the screening of large populations exposed to different pathogens will most likely contribute to a more efficient disease control and outbreak containment.

## Challenges of CRISPR/Cas9

One major challenge in CRISPR/Cas9 mediated genome editing is represented by the off-target mutations, especially in human cells, which may result in cell death or cell transformation (Fu et al. 2013). The presence of multiple identical or highly homologous DNA sequences is usually a characteristic of large genomes, thus creating the possibility for CRISPR/Cas9 to cleave and modify other sequences besides the desired ones. In order to overcome this threat, efforts are made by the research community to reduce and/or remove off-target mutations that occur in the process of genome-editing with the CRISPR/Cas9 systems (Cong et al. 2013, Fu et al. 2013, Hsu et al. 2013, Xiao et al. 2014). In theory, CRISPR/Cas9 technology can be employed in editing a DNA sequence by using an engineered gRNA. But in order to identify the target sequence, besides the complementarity of the sequence with the gRNA, this system compels the presence of PAM sequence downstream of the target site (Jinek et al. 2012). PAM sequences are variable among Cas9 proteins (NGG in *S. pyrogenes;* NGGNG or NNAGAAW in *S. thermophiles*; NNNNGATT in *N. meningitides*) (Gasiunas et al. 2012, Hou et al. 2013, Deltcheva et al. 2011, Garneau et al. 2010, Karvelis et al. 2013, Zhang et al. 2013). The pitfall of PAM-dependence is that it limits the targetable *loci* in the DNA (Zhang et al. 2014). CRISPR/Cas9 mediated genome engineering is also dependent on the gRNA production. The recommendation for the selection of the desired gRNAs is to identify a target site that is predicted to have a limited number of off-target sequences. In this respect, several online tools simplify the identification and selection of an appropriate target area (http://crisp.mit.edu/). Given the posttranscriptional modifications and processing followed by the RNA polymerase II transcription of the mRNA, using this enzyme to produce gRNA becomes difficult. In the present time, for the production of gRNAs in vivo, RNA polymerase III, U3 and U6 snRNA (Small nuclear RNA) promoters are generally used. Lack of commercially available RNA polymerase III and the fact that U3 and U6 are currently used as housekeeping genes due to their ubiquitous expression, it hinders the production of tissue/cell-specific gRNA (Gao and Zhao 2014). It has been suggested that using truncated gRNAs at the 5' end (2-3 nt) may increase specificity, resulting in less off-target effects. Still, this approach causes a reduction in the efficiency of genome editing (Fu et al. 2014). Another challenge refers to the fact that efficient delivery of the CRISPR/Cas9 complex to the targeted organs, tissues or cells, which would allow it to display the most effective genome editing capabilities, is usually hindered by various factors that characterize the different parts of the system, such as the relatively large dimension of the Cas9 component, or the anionic features of the sgRNA (Luther et al. 2018, Zuris et al. 2015).

## Conclusions

With the discovery and the development of CRISPR/Cas9 technology, biomedical research has been revolutionized, allowing targeted and specific genome modifications in the majority of cell types and organisms. The rapid evolution and broad use in biomedical applications reveals its efficiency and utility, even though the full potential of CRISPR/Cas9 has not yet been tackled. CRISPR/Cas9 is a genome-editing tool used to introduce double-strand breaks at specific sites in the genome. The applicability of this tool went from single-gene editing to multiplexed genome-engineering, gene expression regulation and genome-wide screens. Besides its utility in gene function analysis, CRISPR/Cas9 can be applied for gene therapy, disease diagnosis, targeted drug development, and for the construction of animal models. However, there are still several challenges in CRISPR/Cas9 system that have to be addressed, such as: delivery methods (must be optimized to ensure the efficiency of this method), off-target mutations (efficiency must be coupled with specificity), and various ethical issues. Nevertheless, future perspectives regarding CRISPR/Cas9 technology involves the transfer of the biomedical researcher's knowledge to the clinicians to use it as a new diagnostic technique or for the development of new therapies for genetic diseases.

## References

Adli, M. 2018. The CRISPR tool kit for genome editing and beyond. Nature Communications, 9(1): 1911.

Allen, A. G., Chung, C. H., Atkins, A., Dampier, W., Khalili, K., Nonnemacher, M. R., et al. 2018. Gene editing of HIV-1 co-receptors to prevent and/or cure virus infection. Frontiers in Microbiology, 9: 2940.

Amitai, G. and Sorek, R. 2016. CRISPR-Cas adaptation: Insights into the mechanism of action. Nature Reviews Microbiology, 14(2): 67-76.

Annunziato, S., Lutz, C., Henneman, L., Bhin, J., Wong, K., Siteur, B., et al. 2020. In situ CRISPR-Cas9 base editing for the development of genetically engineered mouse models of breast cancer. EMBO Journal, 39(5): e102169.

Azangou-Khyavy, M., Ghasemi, M., Khanali, J., Boroomand-Saboor, M., Jamalkhah, M., Soleimani, M., et al. 2020. CRISPR/Cas: From tumor gene editing to T cell-based immunotherapy of cancer. Frontiers in Immunology, 11: 2062.

Barrangou, R., Fremaux, C., Deveau, H., Richards, M., Boyaval, P., Moineau, S., et al. 2007. CRISPR provides acquired resistance against viruses in prokaryotes. Science, 315 (5819): 1709-1712.

Barrangou, R. and Marraffini, L. A. 2014. CRISPR-Cas systems: Prokaryotes upgrade to adaptive immunity. Molecular Cell, 54(2): 234-244.

Berindan-Neagoe, I., Monroig Pdel, C., Pasculli, B. and Calin, G. A. 2014. MicroRNAome genome: A treasure for cancer diagnosis and therapy. CA: A Cancer Journal for Clinicians, 64(5): 311-336.

Bialk, P., Wang, Y., Banas, K. and Kmiec, E. B. 2018. Functional gene knockout of NRF2 increases chemosensitivity of human lung cancer A549 cells in vitro and in a xenograft mouse model. Molecular Therapy – Oncolytics, 11: 75-89.

Bikard, D., Euler, C. W., Jiang, W., Nussenzweig, P. M., Goldberg, G. W., Duportet, X., et al. 2014. Exploiting CRISPR-Cas nucleases to produce sequence-specific antimicrobials. Nature Biotechnology, 32(11): 1146-1150.

Bikard, D., Jiang, W., Samai, P., Hochschild, A., Zhang, F. and Marraffini, L. A. 2013. Programmable repression and activation of bacterial gene expression using an engineered CRISPR-Cas system. Nucleic Acids Research, 41(15): 7429-7437.

Bolotin, A., Quinquis, B., Sorokin, A. and Ehrlich, S. D. 2005. Clustered regularly interspaced short palindrome repeats (CRISPRs) have spacers of extra chromosomal origin. Microbiology (Reading), 151(Pt 8): 2551-2561.

Broughton, J. P., Deng, X., Yu, G., Fasching, C. L., Servellita, V., Singh, J., et al. 2020. CRISPR-Cas12-based detection of SARS-CoV-2. Nature Biotechnology, 38(7): 870-874.

Brouns, S. J., Jore, M. M., Lundgren, M., Westra, E. R., Slijkhuis, R. J., Snijders, A. P., et al. 2008. Small CRISPR RNAs guide antiviral defense in prokaryotes. Science, 321(5891): 960-964.

Charpentier, E. and Doudna, J. A. 2013. Biotechnology: Rewriting a genome. Nature, 495(7439): 50-51.

Chen, B., Gilbert, L. A., Cimini, B. A., Schnitzbauer, J., Zhang, W., Li, G. W., et al. 2013. Dynamic imaging of genomic loci in living human cells by an optimized CRISPR/Cas system. Cell, 155(7): 1479-1491.

Chen, J. S., Ma, E., Harrington, L. B., Da Costa, M., Tian, X., Palefsky, J. M., et al. 2018. CRISPR-Cas12a target binding unleashes indiscriminate single-stranded DNase activity. Science 360(6387): 436-439.

Chira, S., Gulei, D., Hajitou, A. and Berindan-Neagoe, I. 2018. Restoring the p53 'Guardian' phenotype in p53-deficient tumor cells with CRISPR/Cas9. Trends in Biotechnology, 36(7): 653-660.

Chira, S., Gulei, D., Hajitou, A., Zimta, A. A., Cordelier, P. and Berindan-Neagoe, I. 2017. CRISPR/Cas9: Transcending the reality of genome editing. Molecular Therapy – Nucleic Acids, 7: 211-222.

Chiu, C. 2018. Cutting-edge infectious disease diagnostics with CRISPR. Cell Host Microbe, 23(6): 702-704.

Cho, S. W., Kim, S., Kim, J. M. and Kim, J. S. 2013. Targeted genome engineering in human cells with the Cas9 RNA-guided endonuclease. Nature Biotechnology, 31(3): 230-232.

Chylinski, K., Le Rhun, A. and Charpentier, E. 2013. The tracrRNA and Cas9 families of type II CRISPR-Cas immunity systems. RNA Biology, 10(5): 726-737.

Cong, L., Ran, F. A., Cox, D., Lin, S., Barretto, R., Habib, N., et al. 2013. Multiplex genome engineering using CRISPR/Cas systems. Science 339(6121): 819-823.

Dampier, W., Nonnemacher, M. R., Sullivan, N. T., Jacobson, J. M. and Wigdahl, B. 2014. HIV Excision utilizing CRISPR/Cas9 technology: Attacking the proviral quasispecies in reservoirs to achieve a cure. MedCrave online Journal of Immunology, 1(4).

Deltcheva, E., Chylinski, K., Sharma, C. M., Gonzales, K., Chao, Y., Pirzada, Z. A., et al. 2011. CRISPR RNA maturation by trans-encoded small RNA and host factor RNase III. Nature, 471(7340): 602-607.

Deng, L., Luo, M., Velikovsky, A. and Mariuzza, R. A. 2013. Structural insights into the evolution of the adaptive immune system. Annual Review of Biophysics, 42: 191-215.

Doench, J. G., Hartenian, E., Graham, D. B., Tothova, Z., Hegde, M., Smith, I., et al. 2014. Rational design of highly active sgRNAs for CRISPR-Cas9-mediated gene inactivation. Nature Biotechnology, 32(12): 1262-1267. doi: 10.1038/nbt.3026.

Doudna, J. A. and Charpentier, E. 2014. Genome editing. The new frontier of genome engineering with CRISPR-Cas9. Science, 346(6213): 1258096.

Dove, S. L. and Hochschild, A. 1998. Conversion of the omega subunit of Escherichia coli RNA polymerase into a transcriptional activator or an activation target. Genes & Development, 12(5): 745-54.

Duchartre, Y., Kim, Y. M. and Kahn, M. 2016. The Wnt signaling pathway in cancer. Critical Reviews in Oncology/Hematology, 99: 141-149.

Fairclough, R. J., Wood, M. J. and Davies, K. E. 2013. Therapy for Duchenne muscular dystrophy: Renewed optimism from genetic approaches. Nature Reviews Genetics, 14(6): 373-378.

Fonfara, I., Le Rhun, A., Chylinski, K., Makarova, K. S., Lecrivain, A. L., Bzdrenga, J., et al. 2014. Phylogeny of Cas9 determines functional exchangeability of dual-RNA and Cas9 among orthologous type II CRISPR-Cas systems. Nucleic Acids Research, 42(4): 2577-2590.

Fu, Y., Foden, J. A., Khayter, C., Maeder, M. L., Reyon, D., Joung, J. K., et al. 2013. High-frequency off-target mutagenesis induced by CRISPR-Cas nucleases in human cells. Nature Biotechnology, 31(9): 822-826.

Fu, Y., Sander, J. D., Reyon, D., Cascio, V. M. and Joung, J. K. 2014. Improving CRISPR-Cas nuclease specificity using truncated guide RNAs. Nature Biotechnology, 32(3): 279-284.

Gao, Y. and Zhao, Y. 2014. Self-processing of ribozyme-flanked RNAs into guide RNAs in vitro and in vivo for CRISPR-mediated genome editing. Journal of Integrative Plant Biology, 56(4): 343-349.

Garmendia, I., Pajares, M. J., Hermida-Prado, F., Ajona, D., Bertolo, C., Sainz, C., et al. 2019. YES1 drives lung cancer growth and progression and predicts sensitivity to dasatinib. American Journal of Respiratory and Critical Care Medicine, 200(7): 888-899.

Garneau, J. E., Dupuis, M. E., Villion, M., Romero, D. A., Barrangou, R., Boyaval, P., et al. 2010. The CRISPR/Cas bacterial immune system cleaves bacteriophage and plasmid DNA. Nature, 468(7320): 67-71.

Gasiunas, G., Barrangou, R., Horvath, P. and Siksnys. V. 2012. Cas9-crRNA ribonucleoprotein complex mediates specific DNA cleavage for adaptive immunity in bacteria. Proceedings of the National Academy of Sciences USA, 109(39): E2579-2586.

Gootenberg, J. S., Abudayyeh, O. O., Kellner, M. J., Joung, J., Collins, J. J. and Zhang, F. 2018. Multiplexed and portable nucleic acid detection platform with Cas13, Cas12a, and Csm6. Science, 360(6387): 439-444.

Haapaniemi, E., Botla, S., Persson, J., Schmierer, B. and Taipale, J. 2018. CRISPR-Cas9 genome editing induces a p53-mediated DNA damage response. Nature Medicine, 24(7): 927-930.

Haft, D. H., Selengut, J., Mongodin, E. F. and Nelson, K. E. 2005. A guild of 45 CRISPR-associated (Cas) protein families and multiple CRISPR/Cas subtypes exist in prokaryotic genomes. PLOS Computational Biology, 1(6): e60.

Hale, C. R., Zhao, P., Olson, S., Duff, M. O., Graveley, B. R., Wells, L., et al. 2009. RNA-guided RNA cleavage by a CRISPR RNA-Cas protein complex. Cell, 139(5): 945-956.

Hanahan, D. and Weinberg, R. A. 2011. Hallmarks of cancer: The next generation. Cell, 144(5): 646-674.

Horvath, P. and Barrangou, R. 2013. RNA-guided genome editing a la carte. Cell Research, 23(6): 733-734.

Hou, Z., Zhang, Y., Propson, N. E., Howden, S. E., Chu, L. F., Sontheimer, E. J., et al. 2013. Efficient genome engineering in human pluripotent stem cells using Cas9 from Neisseria meningitidis. Proceedings of the National Academy of Sciences USA, 110(39): 15644-15649.

Hsu, P. D., Lander, E. S. and Zhang, F. 2014. Development and applications of CRISPR-Cas9 for genome engineering. Cell. 157(6): 1262-1278.

Hsu, P. D., Scott, D. A., Weinstein, J. A., Ran, F. A., Konermann, S., Agarwala, V., et al. 2013. DNA targeting specificity of RNA-guided Cas9 nucleases. Nature Biotechnology, 31(9): 827-832.

Hu, W., Kaminski, R., Yang, F., Zhang, Y., Cosentino, L., Li, F., et al. 2014. RNA-directed gene editing specifically eradicates latent and prevents new HIV-1 infection. Proceedings of the National Academy of Sciences USA, 111(31): 11461-11466.

Ishino, Y., Shinagawa, H., Makino, K., Amemura, M. and Nakata. A. 1987. Nucleotide sequence of the iap gene, responsible for alkaline phosphatase isozyme conversion in Escherichia coli, and identification of the gene product. Journal of Bacteriology, 169(12): 5429-5433.

Izumi, D., Toden, S., Ureta, E., Ishimoto, T., Baba, H. and Goel, A. 2019. TIAM1 promotes chemoresistance and tumor invasiveness in colorectal cancer. Cell Death & Disease, 10(4): 267.

Janik, E., Niemcewicz, M., Ceremuga, M., Krzowski, L., Saluk-Bijak, J. and Bijak, M. 2020. Various aspects of a gene editing system-CRISPR-Cas9. International Journal of Molecular Sciences, 21(24).

Jansen, R., Embden, J. D., Gaastra, W. and Schouls, L. M. 2002. Identification of genes that are associated with DNA repeats in prokaryotes. Molecular Microbiology, 43(6): 1565-1575.

Jin, L. F. and Li, J. S. 2016. Generation of genetically modified mice using CRISPR/Cas9 and haploid embryonic stem cell systems. Dongwuxue Yanjiu, 37(4): 205-213.

Jinek, M., Chylinski, K., Fonfara, I., Hauer, M., Doudna, J. A. and Charpentier, E. 2012. A programmable dual-RNA-guided DNA endonuclease in adaptive bacterial immunity. Science. 337(6096): 816-821.

Jones, P. A. and Baylin, S. B. 2007. The epigenomics of cancer. Cell. 128(4): 683-692.

Kaminski, R., Bella, R., Yin, C., Otte, J., Ferrante, P., Gendelman, H. E., et al. 2016a. Excision of HIV-1 DNA by gene editing: A proof-of-concept in vivo study. Gene Therapy, 23(8-9): 690-695.

Kaminski, R., Chen, Y., Fischer, T., Tedaldi, E., Napoli, A., Zhang, Y., et al. 2016b. Elimination of HIV-1 genomes from human T-lymphoid cells by CRISPR/Cas9 gene editing. Scientific Reports, 6: 22555.

Karvelis, T., Gasiunas, G., Miksys, A., Barrangou, R., Horvath, P. and Siksnys, V. 2013. crRNA and tracrRNA guide Cas9-mediated DNA interference in Streptococcus thermophilus. RNA Biology, 10(5): 841-851.

Koike-Yusa, H., Li, Y., Tan, E. P., Velasco-Herrera Mdel, C. and Yusa, K. 2014. Genome-wide recessive genetic screening in mammalian cells with a lentiviral CRISPR-guide RNA library. Nature Biotechnology, 32(3): 267-273.

Komor, A. C., Badran, A. H. and Liu, D. R. 2017. CRISPR-based technologies for the manipulation of eukaryotic genomes. Cell, 168(1-2): 20-36.

Konermann, S., Brigham, M. D., Trevino, A. E., Joung, J., Abudayyeh, O. O., Barcena, C. et al. 2015. Genome-scale transcriptional activation by an engineered CRISPR-Cas9 complex. Nature, 517(7536): 583-588.

Konermann, S., Brigham, M. D., Trevino, A., Hsu, P. D., Heidenreich, M., Cong, L., et al. 2013. Optical control of mammalian endogenous transcription and epigenetic states. Nature, 500(7463): 472-476.

Kunin, V., R. Sorek and P. Hugenholtz. 2007. Evolutionary conservation of sequence and secondary structures in CRISPR repeats. Genome Biology, 8(4): R61.

Labrie, S. J., Samson, J. E. and Moineau, S. 2010. Bacteriophage resistance mechanisms. Nature Reviews Microbiology, 8(5): 317-327.

Lane, D. P. 1992. Cancer. p53, guardian of the genome. Nature, 358(6381): 15-16.

Larson, M. H., Gilbert, L. A., Wang, X., Lim, W. A., Weissman, J. S. and Qi, L. S. 2013. CRISPR interference (CRISPRi) for sequence-specific control of gene expression. Nature Protocols, 8(11): 2180-2196.

Ledford, H. 2020. CRISPR treatment inserted directly into the body for first time. Nature, 579(7798): 185.

Lee, C. 2019. CRISPR/Cas9-based antiviral strategy: Current status and the potential challenge. Molecules, 24(7).

Lieber, M. R. 2010. The mechanism of double-strand DNA break repair by the nonhomologous DNA end-joining pathway. Annual Review of Biochemistry, 79: 181-211.

Long, C., Amoasii, L., Mireault, A. A., McAnally, J. R., Li, H., Sanchez-Ortiz, E. et al. 2016. Postnatal genome editing partially restores dystrophin expression in a mouse model of muscular dystrophy. Science, 351(6271): 400-403.

Long, C., McAnally, J. R., Shelton, J. M., Mireault, A. A., Bassel-Duby, R. and Olson, E. N. 2014. Prevention of muscular dystrophy in mice by CRISPR/Cas9-mediated editing of germline DNA. Science, 345(6201): 1184-1188.

Lu, Y., Xue, J., Deng, T., Zhou, X., Yu, K., Deng, L. et al. 2020. Safety and feasibility of CRISPR-edited T cells in patients with refractory non-small-cell lung cancer. Nature Medicine, 26(5): 732-740.

Luther, D. C., Lee, Y. W., Nagaraj, H., Scaletti, F. and Rotello, V. M. 2018. Delivery approaches for CRISPR/Cas9 therapeutics in vivo: Advances and challenges. Expert Opinion on Drug Delivery, 15(9): 905-913.

Maeder, M. L., Linder, S. J., Cascio, V. M., Fu, Y., Ho, Q. H. and Joung, J. K. 2013. CRISPR RNA-guided activation of endogenous human genes. Nature Methods, 10(10): 977-979.

Makarova, K. S., Aravind, L., Wolf, Y. I. and Koonin, E. V. 2011. Unification of Cas protein families and a simple scenario for the origin and evolution of CRISPR-Cas systems. Biology Direct, 6: 38.

Makarova, K. S., Grishin, N. V., Shabalina, S. A., Wolf, Y. I. and Koonin, E. V. 2006. A putative RNA-interference-based immune system in prokaryotes: Computational analysis of the predicted enzymatic machinery, functional analogies with eukaryotic RNAi, and hypothetical mechanisms of action. Biology Direct, 1: 7. doi: 10.1186/1745-6150-1-7.

Makarova, K. S. and Koonin, E. V. 2015. Annotation and classification of CRISPR-Cas systems. Methods in Molecular Biology, 1311: 47-75.

Makarova, K. S., Wolf, Y. I., Iranzo, J., Shmakov, S. A., Alkhnbashi, O. S., Brouns, S. J. J., et al. 2020. Evolutionary classification of CRISPR-Cas systems: A burst of class 2 and derived variants. Nature Reviews Microbiology, 18(2): 67-83.

Mali, P., Aach, J., Stranges, P. B., Esvelt, K. M., Moosburner, M., Kosuri, S., et al. 2013a. CAS9 transcriptional activators for target specificity screening and paired nickases for cooperative genome engineering. Nature Biotechnology, 31(9): 833-838.

Mali, P., Yang, L., Esvelt, K. M., Aach, J., Guell, M., DiCarlo, J. E.. et al. 2013b. RNA-guided human genome engineering via Cas9. Science, 339(6121): 823-826.

Malliri, A., Rygiel, T. P., van der Kammen, R. A., Song, J. Y., Engers, R., Hurlstone, A. F. et al. 2006. The rac activator Tiam1 is a Wnt-responsive gene that modifies intestinal tumor development. Journal of Biological Chemistry, 281(1): 543-548.

Marraffini, L. A. and Sontheimer, E. J. 2008. CRISPR interference limits horizontal gene transfer in staphylococci by targeting DNA. Science, 322(5909): 1843-1845.

Matsuoka, T. and Yashiro, M. 2018. Biomarkers of gastric cancer: Current topics and future perspective. World Journal of Gastroenterology, 24(26): 2818-2832.

Mirgayazova, R., Khadiullina, R., Chasov, V., Mingaleeva, R., Miftakhova, R., Rizvanov, A., et al. 2020. Therapeutic editing of the TP53 gene: Is CRISPR/Cas9 an option? Genes (Basel) 11(6).

Mojica, F. J., Diez-Villasenor, C., Garcia-Martinez, J. and Soria, E. 2005. Intervening sequences of regularly spaced prokaryotic repeats derive from foreign genetic elements. Journal of Molecular Evolution, 60(2): 174-182.

Mojica, F. J., Diez-Villasenor, C., Soria, E. and Juez, G. 2000. Biological significance of a family of regularly spaced repeats in the genomes of archaea, bacteria and mitochondria. Molecular Microbiology, 36(1): 244-246.

Mojica, F. J., Juez, G. and Rodriguez-Valera, F. 1993. Transcription at different salinities of Haloferax mediterranei sequences adjacent to partially modified PstI sites. Molecular Microbiology, 9(3): 613-621.

Myhrvold, C., Freije, C. A., Gootenberg, J. S., Abudayyeh, O. O., Metsky, H. C., Durbin, A. F., et al. 2018. Field-deployable viral diagnostics using CRISPR-Cas13. Science, 360(6387): 444-448.

Naldini, L. 2015. Gene therapy returns to centre stage. Nature, 526(7573): 351-360.

Nelson, C. E., Hakim, C. H., Ousterout, D. G., Thakore, P. I., Moreb, E. A., Castellanos Rivera, R. M., et al. 2016. In vivo genome editing improves muscle function in a mouse model of Duchenne muscular dystrophy. Science, 351(6271): 403-407.

Perez-Pinera, P., Kocak, D. D., Vockley, C. M., Adler, A. F., Kabadi, A. M., Polstein, L. R., et al. 2013. RNA-guided gene activation by CRISPR-Cas9-based transcription factors. Nature Methods 10(10): 973-976.

Perez Rojo, F., Nyman, R. K. M., Johnson, A. A. T., Navarro, M. P., Ryan, M. H., Erskine, W., et al. 2018. CRISPR-Cas systems: Ushering in the new genome editing era. Bioengineered, 9(1): 214-221.

Politi, K., Ayeni, D. and Lynch, T. 2015. The next wave of EGFR tyrosine kinase inhibitors enter the clinic. Cancer Cell, 27(6): 751-753.

Pyzocha, N. K. and Chen, S. 2018. Diverse Class 2 CRISPR-Cas effector proteins for genome engineering applications. ACS Chemical Biology, 13(2): 347-356.

Qi, L. S., Larson, M. H., Gilbert, L. A., Doudna, J. A., Weissman, J. S., Arkin, A. P., et al. 2021. Repurposing CRISPR as an RNA-guided platform for sequence-specific control of gene expression. Cell, 184(3): 844.

Qian, P., De Kumar, B., He, X. C., Nolte, C., Gogol, M., Ahn, Y., et al. 2018. Retinoid-sensitive epigenetic regulation of the hoxb cluster maintains normal hematopoiesis and inhibits leukemogenesis. Cell Stem Cell, 22(5): 740-754 e7.

Qu, X., Wang, P., Ding, D., Li, L., Wang, H., Ma, L., et al. 2013. Zinc-finger-nucleases mediate specific and efficient excision of HIV-1 proviral DNA from infected and latently infected human T cells. Nucleic Acids Research, 41(16): 7771-7782.

Ran, F. A., Hsu, P. D., Lin, C. Y., Gootenberg, J. S., Konermann, S., Trevino, A. E., et al. 2013. Double nicking by RNA-guided CRISPR Cas9 for enhanced genome editing specificity. Cell, 154(6): 1380-1389.

San Filippo, J., Sung, P. and Klein, H. 2008. Mechanism of eukaryotic homologous recombination. Annual Review of Biochemistry, 77: 229-257.

Sanchez-Rivera, F. J. and Jacks, T. 2015. Applications of the CRISPR-Cas9 system in cancer biology. Nature Reviews Cancer, 15(7): 387-395.

Schwank, G., Koo, B. K., Sasselli, V., Dekkers, J. F., Heo, I., Demircan, T., et al. 2013. Functional repair of CFTR by CRISPR/Cas9 in intestinal stem cell organoids of cystic fibrosis patients. Cell Stem Cell 13(6): 653-658.

Shalem, O., Sanjana, N. E. and Zhang, F. 2015. High-throughput functional genomics using CRISPR-Cas9. Nature Reviews Genetics, 16(5): 299-311.

Shmakov, S., Smargon, A., Scott, D., Cox, D., Pyzocha, N., Yan, W., et al. 2017. Diversity and evolution of class 2 CRISPR-Cas systems. Nature Reviews Microbiology, 15(3): 169-182.

Sorek, R., Lawrence, C. M. and Wiedenheft, B. 2013. CRISPR-mediated adaptive immune systems in bacteria and archaea. Annual Review of Biochemistry, 82: 237-266.

Strich, J. R. and Chertow, D. S. 2019. CRISPR-Cas biology and its application to infectious diseases. Journal of Clinical Microbiology, 57(4).

Swiech, L., Heidenreich, M., Banerjee, A., Habib, N., Li, Y., Trombetta, J., et al. 2015. In vivo interrogation of gene function in the mammalian brain using CRISPR-Cas9. Nature Biotechnology, 33(1): 102-106.

Tamulaitis, G., Venclovas, C. and Siksnys, V. 2017. Type III CRISPR-Cas immunity: Major differences brushed aside. Trends in Microbiology, 25(1): 49-61.

Tang, H. and Shrager, J. B. 2016. CRISPR/Cas-mediated genome editing to treat EGFR-mutant lung cancer: A personalized molecular surgical therapy. EMBO Molecular Medicine, 8(2): 83-85.

van der Oost, J. 2013. Molecular biology. New tool for genome surgery. Science, 339(6121): 768-770.

van der Oost, J., Jore, M. M., Westra, E. R., Lundgren, M. and Brouns, S. J. 2009. CRISPR-based adaptive and heritable immunity in prokaryotes. Trends in Biochemical Sciences, 34(8): 401-407.

van Soolingen, D., de Haas, P. E., Hermans, P. W., Groenen, P. M. and van Embden, J. D. 1993. Comparison of various repetitive DNA elements as genetic markers for strain differentiation and epidemiology of Mycobacterium tuberculosis. Journal of Clinical Microbiology, 31(8): 1987-1995.

Vojta, A., Dobrinic, P., Tadic, V., Bockor, L., Korac, P., Julg, B., et al. 2016. Repurposing the CRISPR-Cas9 system for targeted DNA methylation. Nucleic Acids Research, 44(12): 5615-5628.

Wang, H., Guo, R., Du, Z., Bai, L., Li, L., Cui, J., et al. 2018. Epigenetic targeting of granulin in hepatoma cells by synthetic CRISPR dCas9 Epi-suppressors. Molecular Therapy – Nucleic Acids, 11: 23-33.

Wang, H., La Russa, M. and Qi, L. S. 2016. CRISPR/Cas9 in genome editing and beyond. Annual Review of Biochemistry, 85: 227-264.

Wang, H., Yang, H., Shivalila, C. S., Dawlaty, M. M., Cheng, A. W., Zhang, F. et al. 2013. One-step generation of mice carrying mutations in multiple genes by CRISPR/Cas-mediated genome engineering. Cell, 153(4): 910-918.

Wang, T., Wei, J. J., Sabatini, D. M. and Lander, E. S. 2014. Genetic screens in human cells using the CRISPR-Cas9 system. Science, 343(6166): 80-84.

Woodman, C. B., Collins, S. I. and Young, L. S. 2007. The natural history of cervical HPV infection: Unresolved issues. Nature Reviews Cancer, 7(1): 11-22.

Wright, A. V., Nunez, J. K. and Doudna, J. A. 2016. Biology and applications of CRISPR systems: Harnessing nature's toolbox for genome engineering. Cell, 164(1-2): 29-44.

Wu, Y., Liang, D., Wang, Y., Bai, M., Tang, W., Bao, S., et al. 2013. Correction of a genetic disease in mouse via use of CRISPR-Cas9. Cell Stem Cell, 13(6): 659-662.

Xiao, A., Cheng, Z., Kong, L., Zhu, Z., Lin, S., Gao, G., et al. 2014. CasOT: A genome-wide Cas9/gRNA off-target searching tool. Bioinformatics, 30(8): 1180-1182.

Xiao, Q., Guo, D. and Chen, S. 2019. Application of CRISPR/Cas9-based gene editing in HIV-1/AIDS therapy. Frontiers in Cellular and Infection Microbiology, 9: 69.

Xiao, Z., Wan, J., Nur, A. A., Dou, P., Mankin, H., Liu, T., et al. 2018. Targeting CD44 by CRISPR-Cas9 in multi-drug resistant osteosarcoma cells. Cellular Physiology and Biochemistry, 51(4): 1879-1893.

Yadav, N., Narang, J., Chhillar, A. K. and Rana, J. S. 2021. CRISPR: A new paradigm of theranostics. Nanomedicine, 33: 102350.

Yang, X., Boehm, J. S., Yang, X., Salehi-Ashtiani, K., Hao, T., Shen, Y., et al. 2011. A public genome-scale lentiviral expression library of human ORFs. Nature Methods 8(8): 659-661.

Yin, H., Song, C. Q., Dorkin, J. R., Zhu, L. J., Li, Y., Wu, Q., et al. 2016. Therapeutic genome editing by combined viral and non-viral delivery of CRISPR system components in vivo. Nature Biotechnology, 34(3): 328-333.

You, L., Tong, R., Li, M., Liu, Y., Xue, J. and Lu, Y. 2019. Advancements and Obstacles of CRISPR-Cas9 technology in translational research. Molecular Therapy – Methods & Clinical Development, 13: 359-370.

Zamborsky, R., Kokavec, M., Harsanyi, S. and Danisovic, L. 2019. Identification of prognostic and predictive osteosarcoma biomarkers. Medical Sciences (Basel), 7(2).

Zetsche, B., Gootenberg, J. S., Abudayyeh, O. O., Slaymaker, I. M., Makarova, K. S., Essletzbichler, P., et al. 2015. Cpf1 is a single RNA-guided endonuclease of a class 2 CRISPR-Cas system. Cell, 163(3): 759-771.

Zhang, F., Wen, Y. and Guo, X. 2014. CRISPR/Cas9 for genome editing: Progress, implications and challenges. Human Molecular Genetics, 23(R1): R40-46.

Zhang, L., Jia, R., Palange, N. J., Satheka, A. C., Togo, J., An, Y., et al. 2015. Large genomic fragment deletions and insertions in mouse using CRISPR/Cas9. PLoS One, 10(3): e0120396.

Zhang, Y., Heidrich, N., Ampattu, B. J., Gunderson, C. W., Seifert, H. S., Schoen, C., et al. 2013. Processing-independent CRISPR RNAs limit natural transformation in Neisseria meningitidis. Molecular Cell, 50(4): 488-503.

Zhou, S. J., Deng, Y. L., Liang, H. F., Jaoude, J. C. and Liu, F. Y. 2017. Hepatitis B virus X protein promotes CREB-mediated activation of miR-3188 and Notch signaling in hepatocellular carcinoma. Cell Death Differ, 24(9): 1577-1587.

Zhou, Y., Zhu, S., Cai, C., Yuan, P., Li, C., Huang, Y., et al. 2014. High-throughput screening of a CRISPR/Cas9 library for functional genomics in human cells. Nature, 509(7501): 487-491.

Zuris, J. A., Thompson, D. B., Shu, Y., Guilinger, J. P., Bessen, J. L., Hu, J. H., et al. 2015. Cationic lipid-mediated delivery of proteins enables efficient protein-based genome editing in vitro and in vivo. Nature Biotechnology, 33(1): 73-80.

# Recent Applications of CRISPR-Cas9 in Genome Mapping and Sequencing

**Dharma Varapula, Lahari Uppuluri and Ming Xiao***

School of Biomedical Engineering, Science and Health Systems, Drexel University, 3141 Chestnut St., Philadelphia, PA, USA 19104

## Introduction

CRISPR (Clustered Regularly Interspersed Short Palindromic Repeats) refers to a bacterial defense mechanism that was identified with its unique repeating DNA sequences as defined in its nomenclature. The enzymes associated with this mechanism, Cas (CRISPR-associated) proteins, have been refashioned into genome editing systems, which are primarily concerned with introducing or deleting desired DNA sequences into an organism's genome. While this has been the primary goal, *in vitro* applications to visualize or detect sequences in a particular genome are also simultaneously gaining attention. This DNA sequence detection is centered around the highly efficient targeting characteristics of the CRISPR-Cas9 system, the same feature that made it popular for genome editing. Across the vast landscape of the genome, a myriad of structural complexities exists that make DNA sequencing and mapping efforts challenging using current technology platforms, sometimes making them expensive or tedious, or time-taking. Some of these limitations and inefficiencies are being overcome by the use of the highly versatile CRISPR-Cas9 system to design new methods and enhance existing protocols. In this chapter, we will briefly describe the mechanism and targeting specificity of Cas9 and summarize some of the new developments and applications that will impact DNA mapping and sequencing, with a special focus on the high-throughput single-molecule telomere characterization method.

### Structure, mechanism, and specificity of CRISPR-Cas9

The CRISPR-Cas9 system is classified as Class 2 type II among several CRISPR-associated bacterial defense systems. It consists of the Cas9 enzyme, crRNA, and tracrRNA. The Cas9 enzyme is a DNA endonuclease that consists of an HNH-like

* Corresponding author: mx44@drexel.edu

nuclease domain and an RuvC-like nuclease domain, each responsible for creating single-strand nicks. Together, crRNA and tracrRNA are termed the 'guide RNA' (gRNA), and they form a ribonucleoprotein (RNP) complex with the Cas9 enzyme. If crRNA and tracrRNA are part of a single molecule, they are termed 'single guide RNA'. The crRNA contains a 'spacer' sequence that is designed to be complementary to the target sequence in the DNA ('protospacer'). This spacer sequence effectively guides the Cas9 protein to the destined location in the genome, subject to the presence of a specific protospacer-adjacent motif (PAM) sequence (Figure 1). The PAM sequence in the DNA is examined by the PAM-interacting domain of the Cas9 protein. Only if the PAM criterion is satisfied will the Cas9 protein enable the examination of the spacer sequence in the crRNA. Hence, the activity of Cas9 is dependent on both the PAM and the protospacer sequences on the DNA substrate. The HNH-like nuclease domain cleaves the DNA strand that is complementary to the spacer sequence in crRNA, while the RuvC-like domain cleaves the opposite DNA strand (Figure 1). Either of the two domains could be deactivated to result in a Cas9 that cleaves only a single strand in the dsDNA substrate ('Cas9 nickase' or 'Cas9n'). The site of cleavage is 3 bases upstream of the PAM sequence. When both HNH

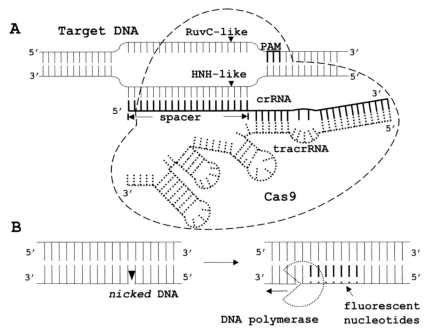

**Figure 1.** CRISPR-Cas9 and fluorescent labeling of DNA. (A) spCas9 complex with crRNA:tracrRNA bound to target DNA; crRNA and tracrRNA when joined together at the first stem loop are referred to as 'single guide RNA' or 'sgRNA' (B) Cas9 nickase-based fluorescent labeling of DNA; Cas9 D10A nickase is used to perform sequence-specific single-strand breaks in DNA ("nicks") targeted either by use of an sgRNA or a crRNA-tracrRNA complex (gRNA). These "nicks" then act as substrates for DNA polymerase-mediated fluorescent nucleotide incorporation, thereby placing multiple fluorophores proximal to a single nick for efficient DNA imaging on a fluorescence microscope system.

and RuvC domains are deactivated, the dead Cas9 (dCas9) enzyme seeks the target sequence and binds at the location without performing any DNA modification. These Cas9 modifications have been exploited in various ways (Gilbert et al. 2013, Ran et al. 2013, Chen et al. 2013).

### The PAM requirement

The most common Cas9 variant is derived from *Streptococcus pyogenes*, SpCas9, and is available from several commercial manufacturers. Any 20-base sequence upstream to the characteristic PAM sequence can be targeted in a DNA molecule using SpCas9. The native PAM motif sequence for the CRISPR-SpCas9 system is 5′-NGG-3′ or NGG, which is widely and frequently distributed across the genome. Although non-canonical PAM, 5′-NAG-3′, is also recognized, it is done so at much lower efficiency (Jiang et al. 2013). The binding of Cas9 to the PAM sequence initiates conformational changes in the Cas9 molecule that promote the base-pairing of the target DNA strand with the gRNA and consequently, the DNA duplex unwinds. The C-terminal domain (CTD) of the Cas9 protein (apo-Cas9) is responsible for recognition of the PAM sequence. The CTD domain is active only when Cas9 is bound to gRNA and is otherwise in a disordered configuration (Jiang and Doudna 2017). This was observed in a single-molecule DNA experiment, where apo-Cas9 (Cas9 without RNA) bound to DNA molecules indiscriminately, that is, without any sequence-specificity. This nonspecific binding was reversed rapidly by introducing a gRNA that has a competing sequence. It was also shown that PAM sequence was sought via random 3D collisions of the Cas9 protein with the DNA molecule. Once a correct PAM sequence was located, the Cas9 enzyme probed further for the complementarity with the spacer sequence (Sternberg et al. 2014). PAM recognition is hence a crucial event and target search is highly sensitive to mutations in the PAM sequence of DNA. The biological function of the PAM sequence is indeed to distinguish non-self-sequences from self-sequences. This sensitivity has been exploited to detect mutations in the genome (Abid et al. 2020b).

In studies that performed base editing, it was found that the NGG PAM motif occurs only once every 16 randomly selected genomic *loci*. This could potentially be a limitation in precise genome editing applications (Hu et al. 2018). To broaden the targeting range, Cas9 variants with modified PAM specificity have been proposed as a solution. Bacterial selection-based evolution was used to imbue different PAM specificities to the SpCas9 variant without affecting the spacer sequence-associated specificity across the human genome. An identified VQR variant showed robust activity specifically towards NGAN and NGCG PAM motifs when compared to the wild-type variant. However, the cleavage efficiencies with all NGAN PAM sites were found to be significantly non-uniform - NGAG>NGAT=NGAA>NGAC (Kleinstiver et al. 2015). This is not ideal and will introduce further complexity in the CRISPR-Cas9 targeting system. Moreover, this method to modify PAM specificities is not freely programmable for any desired PAM motif and more likely requires careful engineering efforts that are also tedious. Another approach has been to engineer SpCas9 variants with minimized PAM dependency such that the frequency of the modified PAM sequence is significantly increased. In one study, the nucleobase-specific interaction between the Cas9 protein and the second nucleobase in the NGG PAM sequence has been substituted with a non-nucleobase-specific interaction. This yielded a variant, coined SpCas9-NG, that recognizes only a 5'-NG-3' PAM sequence albeit with reduced cleavage activity compared to the wild-type SpCas9 (Nishimasu et al. 2018). Weakening the specific interactions between the PAM sequence and Cas9 protein might understandably decrease the cleavage activity. In another attempt, phage-assisted continuous evolution of SpCas9 was performed to modify the Cas9 such that the PAM sequence constraint is diluted. The variant, 'xCas9', was shown to recognize multiple sequences, NG, GAA, and GAT, as PAM with lower off-target spacer-specific activity than the wild-type variant. It was shown that this enhanced performance did not necessarily come at the expense of the cleavage activity guided by the spacer. In fact, xCas9 was shown to provide higher efficiency than the wild-type variant (Hu et al. 2018). Alternate PAM specificities were also attained by use of other naturally available Cas9 proteins derived from different bacteria, summarized in Table 1. PAM sequence for Cas9 is G-rich, but another CRISPR system with Cas9-like activity, Cpf1 or Cas12a, was identified to have a T-rich PAM (TTTV, V is A/G/C) (Zetsche et al. 2015). This might provide more flexible targeting in specific areas in the genome depleted with conventional PAM sequences, such as introns and promoters (Wolter and Puchta 2019). SpCas9 and its variants continue to be the most used CRISPR system because of the relatively robust and well-characterized activity, particularly its high PAM-associated specificity and the generally frequent occurrence of NGG in the genome.

**Table 1.** CRISPR-Cas9 variants and their PAM sequences

| Source | Notation | 5'-PAM-3' | Spacer length | Reference |
|---|---|---|---|---|
| *Streptococcus pyogenes* | SpCas9 | NGG | 20 | (Jinek et al. 2012) |
| *Staphylococcus aureus* | SaCas9 | NNGRR | 21-23 | (Ran et al. 2015) |
| *Neisseria meningitidis* | NmCas9 | NNNNGATT | 21-23 | (Lee et al. 2016) |
| *Campylobacter jejuni* | CjCas9 | NNNNRYAC | 22 | (Kim et al. 2017) |
| *Streptococcus thermophilus* | St1Cas9 | NNAGAAW | 20 | (Müller et al. 2016a) |
| | St3Cas9 | NGGNG | 20 | |

*Binding and cleavage activity of CRISPR-Cas9*

The target search of Cas9 commences with its binding to the gRNA molecule. Once a PAM is identified by the Cas9-gRNA complex, Cas9 initiates DNA-melting adjacent to the PAM sequence. Starting from the region proximal to PAM sequence, the DNA-melt bubble expands away from PAM. As this happens, the gRNA spacer sequence begins to form a heteroduplex with the target DNA strand (protospacer) (Jiang and Doudna 2017). In the single-molecule experiment described earlier (Sternberg et al. 2014), it was noted that the dwell time of each Cas9 binding event during a target search on DNA substrate depended on the extent of sequence complementarity between the spacer and protospacer sequences. The PAM-proximal region of the spacer sequence (first ~10 nt) was found to have an elevated priority in effecting Cas9 activity. The PAM-distal region was found to be more tolerated for mismatches when evaluating complementarity. This led to the PAM-proximal region being termed a 'seed' sequence (Jinek et al. 2012). Observation of the RNA-DNA heteroduplex's structure showed that when Cas9 recognizes a valid PAM sequence, the protein undergoes a sharp conformational change orienting the first base of the protospacer towards the spacer sequencing in the gRNA and away from the complementary DNA strand containing PAM. The PAM-distal region of the heteroduplex was seen to have a less definite (or more flexible) structure compared to the tight PAM-proximal 'seed' region of the heteroduplex. This structural plasticity is afforded by the protein's relatively flexible structure surrounding the PAM-distal region of the heteroduplex in the bound state (Jiang and Doudna 2017). In effect, mismatches in the PAM-distal region are more tolerated than in the PAM-proximal regions (Hsu et al. 2013).

A detailed ChIP analysis of off-target activity was performed using dCas9 and Cas9 separately with a series of gRNA sequences that had varying number and distribution of mismatches. It was found that dCas9 (which can only bind and not cleave DNA) yielded several fold greater number of off-target sequences compared to Cas9 (Kuscu et al. 2014). This suggests that DNA binding is only a precursor to cleavage, and that more stringent complementarity is necessary for initiating cleavage. Complementarity in the 'seed' sequence alone is sufficient to initiate binding but complementarity in the PAM-distal regions is also necessary for triggering cleavage activity.

## Cas9 specificity

CRISPR-Cas9 has been shown to contain significant off-target (or nonspecific) activity, which is one of the major impediments in genome editing applications. Sequences in the template genome that have high sequence identity to that of the target sequence act as substrate to the Cas9-gRNA complex. In DNA mapping and sequencing applications of the CRISPR-Cas9 system, minimizing off-target activity will greatly improve the efficiency of the method used. Several approaches have been proposed to tackle this problem. Since up to 6 base mismatches are tolerated in the 20-base long spacer sequence by the Cas9 protein (Tsai et al. 2015), most genomics applications that allow a choice of gRNA sequences can benefit from careful screening of the gRNA candidates based on the predicted off-target activity. Other approaches involve either the modification of Cas9 protein, the gRNA, or altering the reaction kinetics.

Initial efforts at achieving highly-specific Cas9 variants were developed for the purpose of genome editing and regulation in cells. To achieve this, a double-nicking configuration using a Cas9 nickase (Cas9n) has been proposed to exploit a cooperativity effect between the two nicking *loci* (Ran et al. 2013). Similarly, a fused complex composed of dCas9 and FokI endonuclease has been employed to lower off-target activity (Tsai et al. 2014, Guilinger et al. 2014), as well as a conditionally-active Cas9 (Davis et al. 2015). These methods may not be suitable for non-genome editing applications. For example, prior knowledge of the region of interest may not be available, such as in the case of an unresolved genome gap. For more general applications, Cas9 protein has been rationally engineered to yield high-specificity variants. In one study, the interactions between the non-target strand and the Cas9 protein were weakened to induce a competition between the RNA-DNA heteroduplex hybridization and the DNA-DNA hybridization. This effectively brought in a more stringent requirement for RNA-DNA hybridization and enhanced the specificity of Cas9 without diminishing its efficiency (Slaymaker et al. 2016). In another study, the nonspecific DNA-protein interactions at four locations in the Cas9 protein have been substituted with mutations such that the energy available to perform off-target activity was depleted. This hypothesis indeed resulted in enhanced specificity. Here, the on-target specificity was shown not to be significantly reduced but comparable to the wild-type Ca9 (Kleinstiver et al. 2016). SaCas9 and NmCas9 have been shown to possess lower off-target activity but these have shown lower on-target enzymatic activity than with SpCas9 (Tycko et al. 2016).

The guide RNA has also been modified to obtain higher specificity. In early efforts, the structure of the gRNA was modified at two locations, the poly-U proximal to the recognition sequence and the length of the hairpin (duplex) responsible for Cas9 recognition. This resulted in enhanced specificity without affecting Cas9 on-target activity (Dang et al. 2015). Subsequently, chemical modifications at the 2'-O and on the phosphate backbone of the gRNA were shown to contain the off-target behavior (Hendel et al. 2015, Rahdar et al. 2015). In one method, specificity has been improved by destabilizing the proximal base pairing within the 20-base recognition sequence, thereby changing the hybridization dynamics at on and off target sites (Ryan et al. 2018). A recent method exploits the self-competition of different Cas9-gRNA complexes towards a limited number of target sites. By combining an on-target Cas9-gRNA complex with an off-target one (the gRNA for the off-target one matches at only 16 bases of the spacer), it has been shown that off-target cleavage can be suppressed without affecting on-target activity. The off-target Cas9-gRNA complex serves to block the off-target sites (but do not cleave) from the on-target complex (Rose et al. 2020). The off-target profile of dCas9 is different from that of Cas9, as stated earlier, and the methods to assay it have been published (Wu et al. 2014, Kuscu et al. 2014). In comparison to Cas9 which is said to tolerate up to 3 consecutive mismatches in the PAM-distal spacer sequence, dCas9 can tolerate a much greater number of nine consecutive mismatches (Kuscu et al. 2014). Hence, the gRNA design requirements for dCas9 will be different.

Early CRISPR-Cas9-based efforts for sequence-specific genome visualization applications have centered around cellular imaging (Chen et al. 2013, Ma et al. 2015, Deng et al. 2015) for efficient and accurate imaging and for tracking dynamic cellular processes. A new *in vitro* CRISPR-Cas9-based method to fluorescently label sequences across the whole genome was developed for the purpose of optical mapping (McCaffrey et al. 2015). Briefly, optical mapping is a technique used to generate whole genome maps that act as scaffolds for *de novo* sequence assembly and for detecting large structural variations. Traditionally, it relied on restriction endonucleases or nicking endonucleases that typically recognize 6-8 bp sequence motif. The distribution of these motifs across the genome is, however, non-uniform and regions depleted with these motifs are difficult or not possible to map. Moreover, in clinical applications and research of large structural variants, the exact location of the structural variant breakpoint might not be available with the traditional nicking endonuclease-based labeling. These limitations were overcome by use of a CRISPR-Cas9 nickase-based labeling strategy, described in Figure 1. Repetitive regions in the human genome, which are difficult to characterize using DNA sequencing alone, often lack the aforementioned 6-8 bp sequence motifs making them invisible to optical mapping as well. Using the CRISPR-Cas9n-based labeling method, these repetitive regions have been targeted and characterized. In the following section, we describe the high-throughput single-molecule telomere characterization method, in which the telomere repeat sequences were targeted using the CRISPR-Cas9-based labeling method. This method combines high-resolution single-molecule DNA visualization with the highly efficient and versatile CRISPR-Cas9 system (McCaffrey et al. 2017).

## CRISPR-Cas9-based fluorescent labeling for optical mapping in nanochannels

Optical mapping is an efficient single-molecule technique for genome assembly and structural variant detection. Although originally developed to scaffold shotgun assembly, optical mapping technology produces contig lengths much longer than seen in long-read sequencing and therefore enable long-range *de novo* assembly and long range structural variant detection and delineation, and correction of sequence mis-assembly in complex segmentally duplicated or repetitive regions (Levy-Sakin et al. 2019). Traditionally, for whole human genome mapping, a highly frequent sequence-motif is targeted with a nickase enzyme that can create nicks at the respective target sites. Subsequently, nicks are labeled with a polymerase in presence of fluorescent nucleotides thereby producing distinct 'maps' specific to a sample which is detected by optical mapping technology. Sequence-motif-based mapping has helped to discover a wide range of structural variations; however, the limited set of nickases that can be used for mapping assays restrict the applicability of mapping as an efficient and accurate large SV diagnostic tool (Levy-Sakin et al. 2019). After the DNA molecules are nicked and fluorescently labeled at specific sequence-motifs, they are detected optically as linearized molecules. The single-molecule technique mainly relies on DNA linearization either via DNA combing, DNA stretching via flow force, or electrophoretic stretching in nanoconfinements. Bionano Genomics' optical mapping technology is based on nanochannel confinement and is considered a high-throughput Next Generation genome Mapping (NGM) platform. In Bionano's nanochannel chip, the fluorescently labeled DNA molecules are introduced into the nanochannel array by gradually confining them from microscale to nanoscale dimensions using electrophoretic force. Once inside the nanochannels, the DNA molecules remain linearized and present themselves for imaging on a fluorescent microscope. Several automated iterations of DNA loading and imaging over hundreds of nanochannels are performed making it a high-throughput platform. Output from scanning the nanochannel array is in the form of a collection of raw image files. From there, they can be analyzed, aligned or *de novo* assembled into consensus contigs using Bionano's Refaligner and Assembler tools (Mak et al. 2016).

With CRISPR-Cas9, since any 20-base sequence can be targeted on the genome, Cas9-based genome editing strategies can be repurposed to help overcome the limitations of traditional sequence-motif mapping methods. Since its introduction, Cas9-based optical mapping has been used in several applications including single-molecule telomere characterization (McCaffrey et al. 2017, Abid et al. 2020a), subtelomere characterization (Young et al. 2017, 2020), identification of antibiotic resistance-encoding plasmids in bacteria (Müller et al. 2016b), long-range haplotyping and single nucleotide polymorphism (SNP) analysis (Abid et al. 2020b) etc. These applications are discussed in detail in the following sections.

## CRISPR-Cas9 in telomeric and subtelomeric characterization

*Telomere length measurement and associated biology*

Telomeres are nucleoprotein units present at the tips of eukaryotic chromosomes. Functional telomeres contribute to genomic stability by preventing aberrant recombination or degradation. They protect the linear DNA ends of chromosomes from being recognized as double strand breaks by the cellular damage and repair pathways (Blackburn et al. 2015, Webb et al. 2013). Human telomeres contain (TTAGGG)n tandem repeats with lengths ranging from 0.5–15 kbp (Aubert et al. 2012). These lengths vary among cells, cell types and chromosomes arms within a single cell. Telomeres in somatic cells lose repeats with every cell division. When their lengths decrease beyond certain thresholds (dependent on cell type), telomere dysfunction can occur and replicative senescence and/or cell apoptosis are triggered (Meier et al. 2007, Samassekou et al. 2010, Aubert et al. 2012, Kaul et al. 2012). In germ cells, critically short telomeres can cause unregulated telomerase (a protein which adds repeats at chromosome ends) activity and compromise tumor suppressor pathways thereby favoring tumor formations (Sabatier et al. 2005, Coppé et al. 2010, Davalos et al. 2010, Jaskelioff et al. 2011). There are strong correlations between shorter average telomere lengths and aging or aging-related diseases. Telomere length shortening has been associated with chronic diseases like cancers (Blackburn et al. 2015, Wentzensen et al. 2011), type 2 diabetes (Zhao et al. 2013), Alzheimer's (Zhan et al. 2015), and other conditions like pulmonary fibrosis, gastrointestinal disorders, liver cirrhosis, and neuropsychiatric symptoms (Armanios and Blackburn 2012). Studies indicate that critically short telomeres cause telomere dysfunction. Compared to average telomere length, identity and frequency of short telomeres are better indicators of chromosome stability and cell viability (Samassekou et al. 2010, Vera and Blasco 2012). In addition, there is evidence that the single-telomere lengths biases in cells are arm-specific and haplotype-dependent (McCaffrey et al. 2017). Understandably, there is great interest in using telomere length as a clinical biomarker for estimating disease risk, especially for cardiovascular diseases, cancers, viral respiratory infections, and autoimmune disorders (Cohen et al. 2013, Fasching 2018).

There is also evidence that subtelomeres (the distal 500kbp of each chromosome) are important for maintaining telomere length and stability (Britt-Compton et al. 2006). For a long time, these regions were poorly understood and inadequately characterized due to high segmental duplications, extensive structural variations, and lack of good reference sequences. Consequently, an assay that can identify the lengths, locations, and frequencies of critically short telomeres as well as provide the linked subtelomeric information could be a useful tool in predicting cancer or disease risk-factors and understanding the pathology of aging-related diseases. Today, users have several telomere length measurement methods available to choose from depending on the objective. Each method has advantages and limitations, and features differing from each other. For instance, Terminal Restriction Fragments (TRF) and Quantitative Polymerase Chain Reaction (qPCR) are two widely used techniques to characterize telomeres. TRF varies with the set of restriction enzymes used and needs 1.5 μg of input DNA; it has poor resolution in detecting telomere lengths <2kbp. A commercial kit, TeloTAGGG assay from Roche takes 15-18 hr to perform. Inclusion of subtelomeric fragments in analysis can overestimate average telomere lengths by several kbp (Samassekou et al. 2010, Montpetit et al. 2014). Comparatively, qPCR only needs 25-50 ng input DNA to perform the assay, but it estimates the telomere length as a ratio of total telomere repeats in a sample relative to a single-copy gene (T/S ratio). T/S is a relative measure and does not quantify telomere length in kbp, calibration against a gold standard method (TRF) becomes necessary to interpret T/S in kbp (Cawthon 2002, 2009). However, both TRF and qPCR methods are bulk telomere length measuring methods and estimate average telomere length. To overcome the insufficiency of average telomere length in detailed investigations, single-molecule telomere length assays including, but not limited to, Single Telomere Length Analysis (STELA) (Baird et al. 2003), Universal STELA (Bendix et al. 2010), Telomere Shortest Length Assay (TeSLA) (Lai et al. 2017), Single Telomere Absolute-length Rapid (STAR) (Luo et al. 2020), Quantitative Fluorescence In Situ Hybridization (Q-FISH) (Lansdorp et al. 1996), Flow-FISH (Baerlocher et al. 2006), Cas9-mediated high-throughput single-molecule telomere characterization method using nanochannels (McCaffrey et al. 2017) and Telomere length Combing Assay (TCA) (Kahl et al. 2020) have been developed in the last two decades. Several excellent and exhaustive reviews about telomere length measurements have been published elsewhere and we recommend them to interested readers (Aubert et al. 2012, Montpetit et al. 2014, Uppuluri et al. 2020). One common limitation of most bulk and single telomere length measurement methods is acquiring haplotype-resolved and subtelomere-linked data. This need is addressed by the Cas9-mediated high-throughput telomere characterization method (McCaffrey et al. 2017).

## High-throughput telomere characterization method

Single molecule level telomere characterization can facilitate a better understanding of telomere regulation and dynamics of telomeres with complex and variable subtelomeric regions. For this purpose, a high-throughput telomere characterization assay leveraging Bionano Genomics' nanochannel array and Cas9-mediated labeling strategies was developed to enable visualization and analysis of telomeric repeats. In addition to single-molecule telomere length, subtelomere-linkage and haplotype-resolution can be obtained through this assay (McCaffrey et al. 2017).

To analyze telomeres in a sample, the consensus contig maps of the sample with just Nt.BspQI (a nicking endonuclease) was first defined without any telomere labels (Lam et al. 2012). Next, a two-color or three-color labeling was performed to specifically target telomere repeat sites. Any extra labels found on single molecules were considered telomeres and analyzed further. Figure 2 (A-D) shows a schematic overview of two-color and three-color co-labeling schemes for the single-molecule telomere length assay. Briefly, 300 ng genomic DNA was globally nicked with Nt.BspQI nicking endonuclease and labeled with green fluorophores. A second set of specific labels, in green or red colors, were introduced through labeling of Cas9-mediated nicks. In two-color labeling, as shown in Figure 2A, Nt.BspQI nicking endonuclease along with Cas9 D10A, a mutant variant of Cas9 nuclease with inactivated RuvC domain, are used in a single nick-labeling reaction. Cas9n, directed by a synthetic guide RNA containing UUAGGGUUAGGGUUAGGGUU in its spacer sequence, introduces nicks at telomeric repeats (TTAGGG)n. Simultaneously, Nt.BspQI introduces nicks at GCTCTTCN. In the subsequent labeling reaction, both types of nicks are labeled with green-fluorescent nucleotides using Taq DNA polymerase. In three-color labeling, as shown in Figure 2B, two nick labeling reactions are performed. In the first reaction, Nt.BspQI nicks at GCTCTTCN and Taq DNA

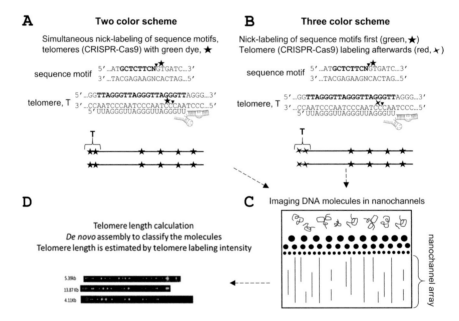

**Figure 2**. (A) Two-color labeling scheme: simultaneous nick labeling of Nt.BspQI sites and Cas9n-gRNA-directed telomeric repeats (marked with 'T') with green dye (5-point star symbol). (B) Three-color labeling scheme: nick-labeling of Nt.BspQI sites with green dye (5-point star symbol), followed by Cas9n-gRNA-directed telomeric repeat labeling with red dye (4-point star symbol). (C) Labeled DNA is loaded into Bionano's nanochannel array chip where molecules uncoil and linearize as they enter the nanochannel array; DNA is imaged at 60X using green, red, and blue lasers. (D) Raw images of molecules (black background) containing bright telomere labels (white dots); label intensities are measured to estimate telomere lengths. (Images A-D adapted with permission from the original (McCaffrey et al. 2017), are licensed under CC BY-NC 4.0 (https://creativecommons.org/licenses/by-nc/4.0/)).

polymerase incorporates green-fluorescent nucleotides. The second reaction includes Cas9n-sgRNA nicking at telomeric repeats and Taq DNA polymerase incorporation of red-fluorescent nucleotides.

After nick-labeling reactions, DNA sample is treated with protease and is stained with YOYO-1 before running on nanochannels where they are imaged using a high-speed scanning microscope. Figure 2D shows example images (black background) of labeled molecules (white dots are labels) following the three-color scheme. Panels with color microscopy photographs are shown in the color version of Figure 3 included in this publication. Red telomeres are observed at the ends of the molecules and green labels represent global Nt.BspQI nick sites. While the three-color scheme enabled easy visualization of telomere repeats from Nt.BspQI nick sites, DNA undergoes an additional modification step here and takes longer duration. As a result, the number of longer DNA molecules are fewer when compared to the two-color scheme. In identifying subtelomeric regions accurately and efficiently, both the two-color schemes and three-color schemes were found to be effective. After scanning,

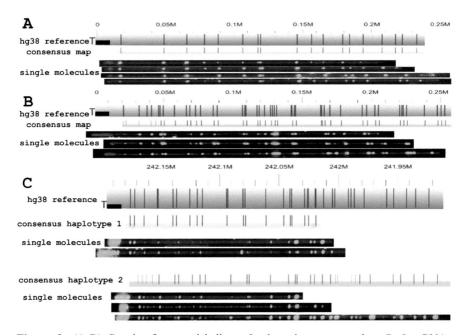

**Figure 3.** (A-D) Results from co-labeling of telomeric repeats using Cas9n-gRNA-directed (red) and of subtelomeres using global nickase, Nt.BspQI (green); hg38 reference is represented as the blue bar with dark ticks marking Nt.BspQI sites (numbers above the hg38 reference indicate genome length in million base-pair (Mbp)); consensus map is seen as yellow horizontal line right below the reference with marks representing Cas9n (green) and Nt.BspQI (blue) sites. (A) shows 12p arm and (B) shows 3p arm. (C) Two different haplotypes of 2q telomeres from IMR90-PD17 cells visualized using two-color labeling scheme. Bright labels seen at the ends of each single molecule are telomeres. (Images A-D adapted with permission from the original (McCaffrey et al. 2017), are licensed under CC BY-NC 4.0 (https://creativecommons.org/licenses/by-nc/4.0/)).

molecules are assembled based on the Nt.BspQI patterns and mapped to the human reference overall, including subtelomeric regions. Individual DNA molecules containing the telomeric and subtelomeric regions, as well as chromosomes are identified and further analyzed. The telomere length is estimated based on the total Cas9n labeling intensity since the Cas9n nicks more frequently and therefore the total number of incorporated fluorophores are higher. This is a reasonable characteristic especially for short telomere length measurement as even a single point emitter turns on several pixels because of photo scattering thereby improving resolution.

Using this method, extensive characterization of telomeres was performed for aging cells (IMR90), telomerase positive cells (UMUC3, LNCaP), and ALT-positive cells (U2OS, SK-MEL-2, and Saos-2). Some of these results are briefly described as follows. Telomeres in a cellular aging model cell line (IMR 90, human fetal lung fibroblasts) were characterized at three different passages (based on population doublings): early or PD17, late or PD45 and senescence or PD61. The average telomere length in early passage was measured to be 7.2kbp and reduced to 4.5kbp in late and 4.0kbp in senescent cells, respectively. Also, with passage number increase, as the cells drew towards senescence, the differences between longest and shortest telomere lengths also decreased, i.e. 13.2 kb vs 3.6 kb at PD17, 9.3 kb vs 2.4 kb at PD45, and 5.5 kb vs 2.4 kb at PD61. However, this decreasing trend was not uniform across all chromosome arms; 50% decrease from PD17 to PD45 was observed in 1p, 3p, 4p, 5p, 6p-1, 6p-2, 8p, 11p, 12p, 1q, 5q, 12q, 14q, 15q, 21q chromosome arms, while 9p, 20p, 2q-1, 17q did not change, and the decrease in lengths was not significant for 2p, 8q, 9q and 11q. Telomeres with <500bp length were observed in late passage and a five-fold increase was observed in frequencies of very short telomeres less than 100bp in length from PD45 to PD61. Despite this increase, the average telomere lengths of PD45 and PD61 are not statistically different. Interestingly, approximately 8q and 14q arms had 30% and 12.5% of all the very short telomeres detected, suggesting subtelomere-specific vulnerabilities. This trend of biased very short telomere distribution was also observed in specific chromosome arms of telomerase-positive cancer cell lines (8q and 9p of UMUC3; 8q of LNCaP).

Cells lacking telomerase rely on DNA repair and recombination pathways, collectively called Alternative Lengthening of Telomeres (ALT) pathways, to maintain telomere lengths. These pathways are exhibited in approximately 15% of osteosarcoma and glioblastoma cancers and show correlation to poor prognosis (Lazzerini-Denchi and Sfeir 2016, Blasco 2007, Cesare and Reddel 2010, Xu and Blackburn 2007). Compared to aging cells, authors found the terminal telomere lengths of the ALT positive cells U2OS, SK-MEL-2, and Saos2 cell to be highly heterogeneous (Abid et al. 2020a). They also found that ALT cells had more frequent recombinant fragments compared to aging cells; U2OS, Saos-2 and SK-MEL-2 had an average of 24%, 19% and 11% of recombinant telomeres. Their distribution was biased and specific to only a few subtelomeres (1q, 3q, 7p, 8q, 11p 18q,19q and 21q of U2OS; 3q and 20q of SK-MEL-2; and 1p, 3q, 8q,17q and 21q of Saos-2, each had over 50% recombinant telomeres). With this assay, authors also detected unusually high signal- free ends, 8% in U2OS cells and SK-MEL-2, and 5% in Saos-2 cell lines; more than half of them were associated with only a few chromosomes. Very large linear extrachromosomal telomere repeat (ECTR) DNA molecules identified with characteristic punctuated labels on telomeric were also detected in all three ALT cell lines. This was interesting because although ECTR were thought to play a role in ALT maintenance mechanisms (Sobinoff and Pickett 2017, Fasching 2018), their potential involvement in generating new telomeres by acting as templates for break-induced repair was a new insight (Abid et al. 2020a).

The single-molecule telomere length assay can identify and measure individual telomeres at very high resolution making it possible to distinguish features such as end telomeres, novel subtelomeric variants, internal telomere like sequences, extrachromosomal telomere repeats, and previously uncharacterized recombined subtelomeres, all in a single assay. The unique barcodes in the physically linked subtelomeric regions per every chromosome arm enabled the accurate identification of 30-35 telomeres simultaneously (McCaffrey et al. 2017). Because of direct imaging of the single telomere molecules, one can positively detect telomere loss using this method, unlike with extant methods for telomere analysis. Resolution of telomere haplotypes is also possible using this assay, making it a highly unique feature among all telomere analysis techniques. Biologically significant ultra-short telomeres ~0.1 kbp were also detected demonstrating the high resolution and accuracy of the method. Unlike Q-FISH, a common telomere visualization technique, the single molecule assay is not limited to metaphase cells and is applicable to various tissue samples too. Due to its high throughput and resolution, and robust technical processes, the single-molecule telomere characterization assay, that is uniquely enabled by Cas9, will potentially find considerable utility in future telomere studies (Uppuluri et al. 2020).

## CRISPR-Cas9 labeling to interrogate individual base, and tag specific genomic region of interest in optical mapping

The main strategy for long-range optical mapping is based on measuring the distances between the short sequence motifs recognized by nicking endonucleases (6-8 bp) on single long DNA molecules. The key information is the pattern of distances between motifs. Current labeling strategies can only detect single-base differences at polymorphisms that happen to coincide with these sequence motifs, which has limited the application of optical mapping technology in SNP detection and haplotype construction in complex genomes.

CRISPR-Cas9-based labeling was exploited to devise a method that can distinguish single-nucleotide variants within the NGG PAM sites, termed 'PAM SNPs'(Abid et al. 2020b). If one of the G bases of a PAM in one genome is a different allele in another genome, it is not expected to be labeled, even if they share the 20-base recognition sequence. Thus, it was predicted that strong differential labeling at gRNA-directed PAM variants could reliably differentiate the single base difference between two genomes over long distances. As a proof of concept, authors sought to differentiate the two *H. influenzae* strains, RR722 and RR3131 by targeting a highly conserved 100kbp genomic stretch with 99% sequence similarity (819-916 kb of RR722 and 884-981 kb of RR3131). With Nt.BspQI, there would be just one extra nick site in RR3131. So, authors designed gRNAs targeting three distinct 20-base recognition/spacer sequences (or '20-mers'), but for each one of the two *H. influenzae* strains lacked an NGG PAM signal due to single nucleotide variation. At the first *locus*, both strains share same recognition sequence but differ by a single base in the PAM sequence (TGG in RR3131, TGA in RR722). Similarly, two other *loci* were differentiable by lack of a PAM sequence (in RR3131) and a single base mismatch in PAM sequences (AGG in RR3131, ACG in RR722), respectively. Labeling by both Nt.BspQI and CRISPR-Cas9 were performed in a single-tube reaction, and the optical mapping data showed >90% labeling efficiency supporting the potential application of PAM SNPs as a strategy for long-range haplotyping in mixed microbial samples and human genomes.

## Multiplexed single guide RNA (sgRNA) optical mapping

In the same report describing the detection of PAM SNPs, the authors also described a Cas9-based multiplex labeling of various *loci* across the whole genome or targeting a specific genomic region, in a single-step reaction (Abid et al. 2020b). By selecting guide RNA for regions of interest, custom maps can be produced and used to differentiate genomes based on map patterns and/or detect structural variations. This is particularly useful in designing different optical map patterns to differentiate similar genomes or conserved sequences between strains or haplotypes.

Demonstrating a proof of concept, authors sought to differentiate the *H. influenzae* strain RR722 from RR3131 based on 162 *loci*. They designed an informatics pipeline to select 162 sgRNA while minimizing off-targets. RR722 genome was selected as a model system to verify the validity of the pipeline. Step-wise, all possible 20-mer sequences that are followed by an NGG sequence on the 3′ end, having a single perfect match within the reference genome were collected. To reduce potential off-targets, sequences containing PAM-adjacent 8-mers with multiple hits to reference were eliminated and only those with single perfect matches were retained. Next, this list was refined by discarding all single-base mismatched 8-mers with more than 5 hits to the reference. Finally, the list was further refined by retaining only the sgRNA sequences with less than 5 hits of a single-base mismatch in the PAM-distal 12-mers.

The 162 sgRNA sequences obtained in the above-described manner were used for a single-tube reaction of sgRNA synthesis and subsequent nick labeling reaction. The authors reported that after assembly with 100X overall coverage, no off-target labels with >30% frequency were observed. Although 90 of 162 sgRNAs had perfect hits in RR3131 strain, only a tiny fraction of short, labeled RR722 molecules aligned to it. Moreover, this alignment was within the highly conserved regions with high sequence similarities (819–916 kb of RR722 vs. 884–981 kb of RR3131 and 1177–1220 kb of RR722 vs. 1211–1254 kb of RR3131) implying that the custom-designed multiplexed sgRNA strategy can be used for bacterial strain identification.

One of the major limitations for scaling up the use of Cas9 for whole genome applications is understandably the requirement for large numbers of unique sgRNA sequences. As noted above, most studies incorporate region-specific targeting typically consisting of a handful number of sgRNA sequences. In studies where a large number of guide RNAs have been used, custom synthesized crRNA and tracrRNAs have been acquired from a manufacturer and pooled together. The sequence of the sgRNAs is selected after performing application-specific *in silico* design work. The sgRNA sequences are further shortlisted based on screening assays performed to verify sequence specificity, or in other words, minimize off-targets. The cost associated with the customized guide RNA synthesis can become a limitation to the targeting multiplicity, since each sgRNA sequence needs to be synthesized separately and pooled together. However, the multiplexed Cas9 labeling method described here (Abid et al. 2020b) utilized T7 RNA transcription for synthesizing the multiplicity of custom guide RNAs in a one-pot reaction, resulting in nearly an order-of-magnitude reduction in the sgRNA synthesis cost.

Although DNA sequencing is more informative than optical maps, and ultralong sequencing read lengths are now possible with long read sequencing technologies like Oxford Nanopore, optical mapping technology continues to be more economic in providing long-range haplotype information. Megabase-long average molecule lengths in mapping as in the case of alternative mapping platforms (Kaykov et al. 2016, Varapula et al. 2019, Kahl et al. 2020) seem promising in the detection of long-range SVs and identifying mis-assemblies.

## Other mapping applications with CRISPR-Cas9

Plasmids are thought to be important in rapid spread of antibiotic resistance (Brolund and Sandegren 2016). Recently, the CRISPR-Cas9 system was used to rapidly identify the antibiotic resistance genes in plasmid DNA using nanochannel array devices and fluorescence imaging (Müller et al. 2016b). In this method, CRISPR-Cas9 targeting is combined with a more global optical mapping approach using YOYO-1/Netropsin staining, to identify both the circular and linear forms of the plasmid. Briefly, wild-type CRISPR-Cas9 was directed to a resistance specific gene and performs cuts into plasmids. The DNA is then stained with YOYO-1 and netropsin before loading into nanochannels. The presence of gene is confirmed by occurrence of unusually high double-strand breaks at a single location. Due to the ability to perform high-quality imaging in single molecule techniques, the process of identification was automated and the various linear, as well as circular plasmids containing the specific antibiotic gene, were identified (Müller et al. 2016b). This automated approach with significantly simplified sample preparation enables tracking of antibiotic resistance within a hospital, something that can have significant outcomes in preventing hospital-acquired infections. The study has targeted a single set of antibiotic resistance genes, but in case multiple gene sets need to be identified this method becomes unsuitable.

# CRISPR-Cas9 in DNA sequencing of targeted regions

Next-generation sequencing (NGS) has enabled large-scale sequencing of human genomes at about $1000 per genome (Schwarze et al. 2020). However, whole genome sequencing (WGS) is still expensive, especially for a large project involving many samples. Moreover, NGS produces massive amounts of data with high complexity, which requires enormous computational power to analyze the WGS data (Ramos et al. 2012). Therefore, the targeted re-sequencing of specific genomic regions that are relevant to a particular disease has many advantages. Not only can it reduce the cost, but also simplify downstream bioinformatics analysis. Many technologies have been developed to enrich specific target genomic regions (Gnirke et al. 2009, Albert et al. 2007, Turner et al. 2009, Mondal et al. 2011). The most successful example is the whole-exome sequencing (WES), which relies on the large-scale hybridization capture in a microarray (Choi et al. 2009). However, most of the technologies capture single-stranded DNA (ssDNA) or short double-stranded DNA (dsDNA). Because CRISPR-Cas9 can target any 20 bp across the genome, it is being increasingly employed in target enrichment and directly interface NGS.

One example that adapted CRISPR-Cas9 to perform solution-based hybridization capture for NGS is called CRISPR-Cap (Lee et al. 2019). Here, a library of Cas9-gRNA complexes is assembled with biotin tags covalently attached to the sgRNA molecules. Each of these complexes would then act as a bait to isolate the fragments sequence-specifically. This is possible because of the property of CRISPR-Cas9 to remain tightly bound to its target sequence even after cleavage. The sgRNA sequences were designed to target the region of interest, and using streptavidin beads these digested sequences were isolated for library construction for NGS. In this case, the target region is well-characterized and smaller in size (few kb). Such implementations are quick and efficient for applications where short-read sequencing is sufficient but are not generally suitable for studying large complex regions where long-read sequencing has a clear advantage. More importantly, several CRISPR-Cas9 based DNA enrichment methods have been developed to capture hundreds of kilobases, up to megabase-long dsDNA. This is critical in recovering large structural variations and reassembling the haplotype-resolved complex genomic region (Young et al. 2017, Levy-Sakin et al. 2019, Huddleston et al. 2014).

The CRISPR-Cas9 system has been employed to improve Transformation-Associated Recombination (TAR) cloning protocol, which can selectively clone large DNA fragments that are ~250 kbp in size (Kouprina and Larionov 2008). The original TAR technique was exploited for long-range haplotyping applications where PCR or NGS-based mutational analysis would not be feasible due to the high level of homology between sequences of gene members and because they reside within large segmental duplications with >95% identity. This technique obviates the need to construct a genomic library of random clones when interested only in specific regions of the genome, most favorably, the complex regions. The original TAR method had a 2% efficiency, equivalent to screening 148 colonies to pick a single gene-positive clone with 95% confidence. The homologous recombination reaction in TAR cloning depends heavily on the location of the targeted sequences relative to the 5' and 3' ends of the fragment greatly affecting the efficiency. Utilizing CRISPR, precise cuts closer to the targeted sequences (complementary to the TAR vector "hooks") on the genome are made possible, due to which the recombination reaction efficiency is improved. The overall efficiency of the CRISPR-Cas9-mediated TAR process was reported to be around 32%, equivalent to screening only 8 colonies to identify a single gene-positive clone with 95% confidence (Lee et al. 2015). CRISPR was also applied in conjunction with Gibson assembly to isolate long (100 kbp) bacterial chromosomal fragments and assemble them into a BAC for cloning purposes (Jiang et al. 2015).

Recently, CRISPR-Cas has been repurposed to function as a viable targeting system to isolate or enrich sequences of much higher scale (100s of kb and few

Mb) in the human genome prior to DNA sequencing. Here, CRISPR-Cas9 is used to excise the target genome region from a large pool of cells without the standard cloning steps plus BAC library formation. Using sgRNA molecules to target the flanking sequences of the target region, fragments that are megabase-sized could be digested and isolated using PFGE. A sufficient quantity of these fragments will then be used as input DNA for the construction of the DNA sequencing library, thus greatly enriching the target sequence content. In the earliest demonstration of this scheme, greater than 100-fold enrichment was achieved targeting gene cluster regions in the mouse genome that are 283 kbp and 2.3 Mbp long (Bennett-Baker and Mueller 2017). This idea was subsequently applied to targeting the BRCA1 breast cancer marker in the human genome (Gabrieli et al. 2018). The BRCA1 gene region including the coding, regulatory and intronic sequence is about 80 kbp in length and contains over 50% repetitive elements that contribute to genetic instability. The authors achieved a 237-fold enrichment of the 200 kb long target region that contained the 80 kb BRCA1 gene body. In this report, the authors combined CRISPR-Cas9-digestion of the target region with long-read DNA sequencing (Oxford Nanopore), which has been shown to be more suitable to analyze the large variants. The CRISPR-Cas9 enrichment plays a complementary role to long-read sequencing by providing greater coverage and hence, improved accuracy (Gabrieli et al. 2018). The purification of the excised target fragments could play a critical role in the achieved enrichment. In one study, a microfluidic device (called 'µLAS') was used to size-specifically select a 30 kbp target in the plant genome after the CRISPR-Cas9 digestion and PFGE processes. This resulted in a further 10-fold improvement in the enrichment factor (Milon et al. 2019). Rapid high-throughput single molecule resolution DNA manipulation methods could potentially make the Cas9 target isolation process more efficient. This scheme of *in vitro* Cas9 digestion of genomic DNA demands a very large amount of starting DNA material, typically 50-100 µg, to yield the sufficient number of target DNA fragments (3-7 ng) for preparing sequencing libraries. Each PFGE lane must contain a large amount of genomic DNA (6-10 µg) load to facilitate discernment of the band(s) corresponding to the target region(s). Several of these excised gel plugs will then need to be combined until the minimum required input DNA amount is achieved. Moreover, the PFGE process has long process times and low throughput. The CRISPR-Cas9 digestion and isolation-based approach can hence be tedious and further advancements, potentially using microfluidic sample processing, are expected.

Another interesting approach towards analyzing large regions of interest replete with relatively low identity segmental duplications has been to merge Cas9 with long-read sequencing library workflows such that isolation of target fragments prior to the pre-amplification step of library construction becomes unnecessary. Two recent papers described this new methodology (Watson et al. 2020, Gilpatrick et al. 2020). Genomic DNA fragments are first dephosphorylated and then cut using Cas9-sgRNA complexes targeted at the flanking sequences of the target region to perform double-strand breaks. After this, the 3' ends of the target fragments are dA-tailed and subsequently, library adaptors (containing sequencing primers) are ligated on both ends of the target fragments. This relatively direct approach was applied to study allelic variations of the VHL and DMD gene markers containing duplications

(Watson et al. 2020). Duplication-rich variants that span long regions are difficult to characterize accurately with short-read sequencing platforms. PCR-based method would be infeasible too for studying regions exceeding 30 kbp in length and those that are rich in GC-content. Using this CRISPR-Cas9-mediated workflow, nucleotide-level resolution was achieved in both the problematic instances. Long-range variants were detected and breakpoints pin-pointed by first sequencing on Oxford Nanopore platform and then performing Sanger sequencing (Watson et al. 2020). The second research group followed a similar process to demonstrate the feasibility of the method for clinical testing and research on the 80 kbp BRCA1 gene body (Gilpatrick et al. 2020). The same method has also been applied to the plant genome, which is generally a more challenging problem in terms of repetitive elements, to assess SNVs successfully (López-Girona et al. 2020). Starting DNA amount of a few μg were shown to be adequate for these assays. This is in contrast to the steep input DNA amount required for the earlier CRISPR-Cas9 approach for isolating long targets via PFGE before library construction. This more recent approach is also relatively rapid and efficient compared to the PFGE scheme, making it more feasible for clinical research and testing methodologies.

# References

Abid, H. Z., McCaffrey, J., Raseley, K., Young, E., Lassahn, K., Varapula, D., et al. 2020a. Single-molecule analysis of subtelomeres and telomeres in Alternative Lengthening of Telomeres (ALT) cells. BMC Genomics, 21(1).

Abid, H. Z., Young, E., McCaffrey, J., Raseley, K., Varapula, D., Wang, H.-Y., et al. 2020b. Customized optical mapping by CRISPR–Cas9 mediated DNA labeling with multiple sgRNAs. Nucleic Acids Research, 49(2): e8.

Albert, T. J., Molla, M. N., Muzny, D. M., Nazareth, L., Wheeler, D., Song, X., et al. 2007. Direct selection of human genomic loci by microarray hybridization. Nature Methods, 4(11): 903-905.

Armanios, M. and Blackburn, E. H., 2012. The telomere syndromes. Nature Reviews Genetics, 13(10): 693-704.

Aubert, G., Hills, M. and Lansdorp, P. M. 2012. Telomere length measurement—Caveats and a critical assessment of the available technologies and tools. Mutation Research/ Fundamental and Molecular Mechanisms of Mutagenesis, 730(1-2): 59-67.

Baerlocher, G. M., Vulto, I., De Jong, G. and Lansdorp, P. M. 2006. Flow cytometry and FISH to measure the average length of telomeres (flow FISH). Nature protocols, 1(5): 2365.

Baird, D. M., Rowson, J., Wynford-Thomas, D. and Kipling, D. 2003. Extensive allelic variation and ultrashort telomeres in senescent human cells. Nature Genetics, 33(2): 203-207.

Bendix, L., Horn, P. B., Jensen, U. B., Rubelj, I. and Kolvraa, S. 2010. The load of short telomeres, estimated by a new method, Universal STELA, correlates with number of senescent cells. Aging Cell, 9(3): 383-397.

Bennett-Baker, P. E. and Mueller, J. L. 2017. CRISPR-mediated isolation of specific megabase segments of genomic DNA. Nucleic Acids Research, 45(19): e165-e165.

Blackburn, E. H., Epel, E. S. and Lin, J. 2015. Human telomere biology: A contributory and interactive factor in aging, disease risks, and protection. Science, 350(6265): 1193-1198.

Blasco, M. A. 2007. Telomere length, stem cells and aging. Nature Chemical Biology, 3(10): 640-649.

Britt-Compton, B., Rowson, J., Locke, M., Mackenzie, I., Kipling, D. and Baird, D. M. 2006. Structural stability and chromosome-specific telomere length is governed by cis-acting determinants in humans. Human Molecular Genetics, 15(5): 725-733.

Brolund, A. and Sandegren, L. 2016. Characterization of ESBL disseminating plasmids. Infectious Diseases, 48(1): 18-25.

Cawthon, R. M. 2002. Telomere measurement by quantitative PCR. Nucleic Acids Research, 30(10): e47-e47.

Cawthon, R. M. 2009. Telomere length measurement by a novel monochrome multiplex quantitative PCR method. Nucleic Acids Research, 37(3): e21-e21.

Cesare, A. J. and Reddel, R. R. 2010. Alternative lengthening of telomeres: Models, mechanisms and implications. Nature Reviews Genetics, 11(5): 319-330.

Chen, B., Gilbert, L. A., Cimini, B. A., Schnitzbauer, J., Zhang, W., Li, G.-W., et al. 2013. Dynamic imaging of genomic loci in living human cells by an optimized CRISPR/Cas system. Cell, 155(7): 1479-1491.

Choi, M., Scholl, U. I., Ji, W., Liu, T., Tikhonova, I. R., Zumbo, P., et al. 2009. Genetic diagnosis by whole exome capture and massively parallel DNA sequencing. Proceedings of the National Academy of Sciences, 106(45): 19096-19101.

Cohen, S., Janicki-Deverts, D., Turner, R. B., Casselbrant, M. L., Li-Korotky, H.-S., Epel, E. S., et al. 2013. Association between telomere length and experimentally induced upper respiratory viral infection in healthy adults. Jama, 309(7): 699-705.

Coppé, J.-P., Desprez, P.-Y., Krtolica, A. and Campisi, J. 2010. The senescence-associated secretory phenotype: The dark side of tumor suppression. Annual Review of Pathology: Mechanisms of Disease, 5: 99-118.

Dang, Y., Jia, G., Choi, J., Ma, H., Anaya, E., Ye, C., et al. 2015. Optimizing sgRNA structure to improve CRISPR-Cas9 knockout efficiency. Genome Biology, 16(1): 280.

Davalos, A. R., Coppe, J.-P., Campisi, J. and Desprez, P.-Y. 2010. Senescent cells as a source of inflammatory factors for tumor progression. Cancer and Metastasis Reviews, 29(2): 273-283.

Davis, K. M., Pattanayak, V., Thompson, D. B., Zuris, J. A. and Liu, D. R. 2015. Small molecule–triggered Cas9 protein with improved genome-editing specificity. Nature Chemical Biology, 11(5): 316-318.

Deng, W., Shi, X., Tjian, R., Lionnet, T. and Singer, R. H. 2015. CASFISH: CRISPR/Cas9-mediated in situ labeling of genomic loci in fixed cells. Proceedings of the National Academy of Sciences, 112(38): 11870-11875.

Fasching, C. L. 2018. Telomere length measurement as a clinical biomarker of aging and disease. Critical Reviews in Clinical Laboratory Sciences, 55(7): 443-465.

Gabrieli, T., Sharim, H., Fridman, D., Arbib, N., Michaeli, Y. and Ebenstein, Y. 2018. Selective nanopore sequencing of human BRCA1 by Cas9-assisted targeting of chromosome segments (CATCH). Nucleic Acids Research, 46(14): e87-e87.

Gilbert, L. A., Larson, M. H., Morsut, L., Liu, Z., Brar, G. A., Torres, S. E., et al. 2013. CRISPR-mediated modular RNA-guided regulation of transcription in eukaryotes. Cell, 154(2): 442-451.

Gilpatrick, T., Lee, I., Graham, J. E., Raimondeau, E., Bowen, R., Heron, A., et al. 2020. Targeted nanopore sequencing with Cas9-guided adapter ligation. Nature Biotechnology, 38(4): 433-438.

Gnirke, A., Melnikov, A., Maguire, J., Rogov, P., LeProust, E. M., Brockman, W., et al. 2009. Solution hybrid selection with ultra-long oligonucleotides for massively parallel targeted sequencing. Nature Biotechnology, 27(2): 182-189.

Guilinger, J. P., Thompson, D. B. and Liu, D. R. 2014. Fusion of catalytically inactive Cas9 to FokI nuclease improves the specificity of genome modification. Nature Biotechnology 32(6): 577.

Hendel, A., Bak, R. O., Clark, J. T., Kennedy, A. B., Ryan, D. E., Roy, S., et al. 2015. Chemically modified guide RNAs enhance CRISPR-Cas genome editing in human primary cells. Nature Biotechnology, 33(9): nbt. 3290.

Hsu, P. D., Scott, D. A., Weinstein, J. A., Ran, F. A., Konermann, S., Agarwala, V., et al. 2013. DNA targeting specificity of RNA-guided Cas9 nucleases. Nature Biotechnology, 31(9): 827-832.

Hu, J. H., Miller, S. M., Geurts, M. H., Tang, W., Chen, L., Sun, N., et al. 2018. Evolved Cas9 variants with broad PAM compatibility and high DNA specificity. Nature, 556(7699): 57-63.

Huddleston, J., Ranade, S., Malig, M., Antonacci, F., Chaisson, M., Hon, L., et al. 2014. Reconstructing complex regions of genomes using long-read sequencing technology. Genome Research, 24(4): 688-696.

Jaskelioff, M., Muller, F. L., Paik, J.-H., Thomas, E., Jiang, S., Adams, A. C., et al. 2011. Telomerase reactivation reverses tissue degeneration in aged telomerase-deficient mice. Nature, 469(7328): 102-106.

Jiang, F. and Doudna, J. A. 2017. CRISPR-Cas9 structures and mechanisms. Annual Review of Biophysics, 46: 505-529.

Jiang, W., Bikard, D., Cox, D., Zhang, F. and Marraffini, L. A. 2013. RNA-guided editing of bacterial genomes using CRISPR-Cas systems. Nature Biotechnology, 31(3): 233-239.

Jiang, W., Zhao, X., Gabrieli, T., Lou, C., Ebenstein, Y. and Zhu, T. F. 2015. Cas9-Assisted Targeting of CHromosome segments CATCH enables one-step targeted cloning of large gene clusters. Nature Communications, 6(1): 8101.

Jinek, M., Chylinski, K., Fonfara, I., Hauer, M., Doudna, J. A. and Charpentier, E. 2012. A programmable dual-RNA–guided DNA endonuclease in adaptive bacterial immunity. Science, 337(6096): 816-821.

Kahl, V. F., Allen, J. A., Nelson, C. B., Sobinoff, A. P., Lee, M., Kilo, T., et al. 2020. Telomere length measurement by molecular combing. Frontiers in Cell and Developmental Biology, 8: 493.

Kaul, Z., Cesare, A. J., Huschtscha, L. I., Neumann, A. A. and Reddel, R. R. 2012. Five dysfunctional telomeres predict onset of senescence in human cells. EMBO Reports, 13(1): 52-59.

Kaykov, A., Taillefumier, T., Bensimon, A. and Nurse, P. 2016. Molecular combing of single DNA molecules on the 10 megabase scale. Scientific Reports, 6(1): 1-9.

Kim, E., Koo, T., Park, S. W., Kim, D., Kim, K., Cho, H.-Y., et al. 2017. In vivo genome editing with a small Cas9 orthologue derived from Campylobacter jejuni. Nature Communications, 8(1): 14500.

Kleinstiver, B. P., Pattanayak, V., Prew, M. S., Tsai, S. Q., Nguyen, N. T., Zheng, Z., et al. 2016. High-fidelity CRISPR–Cas9 nucleases with no detectable genome-wide off-target effects. Nature, 529(7587): 490-495.

Kleinstiver, B. P., Prew, M. S., Tsai, S. Q., Topkar, V. V., Nguyen, N. T., Zheng, Z., et al. 2015. Engineered CRISPR-Cas9 nucleases with altered PAM specificities. Nature, 523(7561): 481-485.

Kouprina, N. and Larionov, V. 2008. Selective isolation of genomic loci from complex genomes by transformation-associated recombination cloning in the yeast Saccharomyces cerevisiae. Nature Protocols, 3(3): 371.

Kuscu, C., Arslan, S., Singh, R., Thorpe, J. and Adli, M. 2014. Genome-wide analysis reveals characteristics of off-target sites bound by the Cas9 endonuclease. Nature Biotechnology, 32(7): 677-683.

Lai, T.-P., Zhang, N., Noh, J., Mender, I., Tedone, E., Huang, E., et al. 2017. A method for measuring the distribution of the shortest telomeres in cells and tissues. Nature Communications, 8(1): 1-14.

Lam, E. T., Hastie, A., Lin, C., Ehrlich, D., Das, S. K., Austin, M. D., et al. 2012. Genome mapping on nanochannel arrays for structural variation analysis and sequence assembly. Nature Biotechnology, 30(8): 771-776.

Lansdorp, P. M., Verwoerd, N. P., Van De Rijke, F. M., Dragowska, V., Little, M.-T., Dirks, R. W., et al. 1996. Heterogeneity in telomere length of human chromosomes. Human Molecular Genetics, 5(5): 685-691.

Lazzerini-Denchi, E. and Sfeir, A. 2016. Stop pulling my strings —What telomeres taught us about the DNA damage response. Nature Reviews Molecular Cell Biology, 17(6): 364-378.

Lee, C. M., Cradick, T. J. and Bao, G. 2016. The Neisseria meningitidis CRISPR-Cas9 system enables specific genome editing in mammalian cells. Molecular Therapy, 24(3): 645-654.

Lee, J., Lim, H., Jang, H., Hwang, B., Lee, J. H., Cho, J., et al. 2019. CRISPR-Cap: Multiplexed double-stranded DNA enrichment based on the CRISPR system. Nucleic Acids Research, 47(1): e1-e1.

Lee, N. C. O., Larionov, V. and Kouprina, N. 2015. Highly efficient CRISPR/Cas9-mediated TAR cloning of genes and chromosomal loci from complex genomes in yeast. Nucleic Acids Research, 43(8): e55-e55.

Levy-Sakin, M., Pastor, S., Mostovoy, Y., Li, L., Leung, A. K., McCaffrey, J. et al. 2019. Genome maps across 26 human populations reveal population-specific patterns of structural variation. Nature Communications, 10(1): 1-14.

López-Girona, E., Davy, M. W., Albert, N. W., Hilario, E., Smart, M. E., Kirk, C., et al. 2020. CRISPR-Cas9 enrichment and long read sequencing for fine mapping in plants. Plant Methods, 16(1): 1-13.

Luo, Y., Viswanathan, R., Hande, M. P., Loh, A. H. P. and Cheow, L.F. 2020. Massively parallel single-molecule telomere length measurement with digital real-time PCR. Science Advances, 6(34): eabb7944.

Ma, H., Naseri, A., Reyes-Gutierrez, P., Wolfe, S. A., Zhang, S. and Pederson, T. 2015. Multicolor CRISPR labeling of chromosomal loci in human cells. Proceedings of the National Academy of Sciences, 112(10): 3002-3007.

Mak, A. C., Lai, Y. Y., Lam, E. T., Kwok, T. P., Leung, A. K., Poon, A., et al. 2016. Genome-wide structural variation detection by genome mapping on nanochannel arrays. Genetics, 202(1): 351-362.

McCaffrey, J., Sibert, J., Zhang, B., Zhang, Y., Hu, W., Riethman, H., et al. 2015. CRISPR-CAS9 D10A nickase target-specific fluorescent labeling of double strand DNA for whole genome mapping and structural variation analysis. Nucleic Acids Research, 44(2): e11-e11.

McCaffrey, J., Young, E., Lassahn, K., Sibert, J., Pastor, S., Riethman, H., et al. 2017. High-throughput single-molecule telomere characterization. Genome Research, 27(11): 1904-1915.

Meier, A., Fiegler, H., Muñoz, P., Ellis, P., Rigler, D., Langford, C., et al. 2007. Spreading of mammalian DNA-damage response factors studied by ChIP-chip at damaged telomeres. The EMBO Journal, 26(11): 2707-2718.

Milon, N., Chantry-Darmon, C., Satge, C., Fustier, M.-A., Cauet, S., Moreau, S., et al. 2019. μLAS technology for DNA isolation coupled to Cas9-assisted targeting for sequencing and assembly of a 30 kb region in plant genome. Nucleic Acids Research, 47(15): 8050-8060.

Mondal, K., Shetty, A. C., Patel, V., Cutler, D. J. and Zwick, M. E. 2011. Targeted sequencing of the human X chromosome exome. Genomics, 98(4): 260-265.

Montpetit, A. J., Alhareeri, A. A., Montpetit, M., Starkweather, A. R., Elmore, L. W., Filler, K., et al. 2014. Telomere length: A review of methods for measurement. Nursing Research, 63(4): 289.

Müller, M., Lee, C. M., Gasiunas, G., Davis, T. H., Cradick, T. J., Siksnys, V., et al. 2016a. Streptococcus thermophilus CRISPR-Cas9 systems enable specific editing of the human genome. Molecular Therapy, 24(3): 636-644.

Müller, V., Rajer, F., Frykholm, K., Nyberg, L. K., Quaderi, S., Fritzsche, J., et al. 2016b. Direct identification of antibiotic resistance genes on single plasmid molecules using CRISPR/ Cas9 in combination with optical DNA mapping. Scientific Reports, 6(1): 37938.

Nishimasu, H., Shi, X., Ishiguro, S., Gao, L., Hirano, S., Okazaki, S., et al. 2018. Engineered CRISPR-Cas9 nuclease with expanded targeting space. Science, 361(6408): 1259-1262.

Rahdar, M., McMahon, M. A., Prakash, T. P., Swayze, E. E., Bennett, C. F. and Cleveland, D.W. 2015. Synthetic CRISPR RNA-Cas9–guided genome editing in human cells. Proceedings of the National Academy of Sciences, 112(51): E7110-E7117.

Ramos, R. T. J., Carneiro, A. R., Azevedo, V., Schneider, M. P., Barh, D. and Silva, A. 2012. Simplifier: A web tool to eliminate redundant NGS contigs. Bioinformation, 8(20): 996-999.

Ran, F. A., Cong, L., Yan, W. X., Scott, D. A., Gootenberg, J. S., Kriz, A. J., et al. 2015. In vivo genome editing using Staphylococcus aureus Cas9. Nature, 520(7546): 186-191.

Ran, F. A., Hsu, P. D., Lin, C.-Y., Gootenberg, J. S., Konermann, S., Trevino, A. E., et al. 2013. Double nicking by RNA-guided CRISPR Cas9 for enhanced genome editing specificity. Cell, 154(6): 1380-1389.

Rose, J. C., Popp, N. A., Richardson, C. D., Stephany, J. J., Mathieu, J., Wei, C. T., et al. 2020. Suppression of unwanted CRISPR-Cas9 editing by co-administration of catalytically inactivating truncated guide RNAs. Nature Communications, 11(1): 2697.

Ryan, D. E., Taussig, D., Steinfeld, I., Phadnis, S. M., Lunstad, B. D., Singh, M., et al. 2018. Improving CRISPR-Cas specificity with chemical modifications in single-guide RNAs. Nucleic Acids Research, 46(2): 792-803.

Sabatier, L., Ricoul, M., Pottier, G. and Murnane, J. P. 2005. The loss of a single telomere can result in instability of multiple chromosomes in a human tumor cell line. Molecular Cancer Research, 3(3): 139-150.

Samassekou, O., Gadji, M., Drouin, R. and Yan, J. 2010. Sizing the ends: Normal length of human telomeres. Annals of Anatomy-Anatomischer Anzeiger, 192(5): 284-291.

Schwarze, K., Buchanan, J., Fermont, J. M., Dreau, H., Tilley, M. W., Taylor, J. M., et al. 2020. The complete costs of genome sequencing: A microcosting study in cancer and rare diseases from a single center in the United Kingdom. Genetics in Medicine, 22(1): 85-94.

Slaymaker, I. M., Gao, L., Zetsche, B., Scott, D. A., Yan, W. X. and Zhang, F. 2016. Rationally engineered Cas9 nucleases with improved specificity. Science, 351(6268): 84-88.

Sobinoff, A. P. and Pickett, H. A. 2017. Alternative lengthening of telomeres: DNA repair pathways converge. Trends in Genetics, 33(12): 921-932.

Sternberg, S. H., Redding, S., Jinek, M., Greene, E. C. and Doudna, J. A. 2014. DNA interrogation by the CRISPR RNA-guided endonuclease Cas9. Nature, 507(7490): 62-67.

Tsai, S. Q., Wyvekens, N., Khayter, C., Foden, J. A., Thapar, V., Reyon, D., et al. 2014. Dimeric CRISPR RNA-guided FokI nucleases for highly specific genome editing. Nature Biotechnology, 32(6): 569.

Tsai, S. Q., Zheng, Z., Nguyen, N. T., Liebers, M., Topkar, V. V., Thapar, V., et al. 2015. GUIDE-seq enables genome-wide profiling of off-target cleavage by CRISPR-Cas nucleases. Nature Biotechnology, 33(2): 187-197.

Turner, E. H., Lee, C., Ng, S. B., Nickerson, D. A. and Shendure, J. 2009. Massively parallel exon capture and library-free resequencing across 16 genomes. Nature Methods, 6(5): 315-316.

Tycko, J., Myer, Vic E. and Hsu, Patrick D. 2016. Methods for optimizing CRISPR-Cas9 genome editing specificity. Molecular Cell, 63(3): 355-370.

Uppuluri, L., Varapula, D., Young, E., Riethman, H. and Xiao, M. 2020. Single-molecule telomere length characterization by optical mapping in nano-channel array: Perspective and review on telomere length measurement. Environmental Toxicology and Pharmacology, 82: 103562.

Varapula, D., LaBouff, E., Raseley, K., Uppuluri, L., Ehrlich, G. D., Noh, M. et al. 2019. A micropatterned substrate for on-surface enzymatic labelling of linearized long DNA molecules. Scientific Reports, 9(1): 1-11.

Vera, E. and Blasco, M. A. 2012. Beyond average: Potential for measurement of short telomeres. Aging (Albany NY), 4(6): 379.

Watson, C. M., Crinnion, L. A., Hewitt, S., Bates, J., Robinson, R., Carr, I. M., et al. 2020. Cas9-based enrichment and single-molecule sequencing for precise characterization of genomic duplications. Laboratory Investigation, 100(1): 135-146.

Webb, C. J., Wu, Y. and Zakian, V. A. 2013. DNA repair at telomeres: Keeping the ends intact. Cold Spring Harbor Perspectives in Biology, 5(6): a012666.

Wentzensen, I. M., Mirabello, L., Pfeiffer, R. M. and Savage, S. A. 2011. The association of telomere length and cancer: A meta-analysis. Cancer Epidemiology and Prevention Biomarkers, 20(6): 1238-1250.

Wolter, F. and Puchta, H. 2019. In planta gene targeting can be enhanced by the use of CRISPR/Cas12a. The Plant Journal, 100(5): 1083-1094.

Wu, X., Scott, D. A., Kriz, A. J., Chiu, A. C., Hsu, P. D., Dadon, D. B., et al. 2014. Genome-wide binding of the CRISPR endonuclease Cas9 in mammalian cells. Nature Biotechnology, 32(7): 670.

Xu, L. and Blackburn, E.H. 2007. Human cancer cells harbor T-stumps, a distinct class of extremely short telomeres. Molecular Cell, 28(2): 315-327.

Young, E., Abid, H. Z., Kwok, P.-Y., Riethman, H. and Xiao, M. 2020. Comprehensive analysis of human subtelomeres by whole genome mapping. PLoS Genetics, 16(1): e1008347.

Young, E., Pastor, S., Rajagopalan, R., McCaffrey, J., Sibert, J., Mak, A. C., et al. 2017. High-throughput single-molecule mapping links subtelomeric variants and long-range haplotypes with specific telomeres. Nucleic Acids Research, 45(9): e73-e73.

Zetsche, B., Jonathan, Omar, Ian, Kira, Essletzbichler, P., et al. 2015. Cpf1 is a single RNA-guided endonuclease of a Class 2 CRISPR-Cas system. Cell, 163(3): 759-771.

Zhan, Y., Song, C., Karlsson, R., Tillander, A., Reynolds, C. A., Pedersen, N. L., et al. 2015. Telomere length shortening and Alzheimer disease—A Mendelian randomization study. JAMA Neurology, 72(10): 1202-1203.

Zhao, J., Miao, K., Wang, H., Ding, H. and Wang, D. W. 2013. Association between telomere length and type 2 diabetes mellitus: A meta-analysis. PloS One, 8(11): e79993.

# Gene Editing and Genetic Disorders: Ethical and Legal Concerns

**Vicente Bellver**[1]* **and Federico de Montalvo**[2]

[1] Departamento de Filosofía del Derecho y Política; Facultad de Derecho, Universitat de València, Campus de los Naranjos; Avenida de los Naranjos s/n 46071-Valencia, Spain

[2] Facultad de Derecho, Universidad Pontificia Comillas; Calle de Alberto Aguilera, 23; 28015 - Madrid, Spain

## Introduction

The development of new, more versatile, gene editing techniques since 2012 opens the possibility of treating many genetic disorders in humans. These interventions can be performed in the somatic line or in the germ line**. There are basically two and very relevant differences between these two types of interventions. First, somatic gene editing is carried out on an existing human subject who, in principle, can give free and informed consent under the same conditions as for any other intervention. In contrast, heritable human genome editing (HHGE) is not performed in an individual who can be asked for consent. HHGE is part of a set of actions aimed at creating a new human being. Second, the genetic changes made by somatic gene editing will

---

** "Human genome editing has been the subject of extensive public discussion in many societies, but often important differences between human genome editing in somatic cells and germ cells have been conflated; also, there has been conflation between genome editing in early embryos and other germline cells in vitro for research (sometimes referred to as germline genome editing) and genome editing of germ cells or embryos for reproduction (sometimes referred to as heritable genome editing). Good governance must specifically consider the challenges inherent in both germline and somatic human genome editing, as well as uses for both research and reproduction". Expert Advisory Committee on Developing Global Standards for Governance and Oversight of Human Genome Editing – WHO, Human Genome Editing: A DRAFT Framework for Governance, 3 July 2020, https://www.who.int/docs/default-source/ethics/governance-framework-for-human-genome-editing-2ndonlineconsult.pdf?ua=1 In this work, we will consider exclusively genome editing of embryos or germ cells for reproduction and consequently, heritable human genome editing (HHGE).

---

* Corresponding author: vicente.bellver@uv.es

only have effects on that particular individual, whereas HHGE will affect the new human being and all his offspring.

Gene editing in humans can have three different purposes: therapeutic, preventive or enhancement. However, determining whether an intervention seeks to avoid a serious congenital disease, prevent a risk of disease or introduce an enhancement in the new individual is not always easy. Gene editing aimed at avoiding a monogenetic disorder such as Duchenne muscular dystrophy is undoubtedly therapeutic. But when gene editing tries to avoid a genetic disease conditioned by environmental factors or lifestyles, it is more difficult to discern whether we are dealing with a therapeutic or preventive action. It is even more difficult to differentiate between preventive and enhancement interventions.

In 2018, the scientist He Jiankui tried to immunize twin girls against the risk of HIV infection through gene editing. This technique could be considered both preventive and for enhancement. Just as the concept of genetic disorder is well outlined, the concept of health is very diffuse. Hence, at times, it is difficult to know whether gene editing seeks to prevent certain health problems, guarantee adequate health conditions, choose characteristics that are considered suitable for the future human being, or introduce some hitherto unprecedented genetic enhancements in humans. The law cannot ignore the existence of these diverse possibilities of interventions on the human genome. In fact, many countries in the world have already established more or less precise regulations on this matter. Also, at the international level, legal rules and recommendations have been approved by intergovernmental organizations such as the Council of Europe and UNESCO. The scientific community, for its part, has published important statements and reports with regulatory proposals addressed to governments, or with self-regulatory proposals for its own activity.

Regulations on human gene editing are usually the result of the interaction between three agents: scientists, bioethicists, and the general public. As well as the public has been more active in other bioethical debates such as those related to abortion, euthanasia, cloning, or human embryonic stem cells, it has hardly been involved in the debate about gene editing these past years. However, the participation of scientists and experts in bioethics dates back to the 1960s. Since then, both have offered very different and even contrary public policy proposals. Furthermore, in recent years, with the development of CRIPR-Cas9 technology, these proposals have multiplied.

This chapter is divided into two parts. The first presents the state of the art on ethical and legal aspects related to human gene editing. We briefly refer to the bioethical debate on this matter, which did not begin with the discovery of CRISPR-Cas9, but with that of the structure of DNA, and continued during the 1990s along with the Human Genome Project. Next, we refer not only to the national and international regulations on HHGE, but also to the attempts to reach a universal agreement among scientists. The reason is because, as stated, "international consensus on such standards is important to avoid the potential for researchers to rationalize the justification or seek out convenient locales for conducting dangerous and unethical experimentation" (Dzau et al. 2018). Although the establishment of international scientific standards cannot replace the adoption of national regulations, this can be a way to feed or even condition such regulations. Since HHGE has a universal and intergenerational impact, having as regulating norms only those approved by each nation does not make sense. So the answer to the question about who should decide on the regulations of this matter is all human beings.

The second part of the work proposes a reflection on the framework that should state any regulation on HHGE. Gene editing is a field in continuous progress, and some of its possible applications generate much hope, but also uncertainties and fears. Thus, it seems reasonable that the regulations on gene editing are based more on principles rather than on rules, which grant greater relevance to the precautionary principle, and may be subject to periodic review.

## The origins of the debate on the ethical aspects related to gene editing in humans

Although the technology for precise gene editing is relatively recent, the debate over its ethical and legal implications dates back to the 1960s, shortly after the discovery of the DNA double helix. Since the early stages of research into DNA, the scientists who were most committed to the social dimension of their work discerned a hitherto unheard-of horizon in the history of humankind: for the first time, human beings would be able to be the master not only of the natural setting in which they lived, but also of their own biological makeup. "We were at an epochal moment, not only for our society or for *Homo sapiens* but for all of life on earth. For the first time in the long course of evolution, for the first time in all time, a species was coming to understand its origins and its inheritance, and with that knowledge would come the ability to alter its inheritance, to determine its own genetic destiny, as well as that of other living species. Through DNA, biology was moving beyond analysis to synthesis" (Sinsheimer 1994).

Robert L. Sinsheimer, one of the "founding fathers" of recombinant DNA and synthetic biology, did not limit himself to stating a new challenge for humankind. In 1966, on the occasion of the 75th anniversary of the California Institute of Technology (CalTech), Sinsheimer declared: "Ours is an age of transition. After two billion years, this is the end of the beginning. It would seem clear, to some achingly clear, that the world, the society, and the man of the future will be far different from that we know. Man is becoming free, not only from the external tyrannies and the caprice of toil and famine and disease, but from the very internal constraints of our animal inheritance, our physical frailties, our emotional anachronisms, our intellectual limits. We must hope for the responsibility and the wisdom and the nobility of spirit to match this ultimate freedom" (Sinsheimer 1966). At this conference, he mentioned some of the most attractive possibilities for human beings offered by science in general and genetics in particular: choosing sex, prolonging life, enhancing intelligence, controlling emotions, altering genetic makeup and, in short, applying intelligence to evolution (Sinsheimer 1966).

Joshua Lederberg, one of the scientists who most intensively reflected and debated the impact of genetic technologies on the future of the human being, posed the crucial question and gave his own answer: "What is the responsibility of each generation for the biological and educational predetermination of its successors? In any event, the central responsibility of the geneticist, qua physician, is to the welfare of his individual patients" (Lederberg 1972). This answer presumes the superiority of individual interests over those of the community, and also that germline genetic interventions are performed on a patient. As shown hereinafter, both considerations are a subject of great controversy: first, because there are reasons to think that individual interests should not necessarily prevail when a decision affects all of humanity, and second, because it could be understood that HHGE is not carried out on an existing subject, but that the technique involves creating the subject.

In this debate about genetic technology and the future of humanity, not only scientists but also some of the most prestigious experts in bioethics participated maintaining radically opposed positions. All assumed that the incorporation of genetic technology into human reproduction made possible the "transition to a wholly new path of evolution" and marked the end of human life as we knew it. But answers to this possibility were completely opposed. Leon Kass affirmed: "It is possible that the nonhuman life that may take our place will be superior, but I think it is most unlikely, and certainly not demonstrable. In either case, we are ourselves human beings; therefore, it is proper for us to have a proprietary interest in our survival, and in our survival as human beings" (Kass 1972). Joseph Fletcher, on the other hand, wrote: "I cannot see how humanity or morality care served by genetic roulette sexually" (Fletcher 1971).

From the beginning, the debate about HHGE was considered in the most radical terms: the creation of human beings better than us as an imperative or as a prohibition. Today this debate remains on the same terms. On the one hand, improving the characteristics of our descendants is seen as an imperative (Savulescu 2001) and, on the other hand, as a betrayal to humanity (Sandel 2009). Little by little, a third option has emerged: authorizing genetic interventions aimed at avoiding serious genetic disorders and prohibiting the others. The following section summarizes the main ethical positions on HHGE and the arguments on which they are based.

## Heritable Human Genetic Editing (HHGE): Ethical reasons and public policy proposals

We can identify many pros and cons with regard to HHGE (van Dijke et al. 2018). All these reasons lead to three models of regulation.

1.  The definitive prohibition of all HHGE interventions. This proposal can be defended from both deontological and consequentialist perspectives. From the deontological perspective, some authors affirm that the guarantee of equality between generations requires that every human being be the result of genetic chance (Habermas 2003). From the consequentialist perspective, it is claimed that the deliberate modification of the human genome will bring with it a myriad of negative individual and social consequences, greater than the specific benefits that could be achieved through HHGE.

Sparrow (2019) stated, "We should assume that gene-editing technology will improve rapidly. However, rapid progress in the development and application of any technology comes at a price: obsolescence. If the genetic enhancements we can achieve get better and better each year, then the enhancements granted to children born in any given year will rapidly go out of date. Sooner or later, every genetically modified child will find him- or herself- to be a "yesterday's child". These negative impacts could be enormous with regard to our individual, social, and philosophical self-understanding" (Sparrow 2019). In trying to escape the Promethean shame that human beings feel in seeing themselves inferior to the technological creations they have created (Anders 2011), all they achieve is to feel permanently obsolete.

We should also accept that the use of HHGE will easily create a new division in humanity between those edited and probably enhanced, and those resulting from the genetic lottery (Silver 1997). If we understand that fragmented societies are undesirable, we will have to admit that promoting HHGE is not a good idea.

For the case of prospective parents at risk of transmitting a genetic condition who wish to avoid doing so and to have genetically related children, they can accomplish this with the existing embryo screening technique called preimplantation genetic diagnosis (PGD). While PGD also raises troubling ethical questions for some authors about what kind of lives we welcome into the world, modifying or introducing traits through genome editing would vastly intensify these concerns (Andorno et al. 2020). Others, on the contrary, affirm HHGE should be considered morally (at least) as acceptable as the selection of genomes on the basis of PGD, so there would be no problem in using one or the other technique depending on the needs of the prospective parents (Hammerstein et al. 2019).

2. The moratorium on HHGE while certain risks cannot be avoided. HHGE can have many negative effects that justify its temporary, but not necessarily permanent, ban. The main risks are the following:

   - There is a practically universal agreement to maintain the prohibition of HHGE until we have sufficient guarantees that the genetically edited human beings, or their closest or most distant descendants in time, are safe from serious harm.
   - HHGE can turn procreation into a manufacturing process in which the child is reduced to a commodity subjected to rigorous quality controls. People with disabilities are seriously worried that the use of these "genetic scissors" will, in the future, cut people like them out of existence without others even noticing (Sufian and Garland-Thomson 2021). This process of reification already occurs when the human embryo is subjected to preimplantation genetic diagnostic tests that are not intended for its own health (Bellver 2013).
   - Although HHGE does not turn the child into a commodity, it can be directed to purposes other than those strictly therapeutic, that is, to what has been called human enhancement. This purpose, which would be framed within the so-called liberal eugenics (Agar 2004), is repudiated by many who justify the use of HHGE only to avoid very serious genetic disorders. Other authors, accepting the legality of some non-therapeutic uses of HHGE, understand that the risks of abuse are great, and their proper regulation and control are difficult; thereby, it is prudent to limit HHGE to exclusively curative purposes.

- If access to HHGE is only possible for people with a good level of personal income, it will be inevitable that this technology contributes to increasing inequalities between human beings. As long as the basic health needs of all human beings are not covered, it is reasonable not to allocate public funds to research and interventions that will only satisfy the demands of those who can afford to pay for them (Bellver 2007).

None of the four risks mentioned leads directly to an outright ban. When HHGE reaches certain safety thresholds, it will not be necessary to maintain a moratorium based on fear of unwanted effects. In turn, three of the risks can be counteracted with regulations that ensure both the well-being of children resulting from HHGE and the equitable access to that technique. With the safety of the technique guaranteed and the establishment of an appropriate regulatory framework, HHGE could be used for noble purposes such as ending very serious genetic diseases (mainly monogenic) not only in an individual, but also in all his offspring (Foht 2015). In any case, it seems that regulations should be adopted by universal agreement since HHGE is a procedure that would affect all future humanity (Baylis 2019).

Many scientists are concerned about the public's reaction to the serious risks of HHGE. They fear that the rejection of this way of gene editing could end up harming the clinical uses of genetically engineered somatic cells. Therefore, they propose the adoption of a moratorium on HHGE. "Key to all discussion and future research is making a clear distinction between genome editing in somatic cells and in germ cells. A voluntary moratorium in the scientific community could be an effective way to discourage human germline modification and raise public awareness of the difference between these two techniques. Legitimate concerns regarding the safety and ethical impacts of germline editing must not impede the significant progress being made in the clinical development of approaches to potentially cure serious debilitating diseases" (Lanphier et al. 2015).

3. The regulation of HHGE as both therapeutic and enhancement interventions. Proponents of this position argue that fear of misuse of this technology is not a reason to abandon it. The way to proceed should be to regulate it to avoid abuse. They reject that human genetic integrity deserves special protection and that predetermining certain genetic characteristics of future human beings goes against their dignity. Rather, they understand that the choice of the genetic characteristics of the children is a prerogative that corresponds to the parents; they assume that the criterion of choice is more valuable than that of the genetic lottery, and they trust that HHGE will contribute to improve the lives of future people, families and the human species as a whole (Harris 2010). They defend the legality of introducing "genetic enhancements" into the human species as long as these changes do not generate situations of domination of some human groups over others, nor do they lead to the homogenization of the species (Stock 2000). Some go as far as to affirm that it is not only lawful but also a duty. When human beings have the possibility of controlling their own evolution, they will have the duty to assume that responsibility and achieve the best living conditions for their descendants and all future humanity (Peters 1997).

They also ensure that prohibiting HHGE will be ineffective because there will always be countries willing to authorize it. Therefore, rather than prohibiting, they propose the adoption of regulations that establish rigorous safety controls and only prohibit the perverse uses of this technology (Stock 2000). Obviously, the argument of the alleged ineffectiveness of a legal norm prohibiting HHGE does not affect at all the legitimacy or illegitimacy of the prohibition (Bellver 2015).

Following the announcement of He Jiankui's experiment in 2018, there was a unanimous reaction of rejection towards his experiment. This unanimity, however, does not mean that there is an agreement on how to proceed with HHGE. We present below the positions expressed in the most relevant reports and statements published for the last three years.

The Nuffield Council on Bioethics released a report in July 2018, prior to Jiankui's announcement, in which it favoured HHGE as long as the safety and dignity of individuals resulting from gene editing was guaranteed. It accepted that HHGE could be used both for therapeutic and enhancement purposes. Although the report insists on the importance of public participation, it does not seem very important since the report adopts a position contrary to the current legal consensus before making a broad citizen consultation (Nuffield Council on Bioethics 2018). This way of proceeding contributes more to conditioning the opinion of the public in a certain sense than to encouraging its participation. The report has been a subject of controversy. Some argue that instead of being simply 'morally permissible', many cases of genome editing should be morally imperative (Gyngell 2019).

In 2020, the National Academy of Medicine, the National Academy of Sciences, and the Royal Society published a report on which they apparently established more restrictive guidelines than those of the Nuffield Council. "Initial uses of heritable human genome editing (HHGE), should a country decide to permit them, should be limited to circumstances that meet all of the following criteria: 1. the use of HHGE is limited to serious monogenic diseases; the Commission defines a serious monogenic disease as one that causes severe morbidity or premature death; 2. the use of HHGE is limited to changing a pathogenic genetic variant known to be responsible for the serious monogenic disease to a sequence that is common in the relevant population and that is known not to be disease-causing; 3. no embryos without the disease-causing genotype will be subjected to the process of genome editing and transfer, to ensure that no individuals resulting from edited embryos were exposed to risks of HHGE without any potential benefit; and 4. the use of HHGE is limited to situations in which prospective parents: (i) have no option of having a genetically related child that does not have a serious monogenic disease, because none of their embryos would be genetically unaffected in the absence of genome editing, or (ii) have extremely poor options, because the expected proportion of unaffected embryos would be unusually low, which the Commission defines as 25 percent or less, and have attempted at least one cycle of preimplantation genetic testing without success" (National Academy of Medicine, National Academy of Sciences, and the Royal Society 2020). All these conditions are required for the initial uses of HHGE. The possibility of removing some of them in the future when the right circumstances exist is not ruled out.

A group of scientists led by Eric Lander and one of the two discoverers of CRISPR-Cas9, the 2020 Nobel Prize winner in Chemistry Emmanuelle Charpentier, released a statement in 2019 calling for a moratorium on the use of HHGE. They are aware of the double weakness of this proposal, because it appeals to the voluntary decision of scientists and can slow down progress in this field of research. However, they consider that it is the best way to regulate this technology at this time, in which its use would be extremely risky and in which it is impossible to establish a regulation of universal scope. They argue that HHGE is a technique that will affect all present and future humanity and, consequently, it must be the whole of humanity who decides what to do with it. This technology should not be pursued if there is no broad agreement with a universal scope that sets out the conditions of its use. Lander proposes that the WHO organizes this framework for universal participation. This is a very different proposal from those of other scientists who, when talking about public participation, actually think that scientists should show others the way to move forward. For example, this was the case of Jeniffer Doudna, discoverer of CRISPR-Cas9 and also a Nobel Prize winner in Chemistry in 2020, when she proposed to "stimulate forums in which experts from the genome-editing and bioethics communities provide information and education for the public about the scientific, ethical, social and legal implications of human-genome modification" (Doudna 2015).

Along the lines of the statement declared by Lander et al. (2019), the Geneva Declaration emphasizes that the decisions affecting all of us must be taken together and not only by small groups of scientists. This Declaration also denounces that, since the International Summit on Human Genetic Editing was held in 2015, policies have not gone into that address. The organizing committee of that meeting asserted that the clinical use of germline editing should not proceed without 'broad societal consensus'. Instead of a sustained commitment and the allocation of significant resources toward this prerequisite, there have been steady efforts to weaken it. The clearest example was the 2018 International Summit on Human Genome Editing. In the meeting in the shadow of the He Jiankui's experiments, this group issued a call for a 'translational pathway to germline editing', with only a cursory mention of the 'attention to societal effects'. The Statement also pays attention to the clarification of misconceptions as that heritable human genome editing is needed to treat or prevent serious genetic diseases. Because of that, heritable human genome editing should be understood not as a medical intervention, but as a way to satisfy parental desires for genetically related children or for children with specific genetic traits (Andorno 2021).

## Legal Aspects of HHGE: International and National Regulations

As we have seen, the debate on the ethical aspects of HHGE dates back to the 1960s. Thirty years later, national and international rules almost unanimously banning HHGE began to be passed. In recent years, two media events have taken place, which has exponentially increased the attention to the regulation that should be applied to HHGE.

The first event occurred in 2015, when the CRISPR-Cas9 technology burst into the world making "cutting and pasting" genes a much easier, cheaper and more effective process than before. The second one took place in November 2018, when the Chinese scientist He Jiankui announced the birth of two genetically modified girls. His intervention did not consist in preventing serious genetic disorders, but in providing the girls with genetic immunity against HIV. This event generated almost unanimous international condemnation because it was considered that the experiment had not been carried out in safe conditions, and because its objective was not to avoid a serious disease but to introduce genetic improvement in offspring. The investigator was prosecuted and convicted by the Chinese authorities (Cyranoski and Ledford 2018).

In light of these two facts, some scientists and bioethical experts began to question the current regulatory framework. On the one hand, they interpreted that the regulations did not contain prohibitions as restricted as previously thought. On the other hand, they proposed legal reforms to allow research in this field and, when the technology was safe, take it to the clinical setting. Finally, they defended the primacy of scientific self-regulation over the regulation of nations and international organizations. Below we refer to both the international and national regulatory frameworks.

The Council of Europe has been the pioneering international organization in regulating matters related to genetic engineering, and the one that has taken the most interest in proposing principles of action that could be universalized.

The first regulation on this matter was approved by the Parliamentary Assembly of the Council of Europe in the form of three recommendations: n. 934/1982, n. 1046/1986 and n. 1100/1989. The first of them indicates: "a) the rights to life and to human dignity protected by Articles 2 and 3 of the European Convention on Human Rights imply the right to inherit a genetic pattern which has not been artificially changed; b) this right should be made explicit in the context of the European Convention on Human Rights; c) the explicit recognition of this right must not impede development of the therapeutic applications of genetic engineering (gene therapy), which holds great promise for the treatment and eradication of certain diseases which are genetically transmitted; d) gene therapy must not be used or experimented with except with the free and informed consent of the person(s) concerned, or in cases of experiment with embryos, foetuses or minors with the free and informed consent of the parent(s) or legal guardian(s)". The Recommendation of including a new right in the European Convention of Human Rights did not happen, but the Convention of Human Rights and Biomedicine (Oviedo Convention 1996) included a relevant mention on this matter in art. 13 that imposes stricter limits to HHGE: "An intervention seeking to modify the human genome may only be undertaken for preventive, diagnostic or therapeutic purposes and only if its aim is not to introduce any modification in the genome of any descendants".

To correctly interpret the meaning of this regulation, it would be appropriate to refer to the Explanatory Report to the Convention (Steering Committee on Bioethics – CDBI 1997). But that text is even more confusing than the wording of art. 13: "The progress of science, in particular in knowledge of the human genome and its application, has raised very positive perspectives, but also questions and even great fears. Whilst developments in this field may lead to great benefit for humanity, misuse of these developments may endanger not only the individual but the species itself. The ultimate fear is of intentional modification of the human genome so as to produce individuals or entire groups endowed with particular characteristics and required qualities" (n. 89). The report then makes two claims that are hardly compatible. First, "Any intervention which aims to modify the human genome must be carried out for preventive, diagnostic or therapeutic purposes. Interventions aimed at modifying genetic characteristics not related to a disease or to an ailment are prohibited" (n. 90). Second, "Interventions seeking to introduce any modification in the genome of any descendants are prohibited" (n. 91).

Andorno considers that, even when presented as "therapeutic", alterations in the germ line are not "therapeutic" in the strict sense, since they do not seek to "cure" a currently existing patient but to alter the gametes that are going to be used for fertilization. It makes no sense to speak of "patient" when there is not even an embryo. But even when what is genetically altered is an embryo, it is forced to argue that HHGE seeks to "cure" that particular embryo. It is clear that embryos are created to later introduce the desired genetic alterations. In fact, it is difficult to separate both events since they are only two stages of the same process. The true objective of this complex procedure is not to "cure" a particular embryo, but to create an embryo with an altered genome in order to satisfy the wishes of a couple willing to have a child with certain qualities (Andorno 2021).

In December 2015, the Committee on Bioethics of the European Council approved an ambiguous Declaration on genome editing technologies. It emphasizes that the Oviedo Convention is the only legally binding international instrument dedicated to the recognition of human rights in the field of biomedicine, and that it includes the principles that can serve as a reference for the debate on gene editing required at the international level (Committee on Bioethics 2015).

UNESCO first referred to HHGE in the Universal Declaration on the Human Genome and Human Rights (1997), which seems to prohibit interventions on the human germ line but in ambiguous terms: "The International Bioethics Committee of UNESCO (…) should give advice concerning the follow-up of this Declaration, in particular regarding the identification of practices that could be against human dignity, such as germ-line interventions".

In October 2015, UNESCO's International Bioethics Committee (IBC) approved a report updating its reflection on the human genome and human rights, in which it deals with the most controversial technological advances that have taken place in recent times in the field of genetic technologies. With regard to the CRISPR-Cas9 genome editing system, the report indicates that the "states and governments, especially in relation to editing the human genome so that genetic modifications would be passed on to future generations, should renounce the possibility of going into it alone within their own legal system" (International Bioethics Committee – UNESCO 2015). Based on the convenience of approving a basic regulation of universal scope, the International Bioethics Committee "recommends a moratorium on genome editing of the human germ line (...) the concerns about the safety of the procedure and its ethical implications are so far prevailing." (International Bioethics Committee – UNESCO 2015).

We have seen that the two international legal instruments regulating HHGE prohibit it, albeit in ambiguous terms. At the national level, laws prohibiting HHGE also dominate, although often in ambiguous terms as well. A recent study identified 106 countries regulating gene editing, of which a large majority (96 out of 106) surveyed have policy documents — legislation, regulations, guidelines, codes, and international treaties — relevant to the use of genome editing to modify early-stage human embryos, gametes, or their precursor cells. Most of these 96 countries do not have policies that specifically address the use of genetically modified *in vitro* embryos in laboratory research (germline genome editing). Of those that do, 23 prohibit this research and 11 explicitly permit it. Seventy-five of the 96 countries prohibit the use of genetically modified *in vitro* embryos to initiate a pregnancy (heritable genome editing). Five of these 75 countries provide exceptions to their prohibitions. No country explicitly permits heritable human genome editing (Baylis et al. 2020).

The international and national regulations of HHGE agree on its prohibition. It is true that the provisions of the Oviedo Convention and the Universal Declaration on the Human Genome and Human Rights are ambiguous, but they do not raise doubts about their objective: to prevent future humans from being the result of genetic editing. This same goal is pursued by most of the national laws that regulate HHGE, although some of them have imprecise formulations because they were approved when gene editing still looked very far away in time. Some authors, interpreting these norms in an analytical and non-comprehensive way, conclude that there is no solid prohibition of HHGE both nationally and internationally and that, therefore, certain uses of HHGE could be authorized (De Miguel and Payán 2019). We understand that this general prohibition exists and that the only thing that proceeds is to foster a public conversation with universal scope to find out whether humanity wants to allow its offspring to be the result of HHGE.

# Is Law enough prepared to address the regulation of gene editing?

## Difficulty and uncertainty as main characteristics of the current context

The regulation of gene editing can be considered a difficult task to address for Law. This is so not only because changes are permanent in this area or because the matter is so technical and scientific, but, above all, because there is a prevalence of difficult cases and there is also uncertainty about the consequences of those techniques.

A difficult case is the one where the solution cannot be found in the normative system itself, understood as a system made up of rules. Difficult cases, or *hard cases* in the traditional terms used by Ronal Dworkin, means, in a metaphoric way, a legal solution based on a decision between two evils (Dworkin 1975). Difficult cases also mean a power of interpretation in the hands of the Court and a main role for legal principles instead of specific rules.

Evans vs the UK, ruled by the European Court of Human Rights, is considered a paradigmatic example of a difficult case (Farnos Amorós 2014). The case concerned a woman who, having been diagnosed with precancerous tumours in both ovaries, decided to cryopreserve her fertilized gametes with those of her husband in case she wants to be a mother in the future and could not be one naturally after the medical treatment to which she is going to be subjected. The husband had accepted it, but after two years their relationship broke down and he had requested that the fertilized eggs be destroyed because he didn't want to be the father, to which she opposed since it was her only opportunity to be a biological mother. The legal debate was between her right to be a mother and his right not to be a father. So there was no intermediate solution, through a proportional reduction of both her and his rights. The solution of the conflict implies a complete sacrifice of one of them. The judges of the minority of the Court (Traja and Mijović) expressly described the case as a real dilemma. In such difficult cases, finding intermediate solutions or compromises is usually impossible. The only way to solve the conflict is through the sacrifice of one of the rights in conflict. In other words, one of the rights would end up taking everything, while the other would be reduced to nothing. The second problem as we have anticipated is uncertainty. As José Esteve Pardo explains, the legal system of modernity was built on certainties to generate security and trust in the economic framework (Esteve Pardo 2015). Law can no longer find in science the certainty to which it aspires to regulate and, therefore, regulation loses its rational ideal based on rules.

In any case, even accepting that the uncertainty has increased recently because of the huge scientific and technological development, it is also true that the absence of certainty defines the existence of human beings since its origins. One of the most significant features of last decades is an acute awareness of the uncertainty in the rule of law (Martínez García 2012). Such is the level of uncertainty in the context of regulating matters, gene editing being a paradigmatic example, that the possibility of predicting with total certainty the result of legal proceedings is even interpreted with real suspicion. A margin of uncertainty is an indicator that the legal system actually works. Absolute certainty will only exist in the context of legal fraud (Martínez García 2012). Gustavo Zagrebelsky affirms that it is not only doubtful that certainty can be a realistic objective today, but it is also doubtful that it is desirable (Zagrebelsky 2009).

Hence, from a legal perspective, the context can be briefly explained through two characteristics: difficulty in resolving legal conflicts through legal rules and uncertainty in the decisions to be made. The legal answer to both characteristics is, as we will explain below, the precautionary principle, which also gives a relevant position to principles in our legal systems over the classical rules.

## Taking principles seriously in the context of gene editing regulation

The dilemmatic character of the conflicts and issues posed by gene editing and the uncertainty about the consequences of the decision gives to the principles a main role in this area. Legal principles play a relevant position because they offer an open solution, not a closed one as the rules does. The solution is not about applying a legal consequence to a specific fact, but to find a legal solution which is not specifically foreseen in the legal system. In a difficult case legal, subsumption (application of the legal rule to a fact) is not possible because there is not a clear rule, a clear legal answer for the conflict. So, interpretation is the only way to solve them. The argument is not about facts, but about law. As David Lyons asserts, in easy cases, court rulings are justified by applying the rules. In difficult cases, it is necessary to act differently from the mere justification based on rules (Lyons 1986). For this reason, the new difficult cases that arise under the advance of biotechnology does not require concrete and closed solutions, but open ones, and therefore, principles instead of rules. We must decide not to abandon the Law in favour of scientific advancement, but to opt for more flexible and agile legislative techniques that allow science and regulation to play together (Junquera de Estéfani 2003). Therefore, law should abandon, in some sense, the forms for positive law and return to natural law in its expression through principles and values. As Robert Alexy points out, positivism becomes non-positivism when it is accepted that the inclusion of moral principles and arguments in law is necessary and not just contingent (Alexy 2016).

As Margarita Beladiez explains, principles are characterized by being a type of legal prescription with a peculiar structure different from the rules, since they are not a legal proposition (factual assumption + legal consequence). Their legal mandate translates into the imposition of a negative duty that prohibits acting against the value established by them. There is not a positive mandate that forces, under the threat of the legal consequence, to act in a certain way (Beladiez Rojo 2010). Principles are a specific type of rules characterized by their fundamentality, vagueness and generality, which contain the imperatives of fairness and justice that define positive morality. Principles fulfill the double function of legalizing morality and moralizing law (Vidal Gil 2016).

For Robert Alexy, principles are norms that order something to be done to the greatest extent possible, within the existing legal and real possibilities. Principles are optimization mandates, characterized by the fact that they can be fulfilled to different degrees and to the extent that their fulfillment not only depends on the real possibilities, but also on the legal ones. Instead, the rules would be definitive mandates or norms that can only be fulfilled or not (Alexy 2008). If a rule is valid, then exactly what it requires must be done, no more, no less. Principles are weighted and the rules are applied. The principles arise as a consequence of hard cases that show the normative and ideological gaps of the order (Vidal Gil 2018).

As Laura Palazzani explains, it is important to recover in Biolaw the structural and specific meaning of legality in the sense that ethical values will play a main role, such as dignity of the human being (Palazzani 2007). In Biolaw, we must look beyond validity and effectiveness, wondering if the legal solution in these matters is fair (Palazzani 2002). The rules, with which the Law has traditionally operated for reasons of legal certainty, are already shown to be obsolete or, at least, insufficient for the resolution of biomedical conflicts. The specificity and concretion of the rule in front of the principle seems to make them not very useful. Therefore, open solutions are needed, because the case to rule is precisely very open, unpredictable.

The reasons that justify a solution are the relevant ones, more than the legal solution. The main objective is to find the ideal arguments for the justification of the legal decision – the criteria of reasonableness that support it (Atienza 2013, Sánchez Hidalgo 2019). However, considering the nature and characteristics of principles, the solution is paradoxical. If science generates uncertainty, should law also generate it? Can we fight against the scientific uncertainty with the legal uncertainty of principles? As Jesús Ignacio Martínez García points out, uncertainty is not reduced in the context of scientific development with certainties but with uncertainties of another type, a determined uncertainty, which is partially bounded, fixed, defined and structured in a certain way (Martínez García 2012). For this reason, through the relevant role of principles, Law approaches Bioethics. Graciano González R. Arnáiz indicates that the paradigm of Bioethics, which emerged in the United States of America a few decades ago, refers to a sociocultural framework common to all versions of applied ethics, whose meeting point is the recognition of moral pluralism (González R. Arnáiz 2016). This is also perfectly predicable of the Law itself.

Law turns now to the traditional formula of Bioethics to solve difficult cases, and not because there has been a positivization or juridification of the ethics, but because the principles are now presented as the only instruments that allow achieving a minimally satisfactory solution in such cases. If Bioethics has gone beyond the traditional methods of deduction, subsumption or induction of the case, the Law also goes beyond formalism (Vidal Gil 2018). Therefore, we could conclude, in a similar way as Toulmin indicated some decades ago that Medicine has saved the life of Ethics (Toulmin 1982), Bioethics can save the life of Law, freeing it from the reductionist tyranny of the rules in the framework of biotechnological uncertainty. It can be seen be that an approximation between Law and Bioethics, is not material, but methodological (Atienza 1998).

This transformation of the Law due to the demands of the advancement of science and technology is precisely what can save with singularity and specificity the work of the legal professionals in the age of digitization and robotization. Can the machine subsume? Clearly yes, through precise algorithms that allow us to interpret the facts in accordance with the written rule. However, when we talk about principles, is the operation so simple? The machine can offer legal security, but not justice based on the interpretation of principles.

The response of Law to the improvement of science is contrary to what happened at the beginning of the 20th century, when the Law tried to approach the scientific method. The presence of rules in the legal system would allow us to think about the machining of the application of the Law. However, the presence of principles next to the rules makes such a possibility absolutely unfeasible (Zagrebelsky 2009).

The prominence of the principles also has an impact on the division of power itself, since it implies that the Courts must have a main role. As Lucas Murillo de la Cueva points out, opting for general clauses (principles) instead of closed clauses (rules) affects the subject who has to make the decision. Thus, in the case of the rules, it will be the legislator who takes the initiative; in the second, it will be the judge who is called to perform the essential function in the system (Lucas Murillo de la Cueva 2004). The dogma of the position of the judge as *bouche qui prononce les paroles de la loi*, in Montesquieu's classic expression, is abandoned (García de Enterría 1963). Hence, the changes in the way of interpreting the Law entails not only how it has to be interpreted, but also who has to interpret it.

Giving the Court the power to interpret the Law through principles does not mean that this power will be arbitrary. On the contrary, the Court must obey the moral demands of practical reason. The Court will have more activity but not more activism, not a political role (Prieto Sanchís 2014). The judges are not the lords of the Law now in the same sense that the legislator was in the last century. They are more exactly the *guardians* of the structural complexity of Law (Zagrebelsky 2009). The Courts will not decide *extra legem*, but *intra principia*.

Are the Courts prepared to play this position? Do they have sufficient knowledge, beyond Law, in the field of Bioethics?

Gustavo Zagrebelsky highlights the importance of having judges, not only well recruited and responsible, but formed in a new panorama that demands a transformation of the judiciary and an overcoming of the anachronistic model in which judges continue to see themselves as repositories of eminently technical-legal knowledge, valid as such, but far from the expectations that the needs of society put in them (Zagrebelsky 2009).

Finally, this legal model derived from the improvement of science demands not only very well-prepared judges in the field of Bioethics and Biolaw, but also an active participation of committees to give them advise in such difficult matters. There is a paradigmatic of example of this new framework for governance, where Courts and Bioethics Committees are playing together, in the seminal case ruled by the Supreme Court of New Jersey, Karen Quinlan. In its ruling, the Court mentioned the relevant position of the Ethical Committees of Hospitals to help judges to solve these difficult conflicts in the area of Biolaw (the case was about the withdrawal of life support measures for a patient in a coma).

## The precautionary principle as an example of the new position of principles in the edge of gene editing

The precautionary principle can be considered one of the main ethical and legal instruments to regulate the issues posed by the advance of science and, therefore, also by gene editing. The precautionary principle finds its most remote roots in Aristotelian thought.

As the World Commission on the Ethics of Scientific Knowledge and Technology (hereinafter, COMEST) indicated in its Report on the Precautionary Principle (March 2005), this principle is often seen as an integral principle of sustainable development, that is development that meets the needs of the present without compromising the abilities of future generations to meet their needs. By safeguarding against serious and, particularly, irreversible harm to the natural resource base that might jeopardize the capacity of future generations to provide for their own needs, it builds on ethical notions of intra- and inter-generational equity. For COMEST, sustainable development implies that the needs of present generations should be met provided they do not impair the ability of future generations to meet their needs, which implies an ethical balance between present and future generations and therein the precautionary principle plays an essential role.

This principle receives different names in the area of Ethics and Law such as prudence or caution, all of them refer to the same ethical and legal concept. In the context of the European Union (EU), where this principle has been formally enacted by the legislation as the main measure to confront new risks and uncertainty, there are two different terms: the proper precautionary principle, which usually operates in the areas of the environment and biomedicine, and the principle of caution, which usually operates in the area of food and health.

Precautionary principle is directly related to the concept of "risk society" developed, among others, by Ulrich Beck (2006) and to the current context characterized by the term of uncertainty where the unprecedented development of science and technology poses great opportunities for the welfare of our communities but also implies new potential risks, in many cases not sufficiently determined. Uncertainty is a core element of the principle with a double meaning: a) uncertainty about the risks derived from the specific science or technology; and b) uncertainty about the decision-making process implementing that principle. So, potential risks and uncertainty must be present in the decision-making process to apply the principle. In this context of limitations associated with predictability, the precautionary principle is not applicable to any risk situation, but only to those with two main characteristics: firstly, a context of scientific uncertainty, and secondly, the possibility of particularly serious damage that may be uncontrollable and irreversible.

For COMEST, "morally unacceptable harm refers to harm to humans or the environment that is threatening to human life or health, is serious and effectively irreversible, is inequitable to present or future generations, and is imposed without adequate consideration of the human rights of those affected".

Nevertheless, in real-life situations, even if we act upon a determinate probability estimate, we are not fully certain that this estimate is exactly correct, hence there is uncertainty. Almost all decisions are made under uncertainty. As COMEST explains, human life is, has always been, and will always be full of risks. The urge to deal with the risks we face is a basic condition of our existence. So, the precautionary principle should be applied when there is a considerable scientific uncertainty about a risk which could cause a serious and irreversible harm. Some form of scientific analysis is needed because a mere fantasy or crude speculation is not enough.

The word risk refers, often rather vaguely, to situations in which it is possible but not certain that some undesirable event will occur. When there is a risk, there must be something that is unknown or has an unknown outcome. Therefore, knowledge about risk is knowledge about lack of knowledge. This combination of knowledge and lack thereof contributes to making issues of risk complicated from an epistemological point of view. The words risk and uncertainty differ along the subjective—objective dimension. Whereas uncertainty seems to belong to the subjective realm, risk has a strong objective component (Stanford Encyclopedia of Philosophy, word risk).

As COMEST explained in its Report on The Precautionary Principle (March 2005), in today's environment of rapid scientific research and technological development, different ways to apply new knowledge and innovations are constantly being engendered that present us with ever more possibilities and challenges. We stand to benefit from the greater range of options this progress brings. However, with more choice also comes more responsibility. Conscious of our roles as stewards of the world in which we live, notably on behalf of future generations, we must therefore take care in exercising these options. Therefore, precautionary principle is also related to responsibility.

Precautionary principle should not be considered as an obstacle for the development of science and technology. On the contrary, the principle has been precisely developed to allow that development but in a responsible way. As the scientific movement, so-called *Slow Science* (http://slow-science.org), promotes, scientists must, beyond the laboratory or the computer, take their time to reflect on the great questions posed by the relentless advance of science. Science and technology need time to think mainly about social justice and future generations. It is not about giving up scientific progress, but about carrying it out with a thoughtful reflection about its consequences. In this way, the precautionary principle is not shown as a mere tool that Ethics and Law offer us to avoid the uncertain risks that could arise from such advances, but as something that goes beyond, like a true new paradigm that, without renouncing the opportunities offered by science, allow us to serenely assess its consequences for the human being and the environment. The metaphor of modernism that proclaims 'the more is better' would be transformed into the opposite, 'the better the more.' More than imposing a 'no- go' or a 'go-slow', the principle also acts as a stimulant for other innovations and clean technological progress. The principle promotes the development of innovative alternatives for potentially risky technologies.

In this new context, where principles and not rules must play an essential role, the position of the three branches changes. The *Montesquieu paradigm* of a division of powers, where the legislative branch occupies a main position in relation to the executive and judicial branches, has been transformed by the evolution of science and technology. Because principles must be interpreted, not merely subsumed in a factual case described by the legal rule, the position of the Courts will be stellar, which obliges the system to prepare them for this legal and also political relevant role.

In the specific area of Law, the precautionary principle involves the transition from the forecasting model (knowledge of risk and causal links) to that of uncertainty about the risk, because the possible damages cannot be determined and the possible causal link between one and the other must be established, which is then supported by statistical calculations and probabilities. Both models converge, however, in the prevention of a feared damage, which is their common objective. The principle has improved a transformation from post-damage control (civil liability) to the level of a prior damage control (anticipatory risk measures).

The principle also expresses a new legal framework, a new era of Law, where principles are playing a main role in regard to rules. Related to this idea of the huge development of principles as flexible norms which must be interpreted, instead of mere rules submitted to a mere assumption, Law is following in some sense the same way to solve conflicts developed by Bioethics, based on principles. Through the new position and value of the principles for the resolution of conflicts in the legal system, the connection of Law and Bioethics is produced, which has been expressed through the idea of taking principles seriously (Vidal Gil 2018). As COMEST explains, this principle does not offer a predetermined solution to every new problem raised by scientific uncertainty. On the contrary, it is a guiding principle that provides helpful criteria for determining the most reasonable course of action in confronting situations of potential risk.

Precautionary principle means, in the words of Hans Jonas, a sort of *in dubio pro malo*, or in very similar terms, an exception for the traditional paradigm of Law, *pro libertate* (*permissum id ese intellegitur, quod non prohibetur* or, in similar terms, *intellegitur consessum quod non est prohibitum*). The burden of proof then would not be anymore in the hands of the public authorities, which decide not to allow the activity or technology, but in the hands of the individual who wants to get the specific authorization. The main burden of providing evidence for safety rests on the proposers of a new technology or activity. As the Italian Comitato Nazionale per la Bioetica stated, in the absence of the possibility of an objective risk assessment, the absence of scientific evidence of a likelihood of harm should be interpreted as evidence in favour of the impossibility of excluding it (Report on ethical and juridical considerations on the use of biotechnologies 2001).

The principle was initially developed in the field of the environment and hence extends to other fields of uncertainty. So, the postulate in it is based on sustainable development, a development that meets the needs of the present without compromising the capabilities of future generations to meet their needs. By safeguarding against serious and irreversible damages, the principle is protecting future generations to meet their own needs, based on the ethical notions of intra and intergenerational equity. Principle 15 of the Rio Declaration on Environment and Development proclaims the following: "In order to protect the environment, the precautionary approach shall be widely applied by States according to their capabilities... Where there are threats of serious or irreversible damage, lack of full scientific certainty shall not be used as a reason for postponing cost-effective measures to prevent environmental degradation".

One of the main criticisms that this principle has received comes from the risk of opening a way towards arbitrariness, creating a relevant margin of discretion by public authorities, although the principle implies a specific evaluation and management procedure whose main purpose is to protect the rights of people. In this sense, it could be affirmed that the precautionary principle has a certain antinomic character, since, having been created to combat uncertainty, it often ends up producing it as regards the final decision. Because of this fear, COMEST considered it important to clarify at least what the principle is not: To avoid misunderstandings and confusions, it is useful to elaborate on what the precautionary principle is not. The precautionary principle is not based on 'zero risks' but aims to achieve lower or more acceptable risks or hazards. It is not based on anxiety or emotion, but is a rational decision rule, based in ethics, that aims to use the best of the 'systems sciences' of complex processes to make wiser decisions. Finally, like any other principle, the PP in itself is not a decision algorithm and thus cannot guarantee consistency between cases. Just as in legal court cases, each case will be somewhat different, having its own facts, uncertainties, circumstances, and decision-makers, and the element of judgement cannot be eliminated.

Because there is a huge improvement in the area of science and technology, evidence can change year by year, so uncertainty and risks can also change. The implementation of this principle demands also the development of monitoring and learning activities that provide performance data on a continuous basis. In the area of Law, many legal jurisdictions have developed in the context of normative quality and under the new concept of better regulation or smart regulation, a specific methodology which must be implemented in the context of the precautionary principle: the ex post evaluation which allows to reconsider the decision adopted, taking into account the proper evolving nature of uncertainty and risks. We can find an example of this evaluation ex post in France, where there is taking place, precisely now, a period of reflection on the update of the norms of bioethical content under the advance of the Science, formally called the National Consultation on Bioethics (les Etats généraux de la Bioéthique). Within this framework, the opinions of collegiate bodies, experts and the general public are collected. The objective of this public consultation process is to approve a new law on bioethics in 2020 that incorporates the results of the aforementioned consultation. The process responds to the regulatory provision contained in the Bioethics Law of 2011 (Law 2011-814, of July 7), whose article 47 provides that the Law will be reviewed by Parliament within a maximum period of seven years from its entry into force, after evaluation of its application by the Parliamentary Office for the Evaluation of Scientific and Technological Regulations.

Through the ex post evaluation, the duty to evaluate periodically the regulations is promoted in order to verify whether they have fulfilled the objectives pursued and if the cost and charges derived from them were justified and adequately valued. It is intended to introduce teleological rationality in the normative process as the legal system is a mere means to achieve goals.

Some authors have proposed (in order to develop a legal framework conducive to innovation but, at the same time, able to minimize the risks that it may entail) the effective development of the following measures: a) the organization of reflection on the risks in ad-hoc self-control commissions, b) the articulation of decision-making procedures that, in addition to participatory, allow professional or expert reflection, and c) the temporalization of the rules in terms that safeguard their predictability by setting ex post evaluation deadlines (Parejo 2016).

In any case, it is a fundamentally a European principle. Its development has taken place within the framework of the European Union, where the institutions have enshrined this principle with enormous legal potential in the face of the uncertainty of scientific progress and the risks for individuals. Thus, Jim Dratwa points out that the precautionary principle plays a very important constitutional function from two perspectives: as a legal way to legitimize the regulation of matters which affect human lives (biopolitical perspective) and as an instrument to legitimize the institutions of the European Union over Member States (supranational perspective) (Dratwa 2011).

The precautionary principle has been defined as a procedural principle called to enhance the evaluation of uncertain risks and enable the adoption of measures against them even when they are largely unknown. In our new context, there is insufficient or limited conceptual construction of the predictability, in which the precautionary principle plays its role. However, the principle is not applicable to all risk situations, but only to those that present two main characteristics: first, a context of scientific uncertainty, and secondly, the possibility of particularly serious damage that may be uncontrollable and irreversible.

This new panorama causes, for example, that the precautionary principle itself alters the operation of the principle of *pro libertate* that inspires our legal systems. The classic aphorism that what is not prohibited is considered permitted (*permissum id ese intellegitur, quod non prohibetur* or, in similar terms, *intellegitur consessum quod non est prohibitum*) is no longer the only and main premise.

The use of the principle presupposes that: a) potentially dangerous effects derived from a phenomenon, a product or a process have been identified, and b) the risk is not determined with sufficient certainty from a scientific perspective. The application of an approach based on the precautionary principle should start with a scientific evaluation, as complete as possible and, if feasible, identifying at each stage the degree of scientific uncertainty. In this way, recourse to the principle is part of the general framework of risk analysis (which includes, apart from risk assessment, risk management and risk communication) and, more specifically, within the framework of risk analysis, risk management that corresponds to the decision-making phase. Those responsible for the decision must be aware of the degree of uncertainty inherent in the result of the evaluation of the available scientific information, and this is not a scientific decision, but an eminently political one, which requires resolving two questions: first, whether you must act or not and, secondly, if you have decided to act, the measures that result from the application of the principle.

The precautionary principle also alters the burden of proof about the level of uncertainty of the risks that could arise from the specific activity. It will be the proponent of the activity who must prove that sufficient certainty exists to exclude risks. In the words of the National Bioethics Committee of Italy, in its 2001 Report on ethical and legal considerations on the use of biotechnologies, in the absence of the possibility of an objective risk assessment, the absence of scientific proof of a probability of harm must be interpreted as evidence in favour of the impossibility of excluding it. Such reflection coincides with what the Commission of the European Union itself provides in its Communication of February 2, 2000, where it states, on the burden of proof, that although, in most cases, European consumers and associations representing them must demonstrate the risk posed by a procedure or a product once it is placed on the market, in the case of an action taken under the precautionary principle, the producer, manufacturer or importer may be required to demonstrate the absence of danger. This possibility must be examined on a case-by-case basis, so it cannot be broadly extended to all products and marketing processes.

With the play of the principle, achieving sufficient certainty is aspired rather than absolute certainty. We settle for what is beyond reasonable doubt. What is now called reasonable is a way of limiting uncertainty (Martínez García 2012).

One of the main criticisms that this principle has received is that of opening a way towards arbitrariness, and although the principle conforms to a specific evaluation and management procedure whose main purpose is to protect the rights of people, it does generate a relevant margin of discretion for public authorities Thus, it could be affirmed that the precautionary principle has a certain antinomic character, since, having been created to combat uncertainty, it ends up on many occasions by producing it in terms of the final decision.

Therefore, the UNESCO Committee on Technology and Science (COMEST) reminds us in its 2005 Report on the precautionary principle that the application of the principle does not extend to any risk derived from scientific progress, but it is limited to the dangers that are unacceptable. COMEST considers that morally unacceptable damage consists of that inflicted on human beings or the environment that is a threat to human health or life, or serious and effectively irreversible, or unfair to present or future generations, or imposed without having duly taking into account the human rights of those affected. Furthermore, interventions should be proportional to the level of protection and the magnitude of the possible harm, recognizing that it will rarely be possible to reduce the risk to zero, and the total ban cannot be considered a response proportional to a potential risk. That is, what we have previously called the reflexive stance of science, which is not contrary to its advances.

COMEST also considers that recourse to the precautionary principle does not *per se* imply a negative view of progress and innovation, but quite the opposite, since resorting more widely to the principle can stimulate both innovation and scientific activity, by replacing the technologies of the 19th century and the elementary science of the first industrial revolution for clean technologies and the science of the systems of a new industrial revolution. This will perhaps help to achieve a better balance between the benefits of innovations and the risks of these new developments. The precautionary principle encourages the development of innovative alternatives to potentially dangerous technologies.

For COMEST, it is important to clarify both what the precautionary principle is and what it is not to prevent an incorrect use of it from causing precisely what the principle does not seek; to avoid misunderstandings and confusion, it is useful to expand on what the precautionary principle is not. The precautionary principle is not based on 'zero risk' but aims to ensure that there are fewer risks or contingencies or that they are more acceptable. It does not obey anxiety or emotion, but constitutes a rational decision rule, based on ethics, and which aims to use the best of the 'systems sciences' of complex processes to make the most reasonable decisions. Ultimately, like any other principle, the precautionary principle itself is not a decision algorithm and therefore cannot guarantee consistency between cases. As in the cases that are heard before the courts, each case will be somewhat different, since it will have its own facts, points of uncertainty, circumstances, and decision-makers, always having a quota of subjectivity that cannot be eliminated. In the terms in which COMEST tries to clarify the principle, it is necessarily subject to casuistry, without prejudice to the fact that, at least, there must always be certain requirements to avoid a random or arbitrary use of the principle.

## Conclusion

Heritable Human Gene Editing (HHGE) is a subject of ethical debate due to the collateral effects it can produce on individuals, societies and future generations. While some see HHGE as an exceptional opportunity to eliminate serious genetic disorders, others are convinced that there are better alternatives to achieve this goal, and that the use of HHGE represents a radical alteration of the intergenerational relationship between human beings.

International laws prohibit this procedure in a rather ambiguous way and, therefore, some authors interpret that such a prohibition does not really exist. Others understand that these international standards clearly prohibit HHGE and that this prohibition should be maintained. Some others propose a reform of these norms to establish rigorous conditions under which HHGE should be allowed. In any case, it seems appropriate that the decision about something that will affect all future humanity is made with the participation of all present-day humanity and not just some specific scientists or countries.

At the present time, there is a significant uncertainty about the possibilities of HHGE. In situations such as this one, law should resort more to principles than to rules and establish periodic reviews of the approved regulations. In this regard, a special role corresponds to the precautionary principle, defined as a procedural principle aimed at enhancing the evaluation of uncertain risks and enabling the adoption of measures against them even when they are largely unknown.

## Acknowledgments

We would like to thank Esperanza Marín for reviewing the chapter.

## References

Agar, N. 2004. Liberal Eugenics: In Defence of Human Enhancement. Wiley-Blackwell, London, UK.

Alexy, R. 2008. Teoría de los derechos fundamentales. CEPC, Madrid, Spain.

Alexy, R. 2016. La institucionalización de la justicia. Comares, Granada, Spain.

Anders, G., 2011. La obsolescencia del hombre. Vol. I. Sobre el alma en la época de la segunda revolución industrial. Pre-Textos, Valencia.

Andorno, R. 2021. Edición genética en la línea germinal humana: breves reflexiones en torno a la Declaración de Ginebra. *In*: Bellver, V. [ed.]. ¿Editamos humanos? Ética y Derecho ante la edición genética en la línea germinal humana. Tirant lo Blanch, Valencia, Spain (in press).

Andorno, R., Baylis, F., Darnovsky, M., Dickenson, D., Haker, H., Hasson, K., et al. 2020. Geneva Statement on heritable human genome editing: The need for course correction. Trends in Biotechnology, 38: 351-354.

Atienza, M. 1998. Juridificar la bioética. Isonomía. Revista de teoría y Filosofía del Derecho, 8: 75-99.

Atienza, M. 2013. Curso de argumentación jurídica. Trotta, Madrid, Spain.

Baylis, F., Darnovsky, M., Hasson, K. and Krahn, T. M., et al. 2020. Human germ line and heritable genome editing: The global policy landscape. The CRISPR Journal, 3: 365-377.

Baylis, F. 2019. Altered Inheritance: CRISPR and the Ethics of Human Genome Editing. Harvard University Press, Cambridge, MA, USA.

Beck, U. 2006. La sociedad del riesgo. Paidós, Barcelona, Spain.

Beladiez Rojo, M. 2010. Los principios jurídicos. Civitas, Madrid, Spain.

Bellver, V. 2013. El consejo genético antenatal: Derecho y buenos prácticas, pp. 49-92. *In*: Romeo Casabona, C.M. [ed.]. Hacia una nueva medicina: consejo genético, Comares, Granada.

Bellver, V. 2007. Intervenciones genéticas en la línea germinal humana y justicia. pp. 461-486. *In*: Ballesteros, J., et al. [eds.]. Biotecnología y Posthumanismo. Thomson-Aranzadi, Pamplona, Spain.

Bellver, V. 2015. Biotechnology, ethics, and society: The case of genetic manipulation. pp. 123-43. *In*: W. J. González [ed.]. New Perspectives on Technology, Values, and Ethics, Boston Studies in the Philosophy and History of Science, vol. 315. Springer, Cham.

Cyranoski, D. and Ledford, H. 2018. Genome-edited baby claim provokes international outcry. Nature, 563: 607-608.

De Miguel, Í. and Payán, E. 2019. Retos éticos y jurídicos que plantea la edición genética embrionaria a la luz del marco legal vigente en el ámbito europeo: una mirada crítica. Anuario de Filosofía del Derecho, 35: 71-92.

Doudna, 2015. Embryo editing needs scrutiny. Nature, 528: s.6.

Dratwa, J. 2011. Representing Europe with the precautionary principle. *In*: Jasanoff, S. [ed.]. Reframing rights. Bioconstitutionalism in the genetic age. The MIT Press, Cambridge, USA.

Dworkin, R. 1975. Hard cases. Harvard Law Review, 88(6): 1057-1109.

Dzau, V., McNutt, M. and Bai, C. 2018. Wake-up call from Hong Kong. Science, 362: 1215.

Esteve Pardo, J. 2015. Decidir y regular en la incertidumbre. Respuestas y estrategias del Derecho público. pp. 33-46. *In*: I. Darnacutella, M.M. Gardella, J. Esteve Pardo and I.S. Döhmann [eds.]. Estrategias del Derecho ante la incertidumbre y la globalización. Marcial Pons, Madrid, Spain.

Farnós Amorós, E. 2014. ¿A quién pertenecen los embriones? Fecundación humana asistida y crisis de pareja. Anuario de la Facultad de Derecho de la Universidad Autónoma de Madrid, 18: 331-349.

Fletcher, J. 1971. Ethical aspects of genetic controls. New England Journal of Medicine, 285: 783.

Foht, B. 2015. Gene editing: New technology, old moral questions. The New Atlantis, 16: 3-15.

García de Enterría, E. 1963. Reflexiones sobre la Ley y los principios generales del Derecho. Rev. Adm. Públ. AP, 40: 189-222.

González R. Arnaiz, G. 2016. Bioética: un nuevo paradigma. De ética aplicada a ética de la vida digna. Tecnos, Madrid, Spain.

Gyngell, C., Bowman-Smart, H. and Savulescu, J. 2019. Moral reasons to edit the human genome: Picking up from the Nuffield report. Journal of Medical Ethics, 45: 514-523.

Habermas, 2003. The Future of Human Nature. Polity Press. Cambridge, UK.

Hammerstein, A. L., Egger, M. and Biller-Andorno, N. 2019. Is selecting better than modifying? An investigation of arguments against germline gene editing as compared to preimplantation genetic diagnosis. BMC Medical Ethics, 20(1): 83.

Harris, J. 2010. Enhancing Evolution: The Ethical Case for Making Better People. Princeton University Press, Princeton, USA.

International Bioethics Committee (IBC), UNESCO. 2015. Report of the IBC on Updating Its Reflection on the Human Genome and Human Rights, October 5.

Junquera de Estéfani, R. 2003. Interrogantes planteados por la manipulación genética y el proyecto genoma humano a la filosofía jurídica. Anuario de Filosofía del Derecho. 20: 165-188.

Kass, L. 1972. Making babies. The New Biology and the Old Morality. The Public Interest, Winter.

Lanphier, E., Urnov, F., Haecker, S. E., Werner, M. and Smolenski, J. 2015. Don't edit the human germline. Nature, 519: 410-411.

Lander, E., Baylis, F., Zhang, F., Charpentier, E., Berg, P., Bourgain, C., et al. 2019. Adopt a moratorium on heritable genome editing. Nature, 567: 165-168.

Lederberg, J. 1972. Biological Innovation and Genetic Intervention. p. 26. *In*: J. A. Behnke [ed.]. Challenging Biological Problems. Directions Toward Their Solution. Oxford University Press, New York, USA.

Lucas Murillo De La Cueva, P. 2004. Derechos fundamentales y avances tecnológicos. Los riesgos del progreso. Boletín Mexicano de Derecho Comparado, 109: 71-110.

Lyons, D. 1986. Ética y derecho. Ariel, Barcelona, Spain.

Martínez García, J. I. 2012. Derecho e incertidumbres. Anuario de Filosofía del Derecho, 28: 97-118.

National Academy of Medicine, National Academy of Sciences, and the Royal Society. 2020. Heritable Human Genome Editing. The National Academies Press, Washington DC, USA.

Nuffield Council on Bioethics, 2018. Genome Editing and Human Reproduction: Social and Ethical Issues. London, UK.

Palazzani, L. 2002. Introduzione alla biogiuridica. G. Giappichelli, Torino, Italy.

Palazzani, L. 2007. Bioética y derechos humanos. pp. 383-403. *In*: J. Ballesteros, and E. Fernández [eds.]. Biotecnología y posthumanismo. Thomson Aranzadi, Cizur Menor, Spain.

Parejo Alfonso, L. 2016. Estado y Derecho en proceso de cambios. Las nuevas funciones de regulación y garantía del Estado social de soberanía limitada. Tirant lo Blanch, Valencia, Spain.

Peters, 1997. Playing God? Genetic Determinism and Human Freedom. Routledge, Nueva York, USA.

Prieto Sanchís, L. 2014. Presupuestos neoconstitucionalistas de la teoría de la argumentación jurídica. pp. 17-42. *In*: M. Gascón Abellán (ed.). M. Argumentación jurídica. Tirant lo Blanch, Valencia, Spain.

Report on ethical and juridical considerations on the use of biotechnologies 2001. Available from http://bioetica.governo.it/en/opinions/opinions-responses/ethical-and-juridical-considerations-on-the-use-of-biotechnologies/.

Rio Declaration on Environment and Development. 1992. Available from: https://www.un.org/en/development/desa/population/migration/generalassembly/docs/globalcompact/A_CONF.151_26_Vol.I_Declaration.pdf.

Sánchez Hidalgo, A. J. 2019. Epistemología y metodología jurídica. Tirant lo Blanch, Valencia, Spain.

Sandel, M. 2009. The Case against Perfection: Ethics in the Age of Genetic Engineering. Belknap Press, Cambridge (MS), USA.

Savulescu, J. 2001. Procreative Beneficence: Why We Should Select the Best Children. Bioethics, 15: 413-426.

Silver, L. 1997. Remaking Eden: Cloning and Beyond in a Brave New World. Avon Books, New York, USA.

Sinsheimer, R. 1966. The end of the beginning. Engineering and Science, 30(3): 7-10.

Sinsheimer, R. 1994. The Strands of a Life: The Science of DNA and the Art of Education. University of California Press, Berkeley (California), USA.

Sparrow, R. 2019. Yesterday's child: How gene editing for enhancement will produce obsolescence—and why it matters. The American Journal of Bioethics, 19: 6-15.

Steering Committee on Bioethics – CDBI. 1997. Explanatory Report to the Convention for the protection of Human Rights and Dignity of the Human Being with regard to the Application of Biology and Medicine: Convention on Human Rights and Biomedicine, Oviedo, April 4.

Stock, G. 2000. Redesigning Humans. Our Inevitable Future, Houghton Mifflin, New York, USA.

Sufian, S. and Garland-Thomson, R. 2021. The Dark Side of CRISPR. Scientific American. February 16.

Toulmin, S. 1982. How medicine saved the life of Ethics. Perspectives on Biology and Medicine, 25: 736-750.

van Dijke, I., Bosch, L., Bredenoord, A. L., Cornel, M., Repping, S. and Hendriks, S. 2018. The ethics of clinical applications of germline genome modification: A systematic review of reasons. Human Reproduction, 33(9): 1777-1796.

Vidal Gil, E. J. 2016. Los retos actuales de la Bioética ¿Qué hacer con los principios? pp. 79-104. *In*: J. A. Santos, M. Albert and C. Hermida [eds.]. Bioética y nuevos derechos. Comares, Granada, Spain.

Vidal Gil, E. J. 2018. Bioética y Derecho: la positivización de los principios. Anales de la Cátedra Francisco Suárez, 52: 23-41.

Zagrebelsky, G. 2009. El derecho dúctil. Trotta, Madrid, Spain.

# Political, Regulatory and Ethical Considerations of the CRISPR/Cas Genome Editing Technology

**Eduardo Rodriguez Yunta\* and Luis María Vaschetto**

Interdisciplinary Center for Studies on Bioethics, University of Chile,
Diagonal Paraguay 265, Office 606, Santiago, Chile

## Introduction

Genome editing refers to "the practice of making targeted interventions at the molecular level of DNA or RNA functions, deliberately to alter the structural or functional characteristics of biological entities" (Nuffield Council on Bioethics 2016). In gene therapy, there has always existed an interest in modifying the genetic information of organisms for diverse purposes. While the fact that genome modifications have been performed for many decades, only nowadays we are able to achieve site-specific genome editing with high efficiency. The Clustered Regularly Interspaced Short Palindromic Repeats/CRISPR-Associated (CRISPR/Cas) genome editing system is a revolutionary technology that has shown potential for the development of personalized treatments and cure genetically inherited diseases, mainly due to its high precision, relatively simple use and low cost (Zhu 2015, Vaschetto 2018). These characteristics make CRISPR/Cas an almost ideal tool for gene therapy and clinical applications. CRISPR/Cas can be used, for example, for stem cell gene editing applications, correction of mutations at multiple genomic sites (*loci*), generation of virus-resistant cells, etc. (Santos et al. 2016, Meier et al. 2018, Ethics Council of the Max Planck Society 2019).

There is, however, a problem: CRISPR/Cas still needs to be carefully assessed in relation to their safety and efficacy. The translation of genome editing technologies in the clinic involves challenges in term of off-target effects, delivery, toxicity and immunogenicity. With regard to regulatory procedures, CRISPR-Cas is also fundamental to policy-makers for the development of proper oversight frameworks for this emerging technology. The lack of regulation may lead to dangerous situations.

\* Corresponding author: erodriguezchi@gmail.com

Two research issues of special concern are the dual use of CRISPR-Cas and genetic intervention in the germ-line, since unwanted effects might eventually be transmitted to the next generation. This chapter will focus on political and regulatory aspects related to the implementation of the CRISPR-Cas technology in human cells and their ethical implications.

## Genome-editing for research purposes using human embryos

Typical clinical research and drug development procedures involve three main stages: preclinical research, clinical trial, and marketing authorization. Each of these phases need to be regulated to ensure human health safety and meet ethical guidelines. Experimental procedures must be supervised by ethical review committees which assess research proposals to determine the risks on animal subjects and human health. Drug control agencies such as the Food and Drug Administration (FDA) in the United States and the European Medicines Agency (EMA) in the European Union regulate clinical investigations of new medical devices, while considering the safety and efficacy of each method and/or biological product. There are also specialized committees that are in charge of assessing specific research activities such as, for example, research involving stem cells. The US National Academies of Sciences, Engineering, and Medicine (NASEM), the US National Institutes of Health (NIH), and the International Society for Stem Cell Research (ISSCR) have developed ethical recommendations/guidelines concerning the use of human embryonic stem cells (hESCs). In the USA, the NIH prohibit the use of federal funding for research using hESCs. The implementation of CRISPR-Cas system to edit human embryo cells in the laboratory is subject to the debate about the moral status of human embryos (Brokowski and Adli 2019).

The United Kingdom (UK) allows (in certain situations) the creation and utilization of embryos for research purposes until 14 days after fertilization (Kipling 2016). Research using eggs, sperms or, embryos stored outside the human body is regulated by the Human Fertilisation and Embryology Authority (HFEA). The HFEA prohibits the implant of genetically modified embryos in the uterus, although this regulatory agency may allow research using human embryo cells outside the body, including research projects involving gene-editing (Human Fertilisation and Embryology Authority 2018). In the USA, the NASEM convened a committee of experts on Human Gene Editing (2017) whose report stated:

"There are, of course, enduring debates about limitations of the current system, particularly with respect to how it addresses the use of gametes, embryos, and fetal tissue, but the regulations are considered adequate for oversight of basic science research, as evidenced by their longevity. Special considerations may come into play for research involving human gametes and embryos in jurisdictions where such research is permitted; in those cases, the current regulations governing such work will apply to genome-editing research as well. Overall, then, basic laboratory research in human genome editing is already manageable under existing ethical norms and regulatory frameworks at the local, state, and federal levels" (NASEM 2017).

In the USA, clinical trials using genome editing techniques must follow guidelines as specified in preclinical and clinical drug development, including guidelines with regard to the specification of the manufacture, product characterization and potency assays, animal models to demonstrate safety and efficacy, risks and benefits assessment, a valid therapeutic endpoint, a control, and proper statistical analyses (Schacker and Seimetz 2019). The FDA prohibits the use of US federal funds for "research in which a human embryo is intentionally created or modified to include a heritable genetic modification" (Consolidated Appropriations Act of 2016, Public Law 114-113, 2015). In this regard, Greely (2019) indicates that "the U.S. Food and Drug Administration (FDA) has taken the position that any genetically, or otherwise substantially-modified, human embryo is a drug or biological product, the clinical use of which requires FDA approval".

In China, concerns about the use of genome editing technologies on human embryos have been raised, especially after He Jiankui's experiments, who used CRISPR-Cas9 to edit human embryos which subsequently proceeded through pregnancy. China is poised to introduce regulatory measures in order to comply with the following criteria: "Experiments on genes in adults or embryos that endanger human health or violate ethical norms can accordingly be seen as a violation of a person's fundamental rights" (Cyranoski 2019). In China, there already exists ethical review committees in charge of assessing research involving such activities, but it should be indicated that concerns regarding their independence to work freely have been expressed by members of the scientific community.

# Governance and regulations

Policy and ethical concerns have been raised since the emergence of genetic engineering technologies capable of modifying the genome. In 1974, the NIH established a Recombinant DNA Advisory Committee (RAC) aimed at developing biosafety guidelines in recombinant DNA (rDNA) research (Kleinman 2000). In 1975, molecular biologists and genetic engineers gathered in Asilomar established a series of guidelines for regulation of rDNA research. Scientists pronounced in favor of self-regulation, transparency, and open participation in policies and procedures. In this regard, it should be noted that the CRISPR-Cas system raises several technical points to consider when developing regulatory policies on genome editing: 1) a novel mode of action, 2) accessibility, 3) speed of use and uptake, and 4) multiplexing (i.e. editing at multiple genomic sites) (Nutfield Council on Bioethics 2016). In most countries, there are already regulations for genome engineering, but such regulations should be updated to include technical considerations. The United Nations Educational, Scientific and Cultural Organization (UNESCO) considers that it is not clear how genome editing technologies can fit within existing legal and regulatory frameworks, thereby new guidelines may be required (Tuerlings 2019). In the United Kingdom, the Nutfield Council on Bioethics (2016) considers that, in general, worldwide regulations might provide inadequate guidance for genome editing. Moreover, it is also important to note that the high efficacy of the CRISPR-Cas9 system to produce site-specific genetic modifications may become difficult to identify which genome modifications have been induced by the technology and which ones occurred in living organisms throughout their evolution. In consequence, CRISPR-Cas9 genome editing could eventually be confused with naturally occurring mutations (substitutions).

New regulatory frameworks should always be updated to clarify what are the limitations faced by genome editing technologies. As for any other technology, it should be assessed by a participatory approach. Government, science, and society need to reach consensus on how to decide, what and when, and on which principles these decisions are based upon (Tuerlings 2019). Public participation is fundamental in order to capture dissenting views, promote transparency, understanding, fairness, inclusion, and respect the autonomy of individuals to make decisions. However, public participation also has its limitations. For instance, data gathered from polls and surveys may not represent an informed opinion or depend on the mood of the moment. Other important issues include the measures aimed at protecting vulnerable populations and include future generations, such as social justice, solidarity, and the protection of human rights, which may help to address the limitations of public participation in terms of quality in decision-making (Halpern et al. 2019). Conducting a Human Rights Impact Assessment can help to determine the potential risks and benefits of genome editing, and simultaneously tackle both political and socioeconomic factors (Halpern et al. 2019). The lack of suitable public policies may represent a serious threat to population health (especially for individuals already marginalized) or include groups which do not need to be included.

The elimination of biosafety threats related to the use of genome engineering technologies can be an important starting point for updating existing regulations, and create new ones. The human rights provide a useful framework for the development of national and international regulations. An important starting point is the rejection of 'eugenic' practices as a form of biomedical technocracy whose objective is to improve human traits by irreversibly altering the human genome. Eugenicists argue that humans can be 'improved' through the use of technologies aimed at increasing the occurrence of desirable characteristics (Brokowski et al. 2015). Although not legally binding, the UNESCO Universal Declaration on the Human Genome and Human Rights is intended to create a framework within which countries, corporations, and other public and private stakeholder entities can develop detailed legislative proposals, policies, guidelines, and coordinate actions (UNESCO 1997). The UNESCO Declaration defines the human genome as 'the common heritage of humanity' and recommends a moratorium for genetic interventions on the human germ-line due to the potential risks that new technologies face for future generations, highlighting that such practices "could be contrary to human dignity" (article 24). Article 11 states that "practices which are contrary to human dignity, such as reproductive cloning of human being, shall not be permitted". The UNESCO Declaration on Bioethics and Human Rights (UNESCO 2005) establishes the primacy of individual interests over the interest of science and society (Article 3, page 76). The Declaration also establishes the right to protect future generations from the risk of genetic intervention (Article 16, page 78).

In 2015, the UNESCO International Bioethics Committee (IBC) offered several points of view on human genome engineering. In its report, the IBC recommended to the governments of the countries: "Agree on a moratorium on genome engineering of the human germline, at least as long as the safety and efficacy of the procedures are not adequately proven as treatments" (…) "Renounce the possibility of acting alone in relation to engineering the human genome and accept to cooperate on establishing a shared, global standard for this purpose, building on the principles set out in the Universal Declaration on the Human Genome and Human Rights and the Universal Declaration on Bioethics and Human Rights" (UNESCO 2015). It is expected that safety measures related to the use of genome editing technologies should prevail over individual considerations. The report also included the following recommendations:

- Technological/scientific advances in the genomics field entail a global responsibility
- Include public debates on bioethical issues
- The Law of Supply and Demand should not be used to determine acceptance
- Precautionary principles should guide decision making
- The United Nations should play an important role in the development of norms and their implementation
- A clear differentiation between medical and non-medical applications should be always established
- New methodologies should be based on the respect for human rights

In particular, it has been raised that the Precautionary Principle should provide a regulatory framework for genome editing technologies. The Precautionary Principle is an epistemological, philosophical and legal approach based on the uncertain effects derived from scientific progress. In this regard, it is imperative to highlight the need to know risks and benefits; however, risks are difficult to assess when there is scientific uncertainty. Bedau and Triant (2014) indicate that "parties should refrain from actions that might harm the environment, and, second, that the burden of proof for assuring the safety of an action falls on those who propose it". Table 1 highlights some implications for the implementation of the CRISPR-Cas technology in human subjects.

**Table 1.** Implications of CRISPR-Cas genome editing

| Characteristics of the technology | Moral considerations | Regulatory aspects |
|---|---|---|
| A RNA programmable mode of action Accessibility and easy to use Difficulty in detecting genome editing at single base level Possibility to make multiple simultaneous genome modifications at different genomic sites (*loci*) | Care for the welfare of future generations General respect for society's values and cultural identity Legal protection of human rights Public participation Rejection of eugenics Social justice and solidarity | Benefits and risks assessment Equivalent guidelines for drug development Human rights impact assessment Monitoring and review policies Precautionary principle Protection of vulnerable individuals and populations |

# Human germ-line genome editing

Jennifer Anne Doudna – Nobel laureate and leading scientist in the development of the CRISPR-Cas9 system, has emphasized the need to work together with regulatory agencies in order to design appropriate regulatory frameworks and monitor the potential harmful effects of genome editing on germ-line cells (Shao and Pershad 2019). A major concern is associated with the fact that human germ-line genome modifications can be transmitted to following generations. This situation raises problems for the implementation of an informed consent and for who is responsible for potential damages when effects are transmitted across generations (Billings et al. 1999, Frankel and Chapman 2000). Frankel and Chapman (2000) emphasize that "if IGM* is to be pursued, an effective system of public oversight must first be in place".

The International Summit on Human Gene Editing (2015) was co-hosted by the US National Academy of Sciences and National Academy of Medicine, the UK's Royal Society, and the Chinese Academy of Sciences. During this meeting, it has been proposed "hold off (place a moratorium) on editing the human germline genome for reproduction while we work out the technical issues of safety, off-target effects, efficacy, efficiency of the edit, and the development of a clinical grade delivery mechanism for the editing system" (Thomson 2015). The International Summit also discussed ethical aspects related to human germ-line genome editing and agreed to initiate an international forum aimed at facing these challenges, as well as propose suitable regulatory measures (Steven 2015). The Second International Human Genome Editing Summit, which was held in Hong Kong in 2018, also recommended a moratorium on germ-line genome modification, "the scientific understanding and technical requirements for clinical practice remain too uncertain and the risks too great to permit clinical trials of germ-line editing at this time" (NASEM 2019). This moratorium is based on the precautionary principle. Table 2 summarizes some arguments in favor and against human germ-line genome modification.

**Table 2.** Arguments in favor and against human germ-line genome editing

| In favor | Against |
|---|---|
| Risks should not be higher than in sexual reproduction | The chance for unintended off-target effects are high |
| Potential to eliminate genetic diseases and disorders | Eugenic practices |
|  | Concerns related to the moral status of embryos and human dignity |

The first genome-edited human embryos using the CRISPR-Cas technology were reported in 2015 by researchers who described the use of this technique in non-viable human tripronuclear zygotes (Liang 2015). This study showed that off-target effects may represent a real concern. In 2018, He Jiankui reported the use of CRISPR-

---

\*    Inheritable Genetic Modification (IGM)

Cas to genetically modify human embryos and thus prevent HIV transmission. The genetically modified embryos were subsequently implanted into a woman's uterus, resulting in the birth of twin girls (Wachowicz 2019). The Chinese Academy of Medical Sciences and the National Health Commission of China opposed to these experiments, indicating that Jiankui He's work violated existing regulations and ethical principles (Ma et al. 2019). The experiment carried out by Jiankui had serious negative consequences: the risks outweighed the benefits, questionable consent, secrecy inappropriate and disregard of the international ethical guidelines (Greely 2019).

In the UK, the Nuffield Council on Bioethics (NCoB) issued a report concluding that germ-line genome editing could be acceptable under certain circumstances (Dickenson and Darnovsky 2019). The report indicates "(1) interventions are intended to secure, and are consistent with, the welfare of a person who may be born as a consequence; (2) such interventions would uphold principles of social justice and solidarity – by this we mean that such interventions should not produce or exacerbate social division, or marginalize or disadvantage groups in society" (Nuffield Council on Bioethics 2018).

In its report *'Human Genome Editing: Science, Ethics, and Governance'*, the NASEM concluded that heritable human genome editing could be permissible under certain circumstances:

"Germline genome editing is unlikely to be used often enough in the foreseeable future to have a significant effect on the prevalence of these diseases but could provide some families with their best or most acceptable option for averting disease transmission, either because existing technologies, such as prenatal or preimplantation genetic diagnosis, will not work in some cases or because the existing technologies involve discarding affected embryos or using selective abortion following prenatal diagnosis (....) for those who are aware they are at risk of passing on such a mutation, the use of heritable genome editing offers a potential avenue to having genetically related children who are free of the mutation of concern. This form of editing could be done either in gametes (eggs, sperm), in gamete precursors, or in early embryos, but it is important to note that IVF* procedures would be required to generate embryos for subsequent genomic modification. In most cases, PGD could be used to identify unaffected embryos to implant" (NASEM 2017).

Currently, the risks of off-target mutations can be considered high enough to justify the application of CRISPR-Cas in human subjects. Therefore, it seems prudent to wait until the technique is safe before it can be routinely used for therapeutic purposes. With regard to human germ-line genome editing, a major concern is the introduction of alleles with potential unintended side effects, which could be recognized only after generations (Evitt et al. 2015). It is reasonable to suppose that if this technique is safe enough, then no differences should exist between parental consent to therapy to treat a particular disease and consent for genome editing in order to prevent the disease in the next generation. Finally, it can be argued that one person has the right to health protection and have access to appropriate treatments,

_____

* In vitro fertilization (IVF)

while legal guardians have the right to procure such treatments (Nordberg et al. 2018).

## Mitochondrial genome editing

Mitochondria can be considered as 'the energy centers' of eukaryotic cells. These organelles function to generate energy (ATP) through the process of cellular respiration, which is then used to sustain metabolic processes. Mitochondria have become attractive targets for the treatment of metabolic diseases. Moreover, mutations in the mitochondrial genome can also contribute to cancer initiation and progression (Kalsbeek et al. 2017). As for other genetic engineering techniques, mitochondrial genome modifications first need to assess the risks and benefits, which (as mentioned above) is a difficult task. Second, nowadays there is a limitation regarding the development of efficient *in vivo* delivery systems for mitochondrial targeting (Gammage et al. 2018, Ho 2020). In 2015, UK's regulations allowed the use of mitochondrial replacement techniques as part of *in vitro* fertilization treatments in order to prevent transmission of mtDNA diseases (Castro 2016). The regulatory framework may have implications in the debate on germ-line genome editing since mitochondrial replacement may eventually be used to transfer foreign genetic material to the next generation through the maternal lineage (i.e. mitochondrial genomes are maternally inherited). The NASEM highlighted the need to be cautious on the use of mitochondrial replacement techniques (NASEM 2016). In the USA, the FDA has claimed jurisdiction over the approval of mitochondrial replacement therapies (Cohen et al. 2020).

## Epigenome editing and modulation of gene expression

Epigenetics refers to changes in chromatin structure that are associated with the regulation of gene expression. Epigenetic mechanisms include DNA methylation, histone modifications (e.g. acetylation, methylation, etc.) and RNA interference (RNAi) pathways. These three types of epigenetic pathways work together to modulate gene expression patterns. Currently, there is great interest in the development of CRISPR-Cas-based tools for the treatment of diseases associated with the alteration of epigenetic pathways and aberrant gene expression patterns. CRISPR-Cas-based epigenetic/transcriptional modulation is based on the use of a nuclease-deficient Cas9 (dCas9) enzyme that can be fused to catalytic domains of epigenetic enzymes (e.g. DNA methyltransferases, histone acetyltransferases, etc.) and modular transcriptional domains (activators or repressors), respectively (Gilbert et al. 2013, Vaschetto 2018). CRISPR-Cas-based epigenome editing could eventually inhibit cancer progression by activating tumor suppressor genes or by suppressing oncogene expression in cancer cells. In addition, CRISPR-based epigenetic therapies could also have potential for reprogramming somatic cells into a pluripotent state, and for inducing cell differentiation in tissue regeneration applications (Pulecio et al. 2017). It is important to note that although CRISPR-Cas-based epigenetic/transcriptional modulation do not involve the alteration of DNA (nucleotide) sequences, these methods also require the delivery of CRISPR-like components at target cells. In consequence, such tools should also be subject to assessment in order to ensure safety and effectiveness for clinical use. Moreover, the utilization of epigenome editing tools also requires more research in order to completely understand how epigenetic marks/modifications are transmitted across generations.

## Genome editing for 'enhancement' of human traits

The CRISPR-Cas technology has potential for many therapeutic applications. For instance, CRISPR-Cas can be used to mediate immunological responses against infections, induce immunological tolerance to transplanted cells/tissues/organs, genetically modify cancer cells, correct mutations associated with genetic disorders, etc. However, in certain situations, CRISPR-Cas raises a conflict between the use for therapeutic purpose and the enhancement of human traits. With regard to this point, Nordberg et al. (2018) have posed an open question: "Is it morally permissible or perhaps required for one generation of humans to make changes to the genetic profile of their descendants through germline genetic therapy or enhancement, or both?" A line of thought believes that the use of genome engineering technologies should be completely prohibited when they could potentially (and irreversibly) alter 'human nature'. Moreover, there also exists a social justice concern related to the idea that some individuals could use new genetic engineering technologies to the detriment of people that have no access, thereby increasing social differences in health, life expectancy, etc. As starting point, it is important to highlight that scientific community and society still must clearly define what we call 'human enhancement'. Table 3 highlights some arguments in favor and against genome editing for enhancement of human traits.

**Table 3.** Genome editing and human enhancement

| Pros | Against |
|---|---|
| Treatment and prevention of diseases and disorders | Uncertainty about off-target effects |
| | Increase of human inequalities |
| Enhancement of human performance | Concerns related to the alteration of human nature (germ-line genome modification) |

## Xenotransplantation

Xenotransplantation refers to the process of transplantation of organs, tissues or cells from one species to another. The xenotransplantation of animal organs to humans is seen as a solution to overcome the worldwide donor organ shortage. The second World Health Organization (WHO) Global Consultation on Regulatory Requirements for Xenotransplantation Clinical Trials (2011) developed a series of guidelines and recommendations for xenotransplantation: "Investigators must ensure that source animals are bred for the purpose and as safe as possible, using a closed colony of consistently known specific pathogen-free animals housed in a well-controlled pathogen-free environment with high levels of biosecurity". Indeed, the main problems that this technique faces are the risk of infection caused by zoonotic pathogens and rejection by the immune system. Moreover, diverse animal models for the study of human diseases have been developed using CRISPR-Cas technology (Shah et al. 2018). For example, CRISPR-Cas9 has allowed to inactivate endogenous retroviruses from genetically modified pigs (Niu et al. 2017), thereby showing its potential to face the challenges of xenotransplantation. In consequence, it is reasonable to conclude that CRISPR/Cas should be take into account when developing regulatory policies for xenotransplantation.

## Intellectual property rights and patenting

The development of patents in the biology field has historically been a complex and controversial issue, and CRISPR-Cas does not represent an exception to this rule. CRISPR/Cas is naturally found both in bacteria and archaea (prokaryotic organisms) as an adaptive defense system that confers resistance against foreign genetic elements. Afterwards, this system was repurposed for genome editing of eukaryotic cells. The first patents for the CRISPR-Cas technology were submitted by Feng Zhang (US Patent No. 8,697,359) and Emmanuelle Charpentier-Jennifer Doudna (US Patent Application No. 13/842,859). Doudna and Charpentier's patent application discloses the use of CRISPR-Cas9 for genome editing of prokaryotic cells (bacteria), whereas Zhang's patent application claimed specific uses on eukaryotic cells (Smith-Willis and San Martin 2015). The US Patent and Trademark Office (USPTO) awarded the first CRISPR-Cas9 patent rights to Zhang's group due to the ability of this system to edit DNA and modify this technology for the manipulation of animal and human cells, which does not occur naturally (Zhang et al. 2020).

## Dual-use research

In life sciences, dual-use research concerns are mainly associated with the idea that scientific knowledge may eventually be used for the development of bioweapons (Dixon 2019). This situation raises an ethical issue: technological advancements and scientific progresses can be used by unscrupulous persons to cause harm to innocent people (Miller and Selgelid 2007). In consequence, there exists a need to develop regulatory and oversight mechanisms in order to address the challenges of dual use research. Genome editing is a powerful technology which can be put to good uses as well as bad ones, and therefore it can be categorized as 'dual use'. For instance, the CRISPR-Cas9 genome editing system could eventually be used to induce mutations in the genome of pathogenic microorganisms in order to create 'biological weapons'. The simplicity, versatility, and low cost of CRISPR-Cas even exacerbate this disturbing question. The current laws and regulations may underestimate the risks of CRISPR-Cas in dual-use research. Efficient oversight mechanisms must be clearly established at all levels: research should always be monitored by independent government agencies; scientists must inform the authorities about important discoveries involving CRISPR-Cas-based technologies; institutions need to develop guidelines in order to monitor CRISPR research that may fall into the category of dual-use; professionals must be trained contemplating ethical issues; sponsors should emphasize the importance of considering the risks of dual-use research; editors and publishers should incorporate measures to handle the studies involving dual use technologies, etc.

## Conclusion

In the last years, ethical and regulatory aspects related to the implementation of the CRISPR-Cas technology have been discussed, and guidelines have accordingly been developed. However, in most countries, public debate and additional regulatory measures are still required to ensure safe and responsible use of this technology. The public access to information on the potential risks of CRISPR-Cas is fundamental to create consensus and decision-making, as well as to gain awareness about its positive potential. Regulations should always be flexible, and simultaneously ensure precautionary measures. The development of efficient and safe CRISPR-Cas applications framed by clear legislation and wide ethical principles is likely one of the most profound challenges that the modern civilization faces.

## References

Advisory Committee on Developing Global Standards for Governance and Oversight of Human Genome Editing, 2019. Report of the First Meeting.

Bedau, M. A. and Triant, M. 2014. Social and ethical implications of creating artificial cells. *In*: Sandler, R. L. (ed.). Ethics and Emerging Technologies. Palgrave Macmillan, London. https://doi.org/10.1057/9781137349088_37.

Billings, P. R., Hubbard, R. and Newman, S. A. 1999. Human germline gene modification: A dissent. The Lancet, 353(9167): 1873-1875.

Brokowski, C. and Adli, M. 2019. CRISPR ethics: Moral considerations for applications of a powerful tool. Journal of Molecular Biology, 431(1): 88-101.

Brokowski, C., Pollack, M. and Pollack, R. 2015. Cutting eugenics out of CRISPR-Cas9. Ethics in Biology, Engineering and Medicine: An International Journal, 6(3-4): 263-279.

Castro, R. J. 2016. Mitochondrial replacement therapy: The UK and US regulatory landscapes. Journal of Law and the Biosciences, 3(3): 726-735.

Center for Genetics and Society. 2019. About human germline gene-editing. Available from: https:// www.geneticsandsociety.org/internal-content/about-human-germline-gene-editing.

Charpentier, E., et al. 2013 Methods and compositions for RNA-directed target DNA modification and for RNA-directed modulation of transcription. International Patent WO2013176772.

Chinese Academy of Sciences, the Royal Society, U.S. National Academy of Sciences, U.S. National Academy of Medicine. 2015. International Summit on Human Gene Editing: A Global Discussion. Available from: http://www.nationalacademies.org/gene-editing/Gene-Edit-Summit/index.htm.

Cohen, I. G., Adashi, E. Y., Gerke, S., Palacios-González, C. and Ravitsky, V. 2020. The regulation of mitochondrial replacement techniques around the world. Annual Review of Genomics and Human Genetics, 21: 565-586.

Consolidated Appropriations Bill. 2016. Pub. L. 114-13 § 749, 129 Stat. 2242, 2283. Available from: https://www.congress.gov/bill/114th-congress/house-bill/2029/text.

Cyranoski, D. 2019. China set to introduce gene-editing regulation following CRISPR-baby furor. Nature News. Available from https://www.nature.com/articles/d41586-019-01580-1

Dickenson, D. and Darnovsky, M. 2019. Did a permissive scientific culture encourage the 'CRISPR babies' experiment? Nature Biotechnology, 37(4): 355-357.

Dixon, T. 2019. Mapping the potential impact of synthetic biology on Australian foreign policy. Australian Journal of International Affairs, 73(3): 270-288.

Ethics Council of the Max Planck Society. 2019. Discussion paper focusing on the scientific relevance of genome editing and on the ethical, legal and societal issues potentially involved. Available from: https://www.mpg.de/13811476/DP-Genome-Editing-EN-Web.pdf.

Evitt, N. H., Mascharak, S. and Altman, R. B. 2015. Human germline CRISPR-Cas modification: Toward a regulatory framework. The American Journal of Bioethics, 15(12): 25-29.

Food and Drug Administration. 2016. Guidance for Industry: Source Animal, Product, Preclinical, and Clinical Issues Concerning the Use of Xenotransplantation Products in Humans. https://www.fda.gov/media/102126/download.

Frankel, M. S. and Chapman, A. R. 2000. Human inheritable genetic modifications: Assessing scientific, ethical, religious, and policy issues. Washington, DC: American Association for the Advancement of Science. Available from: http://www.aaas.org/spp/dspp/sfrl/germline/main.htm.

Gammage, P. A., Moraes, C. T. and Minczuk, M. 2018. Mitochondrial genome engineering: The revolution may not be CRISPR-ized. Trends in Genetics, 34 (2): 101-110.

Genome editing: An ethical review. Nuffield Council on Bioethics. Available from: http://nuffieldbioethics.org/wp-content/uploads/Genome-editing-an-ethical-review.pdf.

Gilbert, L. A., Larson, M. H., Morsut, L., Liu, Z., Brar, G. A., Torres, S. E., et al. 2013. CRISPR-mediated modular RNA-guided regulation of transcription in eukaryotes. Cell, 154(2): 442-451.

Greely, H. T. 2019. CRISPR'd babies: Human germline genome editing in the 'He Jiankui affair'. Journal of Law and the Biosciences, 6(1): 111-183.

Halpern, J., O'Hara, S. E., Doxzen, K. W., Witkowsky, L. B. and Owen, A. L. 2019. Societal and ethical impacts of germline genome editing: How can we secure human rights? The CRISPR Journal, 2(5): 293-298.

Ho, M. 2020. Development of CRISPR/Cas Systems for Mammalian Mitochondria. Doctoral dissertation, City of Hope's Irell & Manella Graduate School of Biomedical Sciences.

Human Fertilisation and Embryology Authority. 2018. HFEA Approves Licence Application to Use Gene-Editing in Research. Available from: https://www.hfea.gov.uk/about-us/news-and-press-releases/2016-news-and-press-releases/hfea-approves-licence-application-to-use-gene-editing-in-research/.

Kalsbeek, A. M., Chan, E. K., Corcoran, N. M., Hovens, C. M. and Hayes, V. M. 2017. Mitochondrial genome variation and prostate cancer: A review of the mutational landscape and application to clinical management. Oncotarget, 8(41): 71342.

Kipling, J. 2016. The European Landscape for Human Genome Editing: A review of the current state of the regulations and ongoing debates in the EU. Federation of European Academies of Medicine (FEAM). Available from: https://acmedsci.ac.uk/file-download/41517-573f212e2b52a.pdf.

Kleinman, D. L. 2000. Democratizations of science and technology. pp. 125-165. In: Kleinman, D. L. (ed.). Science, Technology, and Democracy. State University of New York Press, Albany, NY, USA.

Liang, P., Xu, Y., Zhang, X., Ding, C., Huang, R., Zhang, Z., et al. 2015. CRISPR/Cas9-mediated gene editing in human tripronuclear zygotes. Protein & Cell, 6(5): 363-372.

Ma, Y., Zhang, L. and Qin, C. 2019. The first genetically gene-edited babies: It's "irresponsible and too early". Animal Models and Experimental Medicine, 2: 1-4.

Meier, R. P., Muller, Y. D., Balaphas, A., Morel, P., Pascual, M., Seebach, J. D., et al. 2018. Xenotransplantation: Back to the future? Transplant International, 31(5): 465-477.

Miller, S. and Selgelid, M. J. 2007. Ethical and philosophical consideration of the dual-use dilemma in the biological sciences. Science and Engineering Ethics, 13(4): 523-580.

National Academies of Sciences, Engineering, and Medicine. 2016. Mitochondrial Replacement Techniques: Ethical, Social, and Policy Considerations. National Academies Press. DOI: 10.17226/21871.

National Academies of Sciences, Engineering, and Medicine. 2017. Human Genome Editing: Science, Ethics, and Governance. National Academies Press.

National Academies of Sciences, Engineering, and Medicine. 2019. Second International Summit on Human Genome Editing: Continuing the Global Discussion. Proceedings of a Workshop in Brief. Washington, DC, The National Academies Press. DOI: 10.17226/25343.

Niu, D., Wei, H. J., Lin, L., George, H., Wang, T., Lee, I. H., et al. 2017. Inactivation of porcine endogenous retrovirus in pigs using CRISPR-Cas9. Science, 357(6357): 1303-1307.

Nordberg, A., Minssen, T., Holm, S., Horst, M., Mortensen, K. and Møller, B. L. 2018. Cutting edges and weaving threads in the gene editing (Я)evolution: Reconciling scientific progress with legal, ethical, and social concerns. Journal of Law and the Biosciences, 5(1): 35-83.

Nuffield Council on Bioethics. 2015. Ideas about naturalness in public and political debates about science, technology and medicine. Available from: https://www.nuffieldbioethics. org/wp-content/uploads/Naturalness-analysis-paper.pdf.

Nuffield Council on Bioethics. 2016. Genome editing: An ethical review. Available from: https://www.nuffieldbioethics.org/publications/genome-editing-an-ethical-review.

Nuffield Council on Bioethics. 2018. Genome editing and human reproduction. http://nufeldbioethics.org/wp-content/uploads/Genome-editing-and-human-reproduction-FINAL-website.pdf

Pulecio, J., Verma, N., Mejía-Ramírez, E., Huangfu, D. and Raya, A. 2017. CRISPR/Cas9-based engineering of the epigenome. Cell Stem Cell, 21(4): 431-447.

Santos, D. P., Kiskinis, E., Eggan, K. and Merkle, F. T. 2016. Comprehensive protocols for CRISPR/Cas9-based gene editing in human pluripotent stem cells. Current Protocols in Stem Cell Biology, 38(1): 5B-6.

Schacker, M. and Seimetz, D. 2019. From fiction to science: Clinical potentials and regulatory considerations of gene editing. Clinical and Translational Medicine, 8(1): 1-16.

Shah, S. Z., Rehman, A., Nasir, H., Asif, A., Tufail, B., Usama, M., et al. 2018. Advances in research on genome editing CRISPR-Cas9 technology. Journal of Ayub Medical College Abbottabad, 31(1): 108-122.

Shao, E. and Pershad, Y. 2019. CRISPR co-inventor Jennifer Doudna talks ethics and biological frontiers. Stanford|McCoy Family Center for Ethics in Society. Available from: https://ethicsinsociety.stanford.edu/buzz-blog/crispr-co-inventor-jennifer-doudna-talks-ethics-and-biological-frontiers.

Smith-Willis, H. and San Martín, B. 2015. Revolutionizing genome editing with CRISPR/Cas9: Patent battles and human embryos. Cell and Gene Therapy Insights, 1(2): 253-262.

Steven, O. (Ed.). 2015. Committee on Science, Technology, and Law; Policy and Global Affairs; National Academies of Sciences, Engineering, and Medicine. International Summit on Gene Editing. Available from: http://www.nap.edu/catalog/21913/international-summit-on-human-geneediting-a-global-discussion.

Thomson, C. 2015. The human germline genome editing debate. Impact Ethics. Available from: http://impactethics.ca/2015/12/04/the-human-germline-genome-editing-debate/.

Tuerlings, E. 2019. WHO expert advisory committee on developing global standards for governance and oversight of human genome editing: Background paper governance 1: Human genome editing.

UNESCO. 1997. Universal Declaration on the Human Genome and Human Rights, adopted by the General Conference on Nov. 11, 1997 and endorsed by the United Nations General Assembly, 53rd session, resolution AIRES/53/152, Dec. 9, 1998.

UNESCO. 2005. Universal Declaration on Bioethics and Human Rights, adopted by the General Conference, Oct. 19, 2005.

UNESCO. 2015. Report of the IBC on Updating Its Reflection on the Human Genome and Human Rights. Available from: http://unesdoc.unesco.org/images/0023/002332/233258E.pdf.

Vaschetto, L. M. 2018. Modulating signaling networks by CRISPR/Cas9-mediated transposable element insertion. Current Genetics, 64(2): 405-412.

Wachowicz, J. 2019. The Patentability of Gene Editing Technologies Such as CRISPR and the Harmonization of Laws Relating to Germline Editing. American University Intellectual Property Brief, 10: 34-45.

World Health Organization. 2011. Second WHO Global Consultation on Regulatory Requirements for Xenotransplantation Clinical Trials. Geneva, Switzerland; October 17-19, 2011.

Zhang, D., Hussain, A., Manghwar, H., Xie, K., Xie, S., Zhao, S., et al. 2020. Genome editing with the CRISPR-Cas system: An art, ethics and global regulatory perspective. Plant Biotechnology Journal, 18(8): 1651-1669.

Zhang, F. 2014. CRISPR-Cas systems and methods for altering expression of gene products. International patent WO2014093661.

Zhu, L. J. 2015. Overview of guide RNA design tools for CRISPR/Cas9 genome editing technology. Frontiers in Biology, 10: 289-296.

# CRISPR/Cas and Gene Therapy: An Overview

Luis María Vaschetto

Calle Pública N° 277, Barrio Liniers de Horizonte, Alta Gracia (Posta Code 5186), Córdoba, Argentina

## Introduction

Gene therapy is a procedure based on the therapeutic delivery of exogenous genetic material (either DNA or RNA) in order to treat genetic disorders and diseases. Traditional gene therapy approaches are based on the use of viral vectors and transgene expression to replace defective proteins, but these methods have raised several concerns associated with insertional mutagenesis, use of replication-deficient viral vectors, high immunogenicity and cytotoxicity effects. The development of the clustered regularly interspaced short palindromic repeats (CRISPR)/-CRISPR-associated enzyme (Cas) genome editing systems represents a promising approach for the development of therapeutic applications aimed at treating genetic disorders. Remarkably, CRISPR/Cas9 can be used to correct defective mutations in a versatile, RNA-programmable manner (Vaschetto 2018). However, this technology also faces several limitations that still hinder its applicability in the clinic.

## Showcase examples and limitations of the CRISPR/Cas technology

### CRISPR/Cas9 in gene therapy: Showcase examples

#### Cystic fibrosis

Cystic fibrosis (CF) is one of the most common genetic disorders which is caused by recessive mutations in the cystic fibrosis transmembrane conductance regulator (CFTR) gene. The CRISPR-Cas9 genome editing system has already been employed to repair mutations that lead to CF both *in vitro* and in induced pluripotent stem cells

E-mail: luisvaschetto@hotmail.com

(iPSC) (Schwank et al. 2013, Firth et al. 2015, Xia et al 2018, Ruan et al 2019, Sanz et al. 2017). For example, in iPSCs derived from CF-patients, Firth et al. (2015) reported that CRISPR-Cas9 was successfully used to correct a deletion in the exon 10 of the CFTR gene (F508del), enabling the restoration of normal gene function.

## Duchenne muscular dystrophy

Duchenne muscular dystrophy (DMD) is a X-linked recessive genetic disease caused by mutations in the dystrophin gene, which encodes a cytoskeletal protein (dystrophin). In skeletal muscle of dogs, a CRISPR-Cas9–mediated genome editing approach allowed to correct a stop mutation in the dystrophin gene (Amoasii et al 2018). The technique was successful to restore dystrophin expression, showing a high efficiency (up 90 percent) 8 weeks after adeno-associated virus (AAV)-mediated CRISPR genome editing. In mice, Zhang et al. (2020) also used a self-complementary AAV vector to deliver CRISPR/Cas9 system components and thus correct MDM mutations.

## Retinal diseases

Retinal cells are an interesting target in CRISPR/Cas gene therapy due to the small volume of retinal tissue and its self-contained nature, both being beneficial features to avoid undesired systemic side effects. In 2019, a clinical trial was launched to assess the efficacy of CRISPR/Cas9 to treat Leber congenital amaurosis type 10 (LCA10). LCA10 is a severe retinal dystrophy caused by mutations in the centrosomal protein (CEP) 290 gene. LCA10 often leads to early loss of vision during the first years of life. In mice and non-human primate cells, Maeder et al. (2019) demonstrated that subretinal injection of the AAV-CRISPR-Cas9 vector targeting the CEP290 gene can restore its normal expression pattern.

## Haematological disorders: β-thalassaemia

β-thalassaemia is an autosomal recessive disorder caused by mutations in the β-globin gene (HBB), a subunit of haemoglobin (the protein in the blood that carries oxygen throughout the body). β-thalassaemia causes a type of severe anaemia associated with the inheritance of mutations in the HBB gene. In iPSCs obtained from thalassemia patients, a strategy using CRISPR/Cas9 combined with the piggyBac transposon system successfully corrected mutations in the HBB gene (Xie et al. 2014).

## Alzheimer's disease

Alzheimer's disease (AD) is a neurodegenerative disease and the most common cause of dementia due to the loss of memory and impairment of cognitive functions. The APOE gene, which encodes for a protein (apolipoprotein E) involved in the metabolism of fats, has been identified as a major risk factor for AD. In mouse astrocyte cells, a CRISPR/Cas9 base editing approach was used to correct the apoE4 allele sequence to the functional (apoE3) variant by modifying only one nucleotide in the defective allele. This strategy showed minimal insertion/deletion (indel) formation at the site of DNA cleavage (Komor et al. 2016).

## Waardenburg syndrome

Waardenburg syndrome (WS) is a group of autosomal dominantly inherited disorders that may lead to congenital hearing loss, anophthalmia and pigmentation deficiency. WS can be categorized into 4 major types based on certain clinical features and underlying genetic causes. The WS type 2 (WS2) is associated with mutations in the microphthalmia-associated transcription factor (MITF). Using a pig as a model for WS2, Yao and colleagues observed successful correction of the MITF gene via CRISPR/Cas9-mediated homology-directed repair (HDR). This approach rescued anophthalmia and hearing loss phenotypes observed in pigs (Yao et al. 2021).

# Limitations of the technology

## Off-target effects

Off-target effects are defined as unwanted mutations in the genome and represent a major challenge to be faced in the design of efficient and safe CRISPR-Cas genome editing methods. Although Cas nucleases are highly effective to introduce double strand breaks (DSBs) at desired genomic *loci*, whole-genome sequencing studies have demonstrated the existence of off-target mutations in target cells and tissues (Iyer et al. 2018, Park et al. 2019, Kim et al 2015, Zhang et al. 2015). CRISPR-Cas-based gene therapy can be categorized into four major strategies: gene knockout, gene knockin/replacement, base editing, and prime editing (Zhang 2021). In this regard, it is important to highlight that CRISPR base editors may exhibit DNA editing activity which is independent of nuclease binding (Collins et al. 2020). The discovery of off-target effects requires the development of efficient screening methods for the off-target mutations derived from the use of this technology. In the future, it is expected that the design of high specific guide RNAs (gRNAs) and the development of highly specific Cas variants will enable to overcome this limitation (Zhang 2021).

## Delivery systems

A variety of approaches for delivering CRISPR/Cas components are currently available. CRISPR-Cas delivery systems are divided into two categories: 1. viral methods: adeno-associated virus (AAV), lentivirus (LV) and adenovirus (AdV) vectors, and 2. non-viral methods: electroporation, microinjection, hydrodynamic delivery, etc. Each method has its pros and cons (for a detailed review see Lino et al. 2018). For instance, in CRISPR/Cas9 AAV vector-mediated gene editing therapy, it is important to have into account that the size limit of gene insertion is approximately 4700 bp, which is a challenge when considering that the size of the *Streptococcus pyogenes* Cas9 (spCas9) gene alone is approximately 4,300 base pairs (bp) (Rosenblum et al. 2020).

## Immunogenicity

The stimulation of counterproductive immune responses has historically been one of the most important problems in gene therapy. With regard to the use of CRISPR-Cas, IgG antibodies against *Staphylococcus aureus* Cas9 (saCas9) and SpCas9 proteins in blood samples obtained from healthy patients have recently been detected (Charlesworth et al. 2019). Different strategies including, among others, immune privileged organs, epitope masking, alteration of antigen presentation pathways, non-immunogenic Cas9 variants and induction of immune tolerance, are being assessed in order to solve the immunogenicity of the Cas9 protein (Mehta and Merkel 2020).

# References

Amoasii, L., Hildyard, J. C., Li, H., Sanchez-Ortiz, E., Mireault, A., Caballero, D., et al. 2018. Gene editing restores dystrophin expression in a canine model of Duchenne muscular dystrophy. Science, 362(6410): 86-91.

Charlesworth, C. T., Deshpande, P. S., Dever, D. P., Camarena, J., Lemgart, V. T., Cromer, M. K., et al. 2019. Identification of preexisting adaptive immunity to Cas9 proteins in humans. Nature Medicine, 25: 249-254.

Collins, S. P. and Beisel, C. L. 2020. Your base editor might be flirting with single (stranded) DNA: Faithful on-target CRISPR base editing without promiscuous deamination. Molecular Cell, 79(5): 703-704.

Firth, A. L., Menon, T., Parker, G. S., Qualls, S. J., Lewis, B. M., Ke, E., et al. 2015. Functional gene correction for cystic fibrosis in lung epithelial cells generated from patient iPSCs. Cell Reports, 12(9): 1385-1390.

Iyer, V., Boroviak, K., Thomas, M., Doe, B., Riva, L., Ryder, E., et al. 2018. No unexpected CRISPR-Cas9 off-target activity revealed by trio sequencing of gene-edited mice. PLoS Genetics, 14(7): e1007503.

Kim, D., Bae, S., Park, J., Kim, E., Kim, S., Yu, H. R., et al. 2015. Digenome-seq: Genome-wide profiling of CRISPR-Cas9 off-target effects in human cells. Nature Methods, 12(3): 237-243.

Komor, A. C., Kim, Y. B., Packer, M. S., Zuris, J. A. and Liu, D. R. (2016). Programmable editing of a target base in genomic DNA without double-stranded DNA cleavage. Nature, 533(7603): 420-424.

Lino, C. A., Harper, J. C., Carney, J. P. and Timlin, J. A. 2018. Delivering CRISPR: A review of the challenges and approaches. Drug Delivery, 25(1): 1234-1257.

Maeder, M. L., Stefanidakis, M., Wilson, C. J., Baral, R., Barrera, L. A., Bounoutas, G. S., et al. 2019. Development of a gene-editing approach to restore vision loss in Leber congenital amaurosis type 10. Nature Medicine, 25(2): 229-233.

Mehta, A. and Merkel, O. M. 2020. Immunogenicity of Cas9 protein. Journal of Pharmaceutical Sciences, 109(1): 62-67.

Park, S. H., Lee, C. M., Dever, D. P., Davis, T. H., Camarena, J., Srifa, W., et al. 2019. Highly efficient editing of the β-globin gene in patient-derived hematopoietic stem and progenitor cells to treat sickle cell disease. Nucleic Acids Research, 47(15): 7955-7972.

Rosenblum, D., Gutkin, A., Dammes, N. and Peer, D. 2020. Progress and challenges towards CRISPR/Cas clinical translation. Advanced Drug Delivery Reviews, 54(155): 176-186.

Ruan, J., Hirai, H., Yang, D., Ma, L., Hou, X., Jiang, H., et al. 2019. Efficient gene editing at major CFTR mutation loci. Molecular Therapy-Nucleic Acids: 16, 73-81.

Sanz, D. J., Hollywood, J. A., Scallan, M. F. and Harrison, P. T. (2017). Cas9/gRNA targeted excision of cystic fibrosis-causing deep-intronic splicing mutations restores normal splicing of CFTR mRNA. PloS One, 12(9): e0184009.

Schwank, G., Koo, B. K., Sasselli, V., Dekkers, J. F., Heo, I., Demircan, T., et al. 2013. Functional repair of CFTR by CRISPR/Cas9 in intestinal stem cell organoids of cystic fibrosis patients. Cell Stem Cell, 13(6): 653-658.

Vaschetto, L. M. 2018. Modulating signaling networks by CRISPR/Cas9-mediated transposable element insertion. Current Genetics, 64(2): 405-412.

Xia, E., Duan, R., Shi, F., Seigel, K. E., Grasemann, H. and Hu, J. 2018. Overcoming the undesirable CRISPR-Cas9 expression in gene correction. Molecular Therapy-Nucleic Acids, 13: 699-709.

Xie, F., Ye, L., Chang, J. C., Beyer, A. I., Wang, J., Muench, M. O., et al. 2014. Seamless gene correction of β-thalassemia mutations in patient-specific iPSCs using CRISPR/Cas9 and piggyBac. Genome Research, 24(9): 1526-1533.

Yao, J., Wang, Y., Cao, C., Song, R., Bi, D., Zhang, H., et al. 2021. CRISPR/Cas9-Mediated correction of MITF homozygous point mutation in a Waardenburg syndrome 2A pig model. Molecular Therapy-Nucleic Acids. DOI: https://doi.org/10.1016/j.omtn.2021.04.009

Zhang, B. 2021. CRISPR/Cas gene therapy. Journal of Cellular Physiology, 236(4): 2459-2481.

Zhang, X. H., Tee, L. Y., Wang, X. G., Huang, Q. S. and Yang, S. H. 2015. Off-target effects in CRISPR/Cas9-mediated genome engineering. Molecular Therapy-Nucleic Acids, 4: e264.

Zhang, Y., Li, H., Min, Y. L., Sanchez-Ortiz, E., Huang, J., Mireault, A. A., et al. 2020. Enhanced CRISPR-Cas9 correction of Duchenne muscular dystrophy in mice by a self-complementary AAV delivery system. Science Advances, 6(8): eaay6812.

# CRISPR and Cancer: From Bench to Clinic

**Muhammad Usama Tariq[1]\* and Amina Chaudhry[2]**

[1] Rutgers, State University of New Jersey, New Brunswick, New Jersey,
   United States of America

[2] Rutgers, State University of New Jersey, Waksman Institute of Microbiology,
   Busch Campus, Rutgers University, Piscataway, New Jersey, United States of America

## Introduction to Cancer

Cancer is a heterogeneous group of diseases which occurs as a consequence of acquisition of mutations in genes controlling pathways involved in growth, differentiation, proliferation and many others. Hanahan and Weinberg have reviewed six major deregulated mechanisms of cancer, namely, sustained proliferative signaling, suppression of growth suppressor genes, activation of invasion and metastasis, replicative immortality, activation of angiogenesis, and escape from death (Hanahan and Weinberg 2011). Each hallmark mechanism is regulated by the mutations in different genes which either get activated or inactivated in different cancers.

Interestingly, the pattern and the trend of accumulation of mutations in cells decides the fate of cancer whether it would become malignant or not. Initiation of cancer occurs because of driver mutations in parent cells (Pon and Marra 2015). In hematologic malignancies, the driver mutations are introduced in stem cells, whereas in solid tumors, these mutations occur in differentiated cells of different tissues. However, if a cancer must become malignant, then, overtime, on top of driver mutations, more mutations are introduced which result in malignancy (Bozic et al. 2010).

Importantly, deregulation of certain pathways is associated with different cancers. For instance, disruption of the Wnt/Beta catenin pathway is common in colorectal cancer (Schatoff et al. 2017). The same pathway is also associated with leukemia, breast cancer, and melanoma (Zhan et al. 2017). The next section of this chapter provides further insights on commonly deregulated pathways in several cancers.

\* Corresponding author: usama.tariq@rutgers.edu

With advanced cell biology and sequencing techniques, the information on deregulated pathways has become readily available and this is important for developing treatments against cancer. In fact, researchers have been investing in developing cures by utilizing such data. However, the impact of such approach is not very far reaching as nearly 9 million people lost their lives to cancer in 2018 alone (Bray et al. 2018). Therefore, better treatments are the need of this time. This chapter discusses the available treatments and limitations associated with them. Later in the chapter, we will discuss about the use of CRISPR, a promising tool in effectively treating cancer.

# Major deregulated pathways in cancer

In humans, certain signal transduction pathways are responsible for enhancing growth and proliferation while some others inhibit them. A few of these pathways are described in this section. Recent technical advancements have revealed several signal transduction pathways related to growth, proliferation, and homeostasis; however, we will focus on the well-established, canonical pathways used for the purpose of cancer therapeutics. We divide these pathways into two categories: i. Growth suppressive ii. Growth stimulatory.

## Growth inhibitory or tumor suppressor pathways

### Tumor growth factor beta (TGF-β) pathway

Takaku et al. reported the deregulation of Tumor Growth Factor Beta (TGFβ) pathway in different types of tumors (Takaku et al. 1998). The canonical TGFβ pathway has three basic components: signal transduction components, inhibitors of SMADs, and epigenetic control. Mechanistically, TGFβ ligand binding results in the activation of the pairs of type I and type II TGFβ receptors by oligomerization which is essential for stimulation of downstream components (Massague 2012).

Upon ligand binding, type I and type II receptors come closer, phosphorylate each other, and create the binding sites in the cytoplasmic tails for Receptor activated SMAD (R-SMAD) proteins. Receptor bound SMADs get phosphorylated by both receptors. This phosphorylation event is necessary for allowing the formation of R-SMAD/co-SMAD complex in which two molecules of R-SMAD are bound with a single molecule of co-SMAD. This activated complex shuttles to the nucleus and binds to the promoter region of different genes to regulate their expression, thereby controlling growth and proliferation (Massague 2012).

TGFβ signaling pathway is known for its growth suppressive traits. The common transcription targets of this pathway include the cell cycle inhibitors such as p15 and p21. Moreover, it downregulates the expression of c-Myc (oncogene) by inhibiting the recruitment of transcriptional machinery on c-Myc promoter Apart from growth suppressive traits, this pathway also targets Snail, Zeb1, and Twist1 which are involved in differentiation, resulting in inhibition of proliferation (Peng et al. 2016).

Downregulation of TGFβ pathway is crucial for tumor initiation, in the context of certain cancer types. After a driver mutation strikes the cell, a primary response of a pre-cancer cell is to allow excessive proliferation. This driver mutation may be introduced in the components of TGFβ signaling to inactivate the cascade (Zeitouni et al. 2016). As a result, the inhibition of this pathway on MYC expression is lost. Moreover, the expression of the cell cycle inhibitors such as CDKN1A and CDKN2B is lost too because TGFβ pathway transcriptionally regulates their expression (Massague 2012). Alternatively, this pathway may get activated in response to oncogenic stress. However, due to continuous accumulation of mutations in cancerous cells, components of TGFβ pathway are also altered. These mutations render the growth suppressive function of this pathway to collapse which allows cancer progression.

The growth suppressive nature of TGFβ signaling is usually associated with early tumorigenesis. However, its role in late-stage tumors is quite complicated. Moreover, the role of this pathway varies with the stage of tumorigenesis, given the context (Bierie and Moses 2006). For instance, it inhibits tumor progression during the initial stages by collaborating with p53, a well-known tumor suppressor. Overtime, when p53 mutations are introduced and p53 becomes mutp53, SMADs interact with it and stabilize cyclins as well as Sharp-1 which are important for allowing metastasis. It has been experimentally shown that loss of p53 in TGFβ dependent cancers results in metastatic inhibition (Adorno et al. 2009). In addition, TGFβ stimulation allows epigenetic modifications such as hyper-methylation of certain epithelial marker genes which promotes epithelial to mesenchymal transformation (EMT), a hallmark pre-requisite for cancer metastasis, though the mechanisms of TGFβ dependent epigenetic alterations are not well understood (Papageorgis et al. 2010).

## Hippo pathway

Hippo pathway regulates stem cell maintenance, proliferation, and cell survival. The role of this pathway is important for cell fate determination and tissue growth during different developmental stages. Accumulating evidence suggests that hippo pathway plays a vital role in tumor suppression by controlling the expression of genes responsible for performing the cues mentioned above. Downregulation of the hippo pathway results in the activation of oncogenes, normally suppressed when the pathway is stimulated (Harvey and Tapon 2007, Harvey and Hariharan 2012, Snigdha et al. 2019).

The components involved in canonical hippo signaling are Mammalian STE20-like (MST) kinases, Large Tumor Suppressors (LSTS), Salvador homologue 1 (SAV1), MOB kinase activators, and YAP and TAZ transcriptional coactivators. Mechanistically, under pathway stimulated conditions, LATS dependent phosphorylation of YAP and TAZ coactivators allows their binding with 14-3-3 protein in the cytoplasm. Consequently, YAP/TAZ are sequestered in the cytosol and are no longer allowed to regulate transcription. Conversely, under pathway unstimulated conditions, the cascade of kinases upstream of YAP/TAZ remains inactive. As a result, the coactivators are free to modulate the activity of different transcription factors such as SMADs, TEADs, TBX5, p73 etc. thereby controlling the processes mentioned above (Hong and Guan 2012).

In this pathway, YAP and TAZ are crucial determinants of the pathway because during oncogenesis, mutations in proteins acting upstream of YAP/TAZ result in unwanted activation of targeted genes. Several reports have indicated that deregulation of hippo pathway is responsible for chemo-resistance in cancer stem cells. This is commonly found in the breast, colon, liver, and several other forms of cancer (Bartucci et al. 2015, Mao et al. 2014, Touil et al. 2014, Xu et al. 2015).

## Tumor necrosis factor pathway

Tumor Necrosis Factor (TNF) is a pro-inflammatory cytokine which regulates cell proliferation, differentiation, apoptosis, modulation of immune responses, and induction of inflammation. TNF family of proteins consists of 19 members which recognize 29 different receptors to activate different signaling cascades (Dostert et al. 2019). TNF ligands are known as double-edged swords because their roles are implicated in physiological (mediating immune responses, morphogenesis, and apoptosis) as well as pathological states of the cell (tumorigenesis, transplant rejection, and viral replication) (Aggarwal 2003). Evidence suggests that TNF alpha is a canonical ligand which mediates three signaling pathways *in vivo*. The third one acts as growth inhibitory pathway by controlling apoptosis; therefore, in this section, we will mainly focus on apoptotic pathway stimulation in response to TNF pathway activation, though TNF also controls growth and proliferation (Rath and Aggarwal 1999).

Mechanistically, binding of TNF alpha ligand to its receptors, such as TNF Receptor 1 (TNFR1) and TNRF2 on cell surface results in the binding of signal transducers to receptor cytoplasmic tails. A primary response to receptor activation is the recruitment of TNF Receptor Associated Death Domain (TRADD) protein to the TNFR1. TRADD recruitment facilitates the binding of Fas Associated Death Domain (FADD), Rip Associated ICH-1/CED-3-homologous protein with a Death Domain (RAIDD), MAPK Activating Death Domain (MADD) and Receptor Interacting Protein (RIP) to the receptor. Each of these signal transducers is involved in controlling different processes. For apoptotic regulation, the recruitment of TRADD and FADD is crucial. Binding of TRADD and FADD to TNFR1 draws Caspase 8 towards the receptor bound complex which causes Caspase activation. This caspase is known as initiator caspase because it turns on the cascade of proteolysis, thereby activating executioner caspases such as Caspase 3, 6, 7 which regulate apoptosis. In addition to caspase activation, initiator Caspase also causes proteolysis of BH3 Interacting Death Domain (BID) converting it into truncated BID (tBID). Resultantly, tBID disrupts outer mitochondrial membrane, releasing pro-apoptotic components such as Cytochrome C in the cytoplasm. After a few additional steps, the cell eventually undergoes apoptosis (Varfolomeev and Ashkenazi 2004, Wajant et al. 2003).

TNF alpha dependent apoptotic pathway is vital to optimal regulation of growth and development. Despite its growth inhibitory role, TNF is upregulated in different cancers such as colorectal, prostate, breast, chronic lymphocytic leukemia, Barrett's adenocarcinoma, and multiple other cancers (Montfort et al. 2019, Wang and Lin 2008). This is mainly because of the alternate cascades regulated following receptor activation. This includes activation of NFkB, protein kinase C and MAPK signaling (Wajant et al. 2003). Importantly, the deregulation of this pathway is linked to almost all contexts of tumorigenesis such as cell transformation, metastasis, angiogenesis, invasion, proliferation etc. (Montfort et al. 2019). It is very important to understand how apoptotic regulation resulting from TNFR activation is compromised in tumor cells. Given the frequent upregulation of TNF ligand, answering this question would allow us to find a better treatment of cancers caused by dysregulated TNF pathway.

In addition to TNF pathway activation in different cancers, suppression of this pathway has also been seen. For instance, literature suggests that mice lacking TNF are more prone to develop 3'-methylcholanthrene induced skin carcinoma as compared to mice with activated TNF signaling (Swann et al. 2008). Moreover, evidences show that TNF activation results in suppression of glioma formation in xenograft mouse models. This suppression occurs only when TNF ligand is expressed by tumor associated macrophages (Nakagawa et al. 2007). Furthermore, TNFR2 allows inhibition of angiogenesis by the production of Nitric oxide (NO). Mechanistically, TNFR2 mediated activation allows NO production which somehow attracts macrophages towards the tumors and allows inhibition of angiogenesis. Gene knockout models for TNFR2 does not show such anti-tumorigenic properties (Zhao et al. 2007).

## Growth stimulatory or oncogenic pathways

### Receptor tyrosine kinases

In multicellular organisms, most signaling cascades are activated by ligand dependent activation of cell surface receptors. There are many types of cell surface receptors that regulate growth and proliferation. Receptor Tyrosine Kinases (RTKs) are one of the most important classes of cell surface receptors involved in these processes. In addition to controlling growth and proliferation, these receptors play an important role in development, morphogenesis, mitotic regulation, cell fate determination, differentiation, cell migration, metabolism, and cell cycle control (Wintheiser and Silberstein 2020). RTKs are classified into several subtypes that are characterized by different receptors. Some crucial RTK subclasses include Epidermal Growth Factor Receptor (EGFR), Fibroblast Growth Factor Receptor (FGFR), Insulin- and Insulin like-Growth Factor Receptors (IR, IGFR), Platelet Derived Growth Factor Receptor (PDGFR) and many others (Lemmon and Schlessinger 2010, Paul and Hristova 2019).

Each RTK contains an extracellular domain, an intracellular domain, and a transmembrane domain that spans the plasma membrane at least once. The cytosolic domain of the receptor contains a kinase domain which is important for its activation. Ligand binding allows for receptor dimerization which facilitates the phosphorylation of intracellular domain of the receptor. Each kinase domain in the dimer cross phosphorylates the other one. The phosphorylated residues act as binding sites for different proteins which activate different signaling cascades. MAPK, PI3K, PLC, and calcium signaling are few of the products of RTK activation. Hence, they control multiple downstream pathways and regulate various cellular functions (Du and Lovly 2018, Maruyama 2014, Verstraete and Savvides 2012).

Different RTKs are frequently upregulated in different cancers. RTK upregulation is mostly associated with excessive proliferation (Butti et al. 2018, Sangwan and Park 2006). For instance, in the subsets of Acute Myeloid Leukemia, mutations in gate keeping residues of FLT3 receptor tyrosine kinase result in its unwanted activation, which also activates downstream signaling pathways that regulate proliferation, including many other functions (Kindler et al. 2010, Kiyoi et al. 2020). Accumulating evidence also suggests that RTK deregulation is associated with metastasis (Regad 2015). In addition, multiple genetic mutations have been reported in EGFR, which render breast cancer extremely aggressive. RTK deregulation is also reported in lymphoblastic lymphoma, multiple myeloma, Chronic Myeloid Leukemia, melanoma, and mast cell leukemia (Yamaoka et al. 2018). In summary, activating or gain of function mutations in RTKs lead to their upregulation and subsequent cancer formation.

### Mitogen activated protein kinase (MAPK) pathway

MAPK is an evolutionarily conserved pathway involved in a wide range of cellular functions including, but not limited to, cell proliferation, growth, differentiation, apoptosis, immune response, and stress response (Cargnello and Roux 2011, Zhang and Liu 2002). MAPK pathway is commonly activated downstream of activated RTKs. However, this pathway can also be stimulated by other types of receptors involved in TGFβ pathway, stress response, cytokine receptors etc. (Morrison 2012). Activated receptors may regulate gene expression, protein-protein interactions, and protein localization. In canonical MAPK pathway, message from the receptor is communicated to effectors through a cascade of phosphorylation events. The four main components of this pathway are Ras, Raf, MEK, and ERK. Except for Ras, the other three proteins are kinases (Plotnikov et al. 2011).

Canonically, activated RTK sequentially recruits GRB2 and SOS proteins. SOS is a Guanine Exchange Factor (GEF) for a small G protein called Ras. SOS recruitment to the activated receptor causes Ras to localize at the plasma membrane. Resultantly, SOS allows the exchange of GDP to GTP in membrane-bound Ras for its activation. Activated Ras binds to Raf kinase and activates it. This Ras-Raf binding allows a conformational change in the structure of Raf kinase which leads to its auto-phosphorylation. Activated Raf then phosphorylates MEK which phosphorylates ERK. Raf, MEK, and ERK kinases are significant regulators of MAPK pathway. To keep these kinases in proximity for efficient phosphorylation, they are brought together by binding to a scaffold protein called KSR. The final protein kinase of this pathway, ERK, controls different physiological functions, the most important of which are proliferation and growth by phosphorylating several proteins including, but not limited to, Cyclin D1, Myc, C-fos, C-jun (Plotnikov et al. 2011).

Upregulation of MAPK pathway is associated with a repertoire of cancers, faulty cell development, and cancer progression (Braicu et al. 2019). Each component of MAPK signaling is susceptible to getting deregulated depending on the type of cancer. Since the pathway has multiple effectors and controls many crucial cellular processes, its disruption also affects several physiological functions (Dhillon et al. 2007). Recent reports indicate that MAPK components are frequently upregulated in cancers. For instance, more than 90% of pancreatic cancers show mutated Ras, while nearly 30% AMLs exhibit deregulated Ras protein. In addition, Raf mutations have also been reported in 66% melanomas and 12% colorectal cancers (Braicu et al. 2019). In FLT3 RTK mutated Acute Myeloid Leukemia, constitutively active FLT3 RTK activates MAPK pathway which influences tumor progression (Kiyoi et al. 2020).

*PI3 kinase pathway*

Another pathway triggered by activated RTK is Phosphoinositide 3 kinase (PI3K). This pathway controls various functions such as metabolism, proliferation, survival, motility, and cellular growth (Martini et al. 2014, Yu and Cui 2016). Active PI3K modifies membrane lipids to regulate and perform different functions. There are 8 isoforms on PI3K which are divided into 3 main classes. Kinases from each class act on membrane lipids differently. Interestingly, this pathway sheds light on the underappreciated role of lipids as a significant regulatory component of cell signaling cascade (Jean and Kiger 2014).

Mechanistically, following RTK activation, both catalytic and regulatory subunits of PI3K are recruited to the phosphorylated receptor tails. The regulatory subunit of PI3K is known as p85, whereas the catalytic subunit is referred to as p110. Binding of these subunits to activated RTK brings it in proximity of the plasma membrane where PI3K phosphorylates 3′ OH group of Phosphatidylinositol lipid. This phosphorylated 3′ OH group acts as a ligand for several proteins which can initiate different downstream signaling cascades. Two main proteins recruited to phosphorylated membrane lipid are Akt kinase and PDK1. These proteins bind the plasma membrane using their PH domains and are brought in vicinity. As a result, PDK1 phosphorylates Threonine at position 308 (T308) of Akt to partially activate it. For complete activation, Akt must be phosphorylated by mTORC2 on Serine at the 473$^{rd}$ residue. Fully activated Akt allows for phosphorylation-dependent inhibition of the negative regulator of mTORC1, that is, TSC1/TSC2. mTORC1 complex phosphorylates 4EBP1 and S6K to regulate translation. Importantly, it also upregulates the expression of GLUT transporters, which leads to increased glycolysis and metabolism (Pauls et al. 2012, Yu and Cui 2016).

There are multiple negative regulators of PI3K pathway. The most important one, whose expression is commonly suppressed in multiple cancers, is PTEN phosphatase that removes the phosphorylation mark on Phosphatidylinositol lipid added by PI3K. Once the phosphorylation has been removed, the downstream pathway is rendered inactive. However, in cancers where PTEN expression is suppressed, the membrane lipid-dependent activation of downstream cascade is overachieved (Chalhoub and Baker 2009). Accumulating evidence indicates that along with having inactivating mutations in PTEN genes, gene encoding p110 subunit of PI3K is frequently upregulated or amplified in prostate, breast, cervical, and lung cancers (Liu et al. 2014). Although mutations in the p85-encoding gene are rare, yet they are sufficient to induce excessive proliferation in cultured cells. Moreover, recent advances also show upregulation of mTORC, Akt, TSC1/TSC2 and LKB1 (involved in the activation of tsTSC1/TSC2) in most aggressive forms of cancer (Guertin and Sabatini 2007).

*Wnt/beta catenin pathway*

During early development, Wnt pathway cross talks with other developmental pathways to mediate survival. Particularly during embryonic development, Wnt pathway controls the homeostasis of self-regenerating tissues and cells (Merrill 2012). In fact, this pathway is equally crucial during adulthood. Interestingly, intestinal epithelium has been established as a robust model for studying the mechanistic details of this pathway (Mah et al. 2016).

Intestinal epithelium is continuously in contact with chemicals, bears other types of insults, and acts as a barrier. Therefore, frequent renewal is essential for its optimal function. This layer is continuously revived by stem cells lying at the base of intestinal crypt cells. These stem cells are replenished by activated Wnt signaling (Flanagan et al. 2018). Continuous proliferation pushes the cells upwards. As these cells reach the villi of epithelium, Notch signaling is activated which results in their differentiation. The Wnt ligand is around 40 kDa in size and accommodates a lipid modification which is important for its secretion from the cells. The ligand in mostly produced and released via paracrine signaling. Its receptor is a heterodimeric protein present on the surface of different cells. The receptor heterodimer consists of Frizzled and LRP5/6 proteins. Ligand binding induces a conformational change causing receptor dimerization. Under unstimulated conditions, a cytoplasmic protein degradation complex comprising of GSK3 Beta, Casein Kinase 1, Adenomatous Polyposis Coli (APC ubiquitin ligase) and Axin downregulates Beta catenin by targeting it for proteasome-mediated degradation. Since Beta Catenin is a transcription regulator involved in the activation of the genes which are important for growth, proliferation, and stem cell renewal, its degradation perturbs all these functions. However, following receptor activation and dimerization, the important components of the degradation complex are recruited to the receptor, which results in the release of inhibition of Beta Catenin. As a result, Beta Catenin can enter the nucleus, bind to the effectors of this cascade, TCF/LEF, which subsequently allow the transcription of target genes (MacDonald et al. 2009).

Upregulation of the components of Wnt pathway is commonly associated with various types of cancers. For instance, mutations in LGR5, APC, and Axin2 genes are very common in colorectal cancer (Nie et al. 2020). Although fusion of VTI1A and TCF genes is a rare event in colorectal cancer, it renders the disease very aggressive. In addition, deregulation of the components of Wnt pathway are also found in renal cancer, melanoma, Wilms tumor, liver cancer, and rarely in breast cancer (Chan and Lo 2018, Kovacs et al. 2016, Zhang et al. 2017, Zirn et al. 2006).

# Sequential deregulation of oncogenes and tumor suppressors

In the previous section, we discussed multiple cellular pathways which can be inactivated or hyper-activated in different cancers. However, deregulation of a single pathway is insufficient to render a cancer aggressive or malignant because a limited number of aberrations may only perturb growth or proliferation, which can be easily detected and corrected by the body's immune system. Such complications are benign in nature. Intriguingly, the next logical query is how aggressive cancers steer clear of the immune responses, metastasize, and eventually cause death.

It is indispensable to evade body's defenses against cancer without the activation of oncogenes and inactivation of tumor suppressors. This feat is a result of collective deregulation of mechanisms involved in controlling cell cycle checkpoints, cell attachment, apoptotic regulation, immune surveillance, and other regulatory mechanisms. Herceg and Hainaut listed the mechanisms involved in such response during tumor progression (Herceg and Hainaut 2007). According to the proposed model, which also reiterates the importance of 'the hallmarks of cancer' (Hanahan and Weinberg 2011), prominent determinants for a cancer to become malignant include: unlimited growth and proliferation in the absence of growth signals, inhibition of cellular brakes (tumor suppressors), metastatic potential, chromosomal instabilities arising from perturbation in cell cycle checkpoints leading to faulty cell cycle, formation of new blood vessels, and escape from apoptosis.

A few specific pathways involved in tumor progression are worth mentioning here. This includes the upregulation of pathways involved in growth and proliferation such as signaling downstream of EGFR which activates MAPK and PI3K; inhibition of tumor suppressors and cell cycle inhibitors such as p53 and CKi, respectively; loss of cell surface bound molecules involved in attachment of cells to different substrates; sustenance of telomeres by overexpressing telomerase; induction of growth factors such VEGF involved in blood vessel formation; and the inhibition of proapoptotic and overexpression of anti-apoptotic proteins. Although these processes are very important for tumor progression, deregulation of different pathways results in different kinds of cancers. Such variations make up the identity of each type of cancer. Therefore, identification and treatment of each type of cancer can be devised accordingly.

Deregulation at multiple levels makes cancers very resilient to treatment. Despite such deregulation, a synchrony exists between different pathways involved in cancers for undergoing aggressive growth and proliferation (Andriani et al. 2012, Giancotti 2014, Mieczkowski et al. 2012). In other words, all these pathways, in one way or the other, are simultaneously required for cancer progression. Therefore, inhibiting a certain pathway breaks the ensemble and suppresses cancer progression. This idea lies at the heart of anti-cancer drugs and inhibitors design. In the following section, various therapeutic strategies are described which have been used for the treatment of multiple cancers. The section also highlights the strengths and limitations associated with each treatment method.

## Available treatment approaches

Cancer is a multifaceted disease that cannot be treated by adopting a straightforward procedure. Even in the face of current scientific progress and understanding of eukaryotic biology, development of a single panacea-like treatment for various forms of cancers is highly ambitious. Therefore, scientists from around the globe are investing in understanding differences between different types of cancers and devising strategies for inhibition accordingly. In this section, we will discuss some approaches used for the treatment of cancer at present. We will start with the basic ones and proceed to some advanced treatments while appraising their competence and impediments.

### Surgery

For non-invasive and locally restrained solid tumors on a certain organ, surgery is a possible treatment. A physician trained to operate on humans may skillfully perform the procedure. During surgery, the skin is incised and the tumor is physically removed along with tissue surrounding the tumor mass (Wyld et al. 2015). This method is efficient in treating cancer patients but it is not 100% effective because even after excision of the whole tumor, it is highly likely that one or more cancerous cells are left behind. Even one cancerous cell can potentially grow into a whole new tumor, a process known as relapse (Sabel et al. 2007, Tohme et al. 2017). Therefore, we cannot rely on surgery alone as cancer treatment and better survival. To make surgery more effective, the patient is prescribed drugs which in most cases are different chemotherapeutic agents administered before or after surgery. In addition to relapse, another drawback of surgery is that it can only be used to treat early stages of solid tumors because a tumor metastasizes during late stages and spreads to different organs and parts of the body. Therefore, it is practically impossible to physically excise the tumor from every organ (Tohme et al. 2017).

## Radiation therapy

Localized or solid tumors can also be treated using ionizing radiations. Using this strategy, highly reactive free radicals are produced in cancer cells which damage the DNA. For a cancerous cell, it is very unlikely to keep growing exponentially as result of excessive DNA damage. Radiation therapy unlike surgery is non-invasive because radiation exposure does not require excision of tumor tissue and skin. It only requires exposure of part of the body to ionizing radiations that bears the tumor (Mohan et al. 2019). The intensity of the radiations can be adjusted to contain and limit the exposure to cancer cells only (Baskar et al. 2012).

However, treatment of metastasized tumors using ionizing radiations is very risky because it requires very high intensity of radiations. At such intensity, the radiations can start damaging the normal tissues which leads to unintended consequences after the treatment (Mohan et al. 2019). Like surgery, relapse after radiation therapy is very likely; therefore, this method is supplemented with chemotherapeutics and other drugs as well. Unfortunately, this strategy is not applicable to hematologic malignancies.

## Chemotherapy

Another strategy used for treating tumors is based on inducing cytotoxicity in cells using chemical agents. Such agents are called chemotherapeutics which are administered either orally or intravenously. Once the drug is in the bloodstream, it spreads throughout the circulatory system and reaches the tumor. Such approach is helpful in treating hematologic malignancies in addition to locally residing solid tumors (Allart-Vorelli et al. 2015, Schirrmacher 2019).

A chemotherapeutic drug usually binds and inhibits multiple protein targets in cells. Consequently, perturbation of several pathways results in cell death and tumor reduction. The possibility of a cancer cell to maintain its growth and aggression after treatment with such an agent is next to none (Zitvogel et al. 2013). Moreover, the likelihood of development of resistance after therapy is reduced as compared to other treatments described above because of two main reasons. Firstly, a chemotherapeutic drug minimizes the probability of cancer relapse. Secondly, since a chemotherapeutic drug targets multiple pathways simultaneously, it is highly unlikely for the cancer cells to develop resistance to the treatment (Lundqvist et al. 2015).

A major drawback of this strategy is that the drug spreads throughout the body and does not discriminate between healthy and cancer cells. Consequently, chemotherapeutics causes excessive toxicity in the body causing devastating outcomes. It causes excruciating pain and adversely affects the quality of life of the patient. In the worst cases, it may even lead to death of the individual receiving the treatment. Commonly used chemotherapeutic agents include DNA damaging agents, alkylating agents, anti-metabolites, multiprotein targeting drugs etc.

## Targeted therapy and next generation anticancer inhibitors

Despite the promise of chemotherapeutics in treating cancers, they are still not safe owing to excessive toxicity in the body. The reason for such response, as described above, is the downregulation of multiple pathways in all cells regardless of their disease status. To overcome this problem, scientists have invested their efforts in designing compounds which can specifically target selective pathways. In fact, recently, the efforts have shifted to designing drugs that inhibit a single protein crucial for cancer progression. The main obstruction in such strategy is the identification of pathways that are integral to a specific cancer. Such information helps design inhibitors against specific proteins. For instance, in FLT3 mutated AML, FLT3 is overexpressed which deregulates the downstream pathways required for AML progression. Designing inhibitors against FLT3 helps control the progression of this cancer, which has been shown by many research groups all over the world (Gebru and Wang 2020).

Interestingly, inhibitors that selectively target specific proteins can be designed computationally and synthesized in the lab. Such approach helps us modify the structure of the inhibitor which selectively binds to the protein of interest leaving other pathways unaffected (Ramirez 2016). Fortunately, certain proteins are solely expressed in cancer cells. Computationally designing the inhibitors against such proteins and using them in patients does not cause toxicity in the body (Wang et al. 2018). Apart from designing these drugs computationally, they can also be isolated from plants and other sources (Dai et al. 2016, Greenwell and Rahman 2015). One such example is paclitaxel based on a natural product, Taxol, extracted from plants of Taxus species. It is a successful anti-cancer drug used against ovarian cancer, esophageal cancer, breast cancer, lung cancer, Kaposi sarcoma, cervical cancer, and pancreatic cancer at present (Miele et al. 2012, Zhu and Chen 2019).

Targeted therapy holds great promise in the treatment of many cancers; however, a major limitation in its use is acquisition of resistance. Acquired resistance can be defined as the ability of cancer to stop responding to the drug(s) being used over time. In other words, initially, a drug or inhibitor may exhibit potent effects in killing cancer and reducing its growth; however, due to increased selection pressure on the cancer cells over time, they may alter the structure of the protein being targeted (so that the drug cannot bind) or shift to another pathway for growth. As a result, the drug becomes ineffective as it targets only a single protein (Floros et al. 2020, Lackner et al. 2012). The example of such drug is MLN-518 (Tandutinib), which specifically targeted FLT3 RTK for the treatment of FLT3-dependent AML. Initially, when Tandutinib was introduced in the market, it showed formidable results in treating FLT3-dependent AML. After constant exposure, the amino acid targeted by Tandutinib in the primary sequence of FLT3 RTK mutated. This alteration rendered the drug ineffective and relapse occurred (Weisberg et al. 2009).

The recent development of next generation anti-cancer drugs paved way for combatting resistance and cancer relapse. These new drugs simultaneously target two proteins and in doing so, they do not cause excessive toxicity because targeting two pathways simultaneously is not very toxic to the body. Moreover, it is highly unlikely for a cancer cell to develop resistance against such treatment because even if the primary sequence of one protein is altered, the other protein would still be inhibited. Therefore, use of these drugs holds great promise (Raghavendra et al. 2018). Unfortunately, evidence shows that in response to the next generation inhibitors, instead of altering the primary sequence of the targeted proteins, cancer activates additional pathways for its progression (Zahreddine and Borden 2013). To overcome this challenge, a thorough understanding of the molecular underpinnings of different cancers is essential because based on that information we can design drugs that can keep cancer from developing resistance and exploiting new pathways.

## CRISPR

With the advent of $21^{st}$ century, industrial development has led to burgeoning opportunities in metropolitans and lured huge populations towards urban settings and expansion of cities; it has also laden our environment with copious amounts of obnoxious substances. Collectively, these chemicals and foci of human population have resulted in environmental damage, which is continuously influencing health in several adverse ways. One of the most harmful consequences is increasing incidence of different types of cancer. The biggest challenge is the ever-evolving nature of cancer, which renders the existing treatments inadequate as they may only reduce the aggressiveness and treat symptoms but cannot eradicate the disease *per se.*

Clustered regularly interspaced short palindromic repeats (CRISPR) is a recently developed technique that is revolutionizing the field of life sciences and holds great promise in disease treatment. It has changed the perspective of treating ferocious diseases. This genome editing technique works better than the related techniques such as TALENs, RNA interference and Zinc finger nucleases, etc. in terms of efficacy, affordability, specificity, and ease of use (Khan 2019, Nerys et al. 2018). Hypothetically, the technique can cure any kind of genetic alteration. Since cancer is also a genetic disease, CRISPR can be used for its treatment. While it was first discovered by Ishino and his colleagues in 1987 in Osaka University, the technique was developed over several years (Ishino et al. 1987). The prestigious Nobel Prize in chemistry for the year 2020 was awarded to Charpentier and Doudna who developed the CRISPR as a tool for genome editing (Jinek et al. 2012).

CRISPR was first identified as an adaptive immune response in bacteria which is used to get rid of unwanted viral DNA from the bacterial genome. Mechanistically, a guide RNA (gRNA), which is complementary to the target sequence (where change is required) in the genome, recognizes and binds to its complementary sequence.

gRNA, because of its complementarity with the target and ability to bind to the Cas nucleases, guides the nuclease towards the target sequence. The gRNA consists of a spacer region and a scaffold carrying Cas protein. Spacer region of gRNA is complementary to the targeted sequence in the genome, while Cas protein introduces breaks in the Protospacer Adjacent Motif (PAM) of target site in the host genome (Jiang and Doudna 2017).

Importantly, the cleavage by Cas nuclease causes double strand breaks which can be repaired using Non-Homologous Ends Joining (NHEJ). It is highly unlikely for the same fragments of DNA to rejoin as a result of NHEJ, and hence, the DNA is altered. If such alteration is introduced in an important exon of a gene, it may introduce a pre-mature stop codon in the Open Reading Frame, or the gene may become non-functional due to translocation. Either way, the introduction of breaks and subsequent NHEJ results in the loss of gene function (Guo et al. 2018).

Interestingly, in addition to the introduction of double strand breaks at the target site, the gene can be modified in several ways. Depending on the desired genomic alteration, a different kind of Cas may be used; however, the gRNA used remains the same. CRISPR can be classified based on these different modifications. The most common type is Type II CRISPR in which Cas9 protein is used to introduce double strand breaks. Moreover, the Cas protein can be fused with functionally important protein domains to carry out several functions. For instance, Cas9 catalytic domain can be fused with DNA Methyl Transferases to epigenetically modify the targeted region instead of introducing a break in the DNA. Similarly, CRISPR can also be used to introduce a gene of interest at the targeted site (Makarova and Koonin 2015).

# CRISPR in cancer

## Use of CRISPR for research purpose (bench part)

To find a treatment against an ever-transforming disease like cancer, which accumulates new mutations at every cell division, CRISPR seems like the most reliable technique at present. Its promise in cancer therapeutics is based on the identification of therapeutic targets in different cancers (Xue et al. 2020). In addition to its use in the treatment of cancer, CRISPR can also be utilized in lab for the identification of novel therapeutic targets. With increasing number of identified targets, the data can be utilized to design precise and high-quality gRNA for efficient growth inhibition of cancer *in vivo*. CRISPR can help delete, express, or repress the gene of interest involved in carcinogenesis and probe the pathway for inhibition (Makarova and Koonin 2015). Here onwards, we will talk mainly about CRISPR/Cas9 which is used for creating knockouts or deleting genes. This technique is the most established one in terms of its genome-editing applications.

During carcinogenesis, along with the deregulation of proteins and non-protein coding genes, non-coding regions of genome also undergo changes crucial for cancer progression (Cuykendall et al. 2017). Such modifications of the genome result in alteration of gene regulatory sequences present in the non-coding regions, called enhancers. These non-coding elements may be located several kilobases away from their target genes. Under physiological conditions, these enhancer sequences can interact with genes of interest owing to spatial rearrangement of chromatin into distinct loop structures to regulate gene expression. However, such gene regulation is perturbed in cancerous cells and the enhancers begin regulating the expression of genes associated with cancer progression (Cuykendall et al. 2017). We can use techniques such as Hi-C or 3-C to extract structural information from chromatin indicating deregulation (Barutcu et al. 2016).

After the identification of enhancer sequences responsible for regulating gene expression in cancer cells, we can design gRNA against such sequences to knock them out. This strategy selectively targets cancer cells. Healthy cells would remain unaffected because the properties of the deregulated enhancers, such as distance from the gene of interest and sequence flanking the enhancer, are expected to be absent from normal cells.

CRISPR-mediated knockouts can also be utilized in combination with different cell and molecular biology techniques to study the role of a specific gene in cancer cells. For instance, mRNA expression can be investigated using RNA-Seq in CRISPR knockout mutants. This technique generates important data about the expression patterns of genes controlled by the gene of interest. This approach can help us uncover molecular underpinnings of pathways deregulated in cancer at transcriptional level. Similarly, western blotting could also be performed on CRISPR based knockouts to study protein expression. Interestingly, looking at protein expression and RNA-Seq data, concomitantly, informs us if the gene of interest is specifically associated with either transcriptional or translational regulation in cancer cells.

Literature indicates that PubMed search for publications using the keyword *'CRISPR'* has continuously yielded more hits from years 2002 to 2018, indicating an inclination of the scientific community towards this technique. In total, there were around 9332 CRISPR-related research articles published during the mentioned span of time. Around 30% of these publications were focused on CRISPR/Cas9, while approximately 9% were associated with cancer, excluding the ones that used CRISPR only for a few experiments. Moreover, around 23% of the search results mentioned the use of CRISPR/Cas9 as a therapeutic strategy, indicating the significance and reliability of this system in therapeutic targeting and cancer treatment (Tian et al. 2019).

To date, CRISPR/Cas9 is regarded as the most efficient and successful approach with outstanding target specificity, which makes it a suitable candidate for cancer cure without limitations that are associated with the available treatments, as mentioned in previous sections. Additionally, CRISPR can also be used to study different stages of carcinogenesis in the lab, as discussed in the next section. We will look at the use of CRISPR in therapeutic targets' identification and validation in detail.

## Identification of therapeutic targets in cancer

In a loss of functional genetic screen for the identification of a deregulated protein in a particular cancer, downregulation of a particular gene results in either excessive proliferation or growth arrest, whereas the downregulation of the rest of the genes in the screen does not impact cancer cells. To identify such a gene, RNA interference (RNAi) based methods are feasible for carrying out genome-wide genetic screens. The number of genes included in the screen depends on the research interest. For instance, if the study aims to investigate the degradation mechanism of an established oncogene by identifying its upstream E3 ubiquitin ligase, we would design an RNAi-based library which contains RNAi constructs for targeting all E3 ubiquitin ligases. This strategy would allow gene identification regardless of their type and function. The most critical and important determinant for performing such a screen is the choice of library design and the read out.

It has been established that RNAi is limited in terms of efficacy and has off-target effects (Sledz and Williams 2005). Therefore, it is not very reliable for the identification of therapeutic targets. Fortunately, discovery of CRISPR/Cas9 has helped overcome hurdles associated with RNAi. Owing to the high efficacy and specificity of CRISPR, it can be used to knockout a gene to study its function. Comparative analysis of CRISPR/Cas9 and RNAi efficacy in literature indicates that the treatment of melanoma cells with vemurafenib and CRISPR/Cas9 library (targeting multiple genes in the genome) led to the identification of genes such as neurofibromin 1, 2 and MED12 having pathological relevance to skin cancer. The same screen, when performed using RNAi, could not identify MED12 and neurofibromin 2 as prominent targets of the melanoma. In fact, the only downregulation of neurofibromin 1 in case of RNAi was around 40%, which was far less than what was observed with CRISPR-based screen (Shalem et al. 2014).

Accumulating evidence also suggests that nucleophosmin (a histone chaperone) was recognized as an oncogene using a CRISPR based genetic screen both *in vitro* and *in vivo* models. For the identification of nucleophosmin, over 500 chromatin interacting genes were deleted (Li et al. 2020). In another study, CRISPR-based genome-wide screen was employed to identify genes involved in developing resistance in FLT3 mutated Acute Myeloid Leukemia against a drug, AC220. The researchers found that knockout of SPRY3 (negative regulator of FGF signaling) and GSK3 (negative regulator of Wnt signaling) was able to confer resistance in these cells as confirmed by reactivation of Wnt and FGF signaling (Hou et al. 2017). CRISPR-based genetic screen was also used to identify the target of STF-118804 drug, which was shown to potently inhibit another version of AML. The screen found out nicotinamide phosphoribosyl transferase (NAMPT) as a target for STF-118804 in AML (Matheny et al. 2013).

## Validation and characterization of identified targets

The identification of therapeutic targets in cancer is one of the most important steps for the development of effective therapies. However, a potent cure cannot be developed until or unless a therapeutic target has been validated functionally. Like its use in the identification of targets, RNAi can also be utilized for validation purposes. Yet, based on its limited efficacy, such application is very ambitious. Notably, CRISPR/Cas9 genome editing tool works perfectly for such confirmation owing to its efficacy and specificity. The prowess of this tool can be employed at multiple stages to devise a potent cure for cancer.

Remarkably, CRISPR/Cas9 based functional validation can be studied in several model systems such as mouse, zebra fish, fruit fly, and nematodes. In a recent study, CRISPR mediated modeling was used to functionally validate the tumor suppressor genes in small cell lung cancer (SCLC). In this study, the authors used an established mouse model for small cell lung cancer. Previously, it was reported that p130 and p107 are responsible for tumor progression in SCLC. However, the utilization of CRISPR to knockout p107 resulted in larger and more metastatic tumors as compared to p130 (Ng et al. 2020). In another study, drug resistant mutations in chronic myelogenous leukemia (CML) cells were validated. The researchers combined CRISPR/Cas9 with DNA sequencing technologies by either creating a knockout of 6-thioguanine-HPRT or introducing the drug resistant form of HPRT through knock in to investigate drug sensitivity (Smurnyy et al. 2014). CRISPR/Cas9 tool is also very important for introducing multiple genetic alterations in organoid cultures for validation of different target genes associated with a particular cancer. For instance, in human intestinal organoid model, CRISPR was used to introduce a panel of mutations in different genes to make an *in vitro* cancer model of colorectal cancer for drug as well as functional validation purposes. These mutations include APC, KRAS, SMAD4, PI3K, and TP53 (Matano et al. 2015). Such manipulations have also been done in cell lines to study cancer formation and progression (Heck et al. 2014).

## Double knockout models for the identification of deregulated mechanisms

Single gene knockout models are very important for understanding the role of a gene in a particular pathway involved in carcinogenesis. However, this gene may not present a holistic image of tumor progression. In order to gain a true perspective, simultaneous knockout of more than one gene may prove more valuable. However, this strategy may cause cell death due to excessive toxicity and impede studying the cancer. Interestingly, targeted knockout of two different genes often results in providing important information required for understanding a pathway or different pathways involved in the advancement of cancer. For instance, it was recently reported that a single knockout of either CXCR4 or CXCR7 (receptors for CXCL12 ligand) in Triple Negative Breast Cancer (TNBC) reduces the growth, proliferation, and migration with increased cell cycle progression time. However, double knock out of both receptors synergistically enhances all the mentioned processes. Interestingly, induction of single knockout cells with CXCL12 ligand results in increased migration of TNBCs; nonetheless, this migration efficiency decreases when double knockout TNBCs are induced with CXCL12 suggesting a synergistic role of both receptors in breast cancer progression (Yang et al. 2019).

Serine arginine protein kinases (SPRKs) phosphorylate splicing proteins to regulate their function, post-transcriptionally. SPRK1 and SPRK2 are frequently upregulated in nasopharyngeal carcinoma (NPC). Numerous SPRK1 and SPRK2 targeting inhibitors have been developed to study the long-term roles of these protein in tumorigenesis; however, these inhibitors have transient effects. Prolonged and repeated exposures allow accumulation of mutations and cell lethality, making it impossible to study the specific roles of SPRK1 and SPRK2. Another recent report indicates that dual knockout of SPRK1 and SPRK2 from NPC cells is vital for studying their roles. This model played an important role in understanding the intricacies of NPC progression. Importantly, it was found out that dual inhibition of the proteins is required for investigation of sustained cancer growth and transformation, which is unlikely to decipher using single knockouts or drug targeted approaches (Prattapong et al. 2020).

Furthermore, downregulation of p53 and PTEN tumor suppressors is commonly found in hepatocellular carcinoma (HCC). However, the precise temporal downregulation of both tumor suppressors for tumor progression was previously unknown. Emerging evidence suggests that dual knockout of p53 and PTEN results in tumor formation as depicted by Glutamine synthase positive immunohistochemistry. Likewise, this dual knockout mouse model also highlighted that p53 and PTEN inactivation is one of the most important and primary events in the initiation of HCC (Liu et al. 2017).

## Use of CRISPR in clinic for cancer treatment

Owing to its simplicity and intrinsic programmability, CRISPR/Cas9 can be a very reliable choice for the treatment of cancer. This promising tool is continuously being improved for its safe action in human body. Therefore, multiple in-human clinical trials are currently underway against different types of cancers. Specific details on clinical trials will be discussed in the next section of this chapter.

To understand the use of CRISPR in humans for the treatment of cancer, let's focus on the pioneering studies in human cells that focused on targeting liver and prostate cancers. A fusion gene called TMEM135-CCDC67 is responsible for cancer progression in the prostate; likewise, another fusion gene, MAN2A1-FER, causes a subset of hepatocellular carcinoma. In the founding studies, the authors utilized virus mediated delivery mechanism. Two viruses, one carrying Cas9 and gRNA which allows breakpoint in the fusion gene and the other carrying a segment of DNA to be introduced at the breakpoint, were used. The cancer cells which were introduced with such genomic modifications could not grow and died (Chen et al. 2017).

One of the hallmarks of cancer, as described in the beginning of the chapter, is its ability to successfully evade the immune system. Although cancer cells act as Antigen Presenting Cells (APC), the immune cells fail to recognize the antigens displayed on cancer cells (Gonzalez et al. 2018). There are several reasons for such escape including, but not limited to, expression of inhibitory receptors on the surface of T-cells, exhaustion of T cells, etc. To cater to this problem, a study utilized an *ex-vivo* approach using CRISPR/Cas9 to express Chimeric Antigen Receptor (CAR) on the surface of T cells taken from Acute Lymphoblastic Leukemia patients positive for CD19 receptor. This study also showed that the CAR T-cell therapy responded well in CD20, CD22, and CD30 positive patients as well. Although relapse was observed in a few patients due to the loss of CAR expression overtime, this could be managed in future by using more efficient T cells or simultaneously targeting multiple tumor antigens (Lee et al. 2015).

Recently, three different studies also utilized *ex-vivo* approach to use CRISPR/Cas9 for the treatment of cancer. The first study was based on extracting T cells from late-stage lung cancer patients, their *ex-vivo* manipulation using CRISPR/Cas9 to downregulate PD1, expansion of modified T cells *in vitro*, and finally, transfusing the CRISPR-modified cells back into the patient. PD1, if expressed on T cells, allows the downregulation of immune system. So, knocking out PD1 from T cells triggers the immune system to eliminate cancer cells. Interestingly, the patients receiving the therapy did not show treatment-related toxicities and off-target effects (Lu et al. 2020). In the second study, *ex-vivo* engineering and subsequent transplantation of CCR5 knockout hematopoietic stem cells were used in patients with ALL and HIV infection. CCR5 is expressed on immune cells and helps cancer to steer clear of immune recognition. In addition, CCR5 is also important for the progression of TNBCs and other types of cancers. *Ex-vivo* deletion of CCR5 using CRISPR-mediated genome-editing persisted in patients even after 19 months of treatment, without the possible adverse effects of treatment (Xu et al. 2019). The third study also utilized the same *ex-vivo* strategy. The researchers used gRNAs to downregulate PD1 as well as endogenous T cell receptor (TCR). Moreover, they expressed cancer targeting TCR in the same cells. The study was conducted in four patients and in all of them, T cell mediated changes persisted for 9 months. Although utilization of multiple gRNAs targeting multiple genes overtime introduced some chromosomal translocations in patients, there were no deleterious outcomes of these translocations (Stadtmauer et al. 2020).

These three studies and other treatments involving CRISPR described above highlight the safe use of CRISPR in humans for the treatment of cancer. There are many clinical trials presently undergoing to assess both the efficacy and safety of CRISPR/Cas9 in humans for the treatment of different cancers. Therefore, now we will talk about some of the clinical trials being conducted.

## Clinical trials utilizing CRISPR for the treatment of cancer

Owing to the preclinical progress of CRISPR based genome editing in treating cancer, the technique has received a green signal to be tested in clinic. For this, many research groups from all over the world have collaborated and started testing this tool in clinic for evaluating its efficacy and safety. Currently, most of the studies are focusing on both safety and effectiveness of the treatment option; however, still many of these clinical investigations are taking only safety under consideration.

To date, most of the clinical trials being conducted utilize immunotherapy, which consists of *ex-vivo* manipulation of immune cells followed by transfusion in the patients (as discussed in the previous section). Direct introduction of the constructs carrying gRNA and Cas9 in the body is also possible; however, it may trigger an unwanted immune response, or the tumor microenvironment might not let the construct access the cancer cells. Therefore, only a handful of studies are focusing on direct introduction of this tool in human body.

The above-mentioned strategy of *ex-vivo* development of immune cells has been utilized in several clinical trials to target advanced esophageal cancer, metastatic non-small lung cancer, and metastatic renal cell carcinoma with clinical trial identifiers, NCT03081715, NCT02793856, and NCT02867332, respectively. The number of aborted clinical trials is less than the ones being conducted. In fact, more trials are being approved with time. For instance, approval has also been given for the evaluation of the same *ex-vivo* approach for safety and efficacy studies in the patients of bladder and prostate cancer. Currently, new trials are being planned, which are utilizing the same approach for the treatment of bladder and prostate cancer with clinical trials identifiers NCT02863913 and NCT02867345, respectively.

As described in the previous part, PD1 KO T cell reintroduction has been used for stimulating immune system to target cancer. In addition to the PD1 only knockout cells, T cells with PD1 as well as T cell receptor deletion have also been made *ex-*

*vivo*. These cells were engineered to express CARs that can recognize mesothelin positive solid tumors. Such engineered CAR T cells were reintroduced in patients for assessing the safety and efficacy. This clinical trial, NCT03545815, is currently underway; therefore, the results have not been published yet.

It is noteworthy that a unique trial, with clinical trial identifier NCT03057912, is evaluating the safety and efficacy of CRISPR/Cas9 mediated genome editing for the treatment of HPV-related cervical intraepithelial neoplasia I by direct introduction of CRISPR/Cas9 plasmids in human body. This study is specifically involved in removing HPV16 related DNA sequences from cancer cells, which would expectedly induce apoptosis in these cells as HPV related proteins are essential for the survival of this cancer. The plasmids carrying gRNA and Cas9 are loaded in a gel matrix and selectively directed to the cancer when administered in the body. This is one of the few studies based on direct introduction of CRISPR plasmids in the body.

The first US-based trial, with identifier NCT03399448, for the treatment of patients with multiple myeloma and metastatic sarcoma was initiated and completed by the University of Pennsylvania. In this trial, instead of using a single or a double knockout, the researchers did a triple knock out of PDC1, TRAC and TRBC genes in combination with the introduction of NY-ESO-1 receptor in patient derived T cells. The focus of this study was to assess the safety of the treatment. Although certain side effects were reported, that could be the result of chemotherapy that patients were receiving prior to T cell transfusion.

In metastatic gastrointestinal epithelium cancer patients, Cytokine induced SH2 protein (CISH) is expressed on the surface of Tumor infiltrating lymphocytes (TILs), which dampens the activity of these cells to suppress cancer. In a recently published clinical trial, identifier: NCT04426669, where patients with metastatic gastrointestinal cancers are still being recruited, researchers are planning to develop TILs having CISH gene deleted to assess the efficacy and safety of this strategy in phase I and II clinical trials.

Another interesting phase 1 trial is underway for the treatment of relapsed or refractory B cell malignancies and is expected to be completed by 2026. In this trial, the researchers will be recruiting 131 patients in total. The trial would evaluate the efficacy and safety of allogeneic CRISPR engineered T cells for targeting CD19 expressing relapsed B cell malignant cells. In this unique study, CTX110 CRISPR system would be used for modifying T cells which would subsequently be transfused in patients following lymphodepleting chemotherapy. The clinical trial identifier for this study is NCT04035434. Another phase 1 study with clinical trial identifier NCT04502446 is using a similar principle of transfusing CD70 targeting T cells in patients with relapsed B cell malignancies who were pre-treated with lymphodepleting chemotherapy. The study will use a CTX130 CRISPR system instead of CTX110. It is estimated that around 45 patients would participate, and the trial would be completed by the year 2027.

## Limitations of CRISPR and improvements

Despite the promise and effectiveness of the CRISPR system, it comes with certain limitations. One of the major limitations is its off-target effects, which can occur at different levels. For one thing, PAM sequence, which is recognized by Cas9 protein for cleavage (few nucleotides upstream of PAM), can be found at multiple sites throughout the genome. Accumulating evidence seconds that Cas9 can recognize PAM sequence at multiple off-target locations throughout the genome (Alkan et al. 2018, Boyle et al. 2017). In doing so, it may produce unwanted breaks in the genome which may lead to undesirable outcomes. Alternatively, the sequence of gRNA complementary to the target site may bind to off-target sites despite the presence of mismatches. In other words, the gRNA can direct the whole CRISPR machinery at multiple sites rather than specifying it towards the target alone. Some research groups have also reported that the gRNA with 5-6 or even more mismatches is capable of binding to off-target sites (Kang et al. 2020, Tsai et al. 2015). To eradicate associated unforeseen consequences, better computational programming is indispensable for designing efficient gRNAs. These programs should be able to predict the number of maximum possible mismatches for a given gRNA before its synthesis and use so that gRNAs can be devised with utmost specificity.

Besides, another drawback of using CRISPR is the Non-Homologous End Joining (NHEJ) event which occurs after cleavage. Once Cas9 introduces a double strand break at the target site, NHEJ machinery recognizes the cut and repairs it. This kind of restoration may lead to chromosomal rearrangement, including translocations and deletions, which may cause aberrations on top of the existing ones and result in deleterious outcomes (da Silva et al. 2019, Waters et al. 2014). Although these problems are not obvious as NHEJ may or may not result in harmful alterations, based on chance alone, NHEJ can be risky for use in humans and pose threats; therefore, testing this technique in the clinic so far is not entirely safe. However, in clinical trials, despite its off-target effects, CRISPR has proven successful with manageable side effects. Nonetheless, modified state-of-the-art systems for genetic manipulation are required which repair the genome without undergoing NHEJ.

In conclusion, looking at the scientific advancement in this field, it is safe to say that CRISPR/Cas system holds great potential and promise in treating cancer. However, it still needs some crucial upgradation to improve specificity and safety before it can be used widely in humans. Hopefully in the future, CRISPR may become the therapy of choice for the treatment of multiple diseases, in addition to several forms of cancer.

# References

Adorno, M., Cordenonsi, M., Montagner, M., Dupont, S., Wong, C., Hann, B., et al. 2009. A mutant-p53/smad complex opposes p63 to empower tgfbeta-induced metastasis. Cell, 137(1): 87-98.

Aggarwal, B. B. 2003. Signalling pathways of the tnf superfamily: A double-edged sword. Nature Reviews Immunology, 3(9): 745-756.

Alkan, F., Wenzel, A., Anthon, C., Havgaard, J. H. and Gorodkin, J. 2018. Crispr-cas9 off-targeting assessment with nucleic acid duplex energy parameters. Genome Biology, 19.

Allart-Vorelli, P., Porro, B., Baguet, F., Michel, A. and Cousson-Gélie, F. 2015. Haematological cancer and quality of life: A systematic literature review. Blood Cancer Journal, 5(4): e305-e305.

Andriani, F., Roz, E., Caserini, R., Conte, D., Pastorino, U., Sozzi, G., et al. 2012. Inactivation of both fhit and p53 cooperate in deregulating proliferation-related pathways in lung cancer. Journal of Thoracic Oncology, 7(4): 631-642.

Bartucci, M., Dattilo, R., Moriconi, C., Pagliuca, A., Mottolese, M., Federici, G., et al. 2015. Taz is required for metastatic activity and chemoresistance of breast cancer stem cells. Oncogene, 34(6): 681-690.

Barutcu, A. R., Fritz, A. J., Zaidi, S. K., van Wijnen, A. J., Lian, J. B., Stein, J.L., et al. 2016. C-ing the genome: A compendium of chromosome conformation capture methods to study higher-order chromatin organization. Journal of Cellular Physiology, 231(1): 31-35.

Baskar, R., Lee, K. A., Yeo, R. and Yeoh, K. W. 2012. Cancer and radiation therapy: Current advances and future directions. International Journal of Medical Sciences, 9(3): 193-199.

Bierie, B. and Moses, H. L. 2006. Tumour microenvironment: Tgfbeta: The molecular jekyll and hyde of cancer. Nature Reviews Cancer, 6(7): 506-520.

Boyle, E. A., Andreasson, J. O. L., Chircus, L. M., Sternberg, S. H., Wu, M. J., Guegler, C. K., et al. 2017. High-throughput biochemical profiling reveals sequence determinants of dcas9 off-target binding and unbinding. Proceedings of the National Academy of Sciences USA, 114(21): 5461-5466.

Bozic, I., Antal, T., Ohtsuki, H., Carter, H., Kim, D., Chen, S. N., et al. 2010. Accumulation of driver and passenger mutations during tumor progression. Proceedings of the National Academy of Sciences USA, 107(43): 18545-18550.

Braicu, C., Buse, M., Busuioc, C., Drula, R., Gulei, D., Raduly, L., et al. 2019. A comprehensive review on mapk: A promising therapeutic target in cancer. Cancers (Basel), 11(10).

Bray, F., Ferlay, J., Soerjomataram, I., Siegel, R. L., Torre, L. A., Jemal, A. 2018. Global cancer statistics 2018: Globocan estimates of incidence and mortality worldwide for 36 cancers in 185 countries. CA: A Cancer Journal for Clinicians, 68(6): 394-424.

Butti, R., Das, S., Gunasekaran, V. P., Yadav, A. S., Kumar, D. and Kundu, G. C. 2018. Receptor tyrosine kinases (rtks) in breast cancer: Signaling, therapeutic implications and challenges. Molecular Cancer, 17(1): 34.

Cargnello, M. and Roux, P. P. 2011. Activation and function of the mapks and their substrates, the mapk-activated protein kinases. Microbiology and Molecular Biology Reviews, 75(1): 50-83.

Chalhoub, N. and Baker, S. J. 2009. Pten and the pi3-kinase pathway in cancer. Annual Review of Pathology-Mechanisms of Disease, 4: 127-150.

Chan, K. K. and Lo, R. C. 2018. Deregulation of frizzled receptors in hepatocellular carcinoma. International Journal of Molecular Sciences, 19(1).

Chen, Z.-H., Yu, Y. P., Zuo, Z.-H., Nelson, J. B., Michalopoulos, G. K., Monga, S,, et al. 2017. Targeting genomic rearrangements in tumor cells through cas9-mediated insertion of a suicide gene. Nature Biotechnology, 35(6): 543-550.

Cuykendall, T. N., Rubin, M. A. and Khurana, E. 2017. Non-coding genetic variation in cancer. Current Opinion in Systems Biology, 1: 9-15.

da Silva, J. F., Salic, S., Wiedner, M., Datlinger, P., Essletzbichler, P., Hanzl, A., et al. 2019. Genome-scale crispr screens are efficient in non-homologous end-joining deficient cells. Scientific Reports – UK. 9.

Dai, S. X., Li, W. X., Han, F. F., Guo, Y. C., Zheng, J. J., Liu, J. Q., et al. 2016. In silico identification of anti-cancer compounds and plants from traditional chinese medicine database. Scientific Reports, 6: 25462.

Dhillon, A. S., Hagan, S., Rath, O. and Kolch, W. 2007. Map kinase signalling pathways in cancer. Oncogene, 26(22): 3279-3290.

Dostert, C., Grusdat, M., Letellier, E. and Brenner, D. 2019. The tnf family of ligands and receptors: Communication modules in the immune system and beyond. Physiological Reviews, 99(1): 115-160.

Du, Z. and Lovly, C. M. 2018. Mechanisms of receptor tyrosine kinase activation in cancer. Molecular Cancer, 17(1): 58.

Flanagan, D. J., Austin, C.R., Vincan, E. and Phesse, T. J. 2018. Wnt signalling in gastrointestinal epithelial stem cells. Genes-Basel, 9(4).

Floros, K. V., Hata, A. N. and Faber, A. C. 2020. Investigating new mechanisms of acquired resistance to targeted therapies: If you hit them harder, do they get up differently? Cancer Research, 80(1): 25-26.

Gebru, M. T. and Wang, H. G. 2020. Therapeutic targeting of flt3 and associated drug resistance in acute myeloid leukemia. Journal of Hematology and Oncology, 13(1): 155.

Giancotti, F. G. 2014. Deregulation of cell signaling in cancer. The Federation of European Biochemical Societies Letters, 588(16): 2558-2570.

Gonzalez, H., Hagerling, C. and Werb, Z. 2018. Roles of the immune system in cancer: From tumor initiation to metastatic progression. Genes and Development, 32(19-20): 1267-1284.

Greenwell, M. and Rahman, P. K. S. M. 2015. Medicinal plants: Their use in anticancer treatment. International Journal of Pharmaceutical Sciences and Research, 6(10): 4103-4112.

Guertin, D. A. and Sabatini, D. M. 2007. Defining the role of mtor in cancer. Cancer Cell, 12(1): 9-22.

Guo, T., Feng, Y. L., Xiao, J. J., Liu, Q., Sun, X. N., Xiang, J. F., et al. 2018. Harnessing accurate non-homologous end joining for efficient precise deletion in CRISPR/Cas9-mediated genome editing. Genome Biology, 19(1): 170.

Hanahan, D. and Weinberg, R. A. 2011. Hallmarks of cancer: The next generation. Cell, 144(5): 646-674.

Harvey, K. and Tapon, N. 2007. The salvador-warts-hippo pathway – An emerging tumour-suppressor network. Nature Reviews Cancer, 7(3): 182-191.

Harvey, K. F. and Hariharan, I. K. 2012. The hippo pathway. Cold Spring Harbor Perspectives in Biology, 4(8): a011288.

Heck, D., Kowalczyk, M. S., Yudovich, D., Belizaire, R., Puram, R. V., McConkey, M. E., et al. 2014. Generation of mouse models of myeloid malignancy with combinatorial genetic lesions using crispr-cas9 genome editing. Nature Biotechnology, 32(9): 941-946.

Herceg, Z. and Hainaut, P. 2007. Genetic and epigenetic alterations as biomarkers for cancer detection, diagnosis and prognosis. Molecular Oncology, 1(1): 26-41.

Hong, W. and Guan, K. L. 2012. The yap and taz transcription co-activators: Key downstream effectors of the mammalian hippo pathway. Seminars in Cell and Developmental Biology, 23(7): 785-793.

Hou, P. P., Wu, C., Wang, Y. C., Qi, R., Bhavanasi, D., Zuo, Z. X., et al. 2017. A genome-wide crispr screen identifies genes critical for resistance to flt3 inhibitor ac220. Cancer Research, 77(16): 4402-4413.

Ishino, Y., Shinagawa, H., Makino, K., Amemura, M. and Nakata, A. 1987. Nucleotide-sequence of the iap gene, responsible for alkaline-phosphatase isozyme conversion in escherichia-coli, and identification of the gene-product. Journal of Bacteriology, 169(12): 5429-5433.

Jean, S. and Kiger, A. A. 2014. Classes of phosphoinositide 3-kinases at a glance. Journal of Cell Science, 127(5): 923-928.

Jiang, F. G. and Doudna, J. A. 2017. Crispr-cas9 structures and mechanisms. Annual Review of Biophysics, 46: 505-529.

Jinek, M., Chylinski, K., Fonfara, I., Hauer, M., Doudna, J. A. and Charpentier, E. 2012. A programmable dual-rna-guided DNA endonuclease in adaptive bacterial immunity. Science, 337(6096): 816-821.

Kang, S. H., Lee, W. J., An, J. H., Lee, J. H., Kim, Y. H., Kim, H., et al. 2020. Prediction-based highly sensitive crispr off-target validation using target-specific DNA enrichment. Nature Communications, 11(1).

Khan, S. H. 2019. Genome-editing technologies: Concept, pros, and cons of various genome-editing techniques and bioethical concerns for clinical application. Molecular Therapy – Nucleic Acids, 16: 326-334.

Kindler, T., Lipka, D. B. and Fischer, T. 2010. Flt3 as a therapeutic target in aml: Still challenging after all these years. Blood, 116(24): 5089-5102.

Kiyoi, H., Kawashima, N. and Ishikawa, Y. 2020. Flt3 mutations in acute myeloid leukemia: Therapeutic paradigm beyond inhibitor development. Cancer Science, 111(2): 312-322.

Kovacs, D., Migliano, E., Muscardin, L., Silipo, V., Catricala, C., Picardo, M., et al. 2016. The role of wnt/beta-catenin signaling pathway in melanoma epithelial-to-mesenchymal-like switching: Evidences from patients-derived cell lines. Oncotarget, 7(28): 43295-43314.

Lackner, M. R., Wilson, T. R. and Settleman, J. 2012. Mechanisms of acquired resistance to targeted cancer therapies. Future Oncology, 8(8): 999-1014.

Lee, D. W., Kochenderfer, J. N., Stetler-Stevenson, M., Cui, Y. Z. K., Delbrook, C., Feldman, S. A., et al. 2015. T cells expressing cd19 chimeric antigen receptors for acute lymphoblastic leukaemia in children and young adults: A phase 1 dose-escalation trial. Lancet, 385(9967): 517-528.

Lemmon, M. A. and Schlessinger, J. 2010. Cell signaling by receptor tyrosine kinases. Cell, 141(7): 1117-1134.

Li, F., Ng, W. L., Luster, T. A., Hu, H., Sviderskiy, V. O., Dowling, C. M., et al. 2020. Epigenetic crispr screens identify npm1 as a therapeutic vulnerability in non-small cell lung cancer. Cancer Research, 80(17): 3556-3567.

Liu, S. J., Knapp, S. and Ahmed, A. A. 2014. The structural basis of pi3k cancer mutations: From mechanism to therapy. Cancer Research, 74(3): 641-646.

Liu, Y. Z., Qi, X. W., Zeng, Z. Z., Wang, L., Wang, J., Zhang, T., et al. et al. 2017. CRISPR/Cas9-mediated p53 and pten dual mutation accelerates hepatocarcinogenesis in adult hepatitis b virus transgenic mice. Scientific Reports – UK, 7.

Lu, Y., Xue, J. X., Deng, T., Zhou, X. J., Yu, K., Deng, L., et al. 2020. Safety and feasibility of crispr-edited t cells in patients with refractory non-small-cell lung cancer (vol. 23, pg 831, 2020). Nature Medicine, 26(7): 1149-1149.

Lundqvist, E. A., Fujiwara, K. and Seoud, M. 2015. Principles of chemotherapy. International Journal of Gynecology & Obstetrics, 131(Suppl 2): S146-149.

MacDonald, B. T., Tamai, K. and He, X. 2009. Wnt/beta-catenin signaling: Components, mechanisms, and diseases. Developmental Cell, 17(1): 9-26.

Mah, A. T., Yan, K. S. and Kuo, C. J. 2016. Wnt pathway regulation of intestinal stem cells. The Journal of Physiology – London, 594(17): 4837-4847.

Makarova, K. S. and Koonin, E. V. 2015. Annotation and classification of crispr-cas systems. Methods in Molecular Biology, 1311: 47-75.

Mao, B., Hu, F., Cheng, J., Wang, P., Xu, M., Yuan, F., et al. 2014. Sirt1 regulates yap2-mediated cell proliferation and chemoresistance in hepatocellular carcinoma. Oncogene, 33(11): 1468-1474.

Martini, M., De Santis, M. C., Braccini, L., Gulluni, F. and Hirsch, E. 2014. PI3K/AKT signaling pathway and cancer: An updated review. Annals of Medicine, 46(6): 372-383.

Maruyama, I. N. 2014. Mechanisms of activation of receptor tyrosine kinases: Monomers or dimers. Cells-Basel, 3(2): 304-330.

Massague, J. 2012. Tgfbeta signalling in context. Nature Reviews Molecular Cell Biology, 13(10): 616-630.

Matano, M., Date, S., Shimokawa, M., Takano, A., Fujii, M., Ohta, Y., et al. 2015. Modeling colorectal cancer using crispr-cas9–mediated engineering of human intestinal organoids. Nature Medicine, 21(3): 256-262.

Matheny, C. J., Wei, M. C., Bassik, M. C., Donnelly, A. J., Kampmann, M. and Iwasaki, M. 2013. Next-generation NAMPT inhibitors identified by sequential high-throughput phenotypic chemical and functional genomic screens. Chemistry & Biology, 20(11): 1352-1363.

Merrill, B. J. 2012. Wnt pathway regulation of embryonic stem cell self-renewal. Cold Spring Harbor Perspectives in Biology, 4(9).

Mieczkowski, J., Swiatek-Machado, K. and Kaminska, B. 2012. Identification of pathway deregulation - gene expression based analysis of consistent signal transduction. Plos One. 7(7).

Miele, M., Mumot, A. M., Zappa, A., Romano, P. and Ottaggio, L. 2012. Hazel and other sources of paclitaxel and related compounds. Phytochemistry Reviews, 11(2-3): 211-225.

Mohan, G., AH, T. P., Jijo, A J, K M SD, Narayanasamy, A. and Vellingiri, B. 2019. Recent advances in radiotherapy and its associated side effects in cancer—A review. The Journal of Basic and Applied Zoology, 80(1): 14.

Montfort, A., Colacios, C., Levade, T., Andrieu-Abadie, N., Meyer, N. and Segui, B. 2019. The tnf paradox in cancer progression and immunotherapy. Frontiers in Immunology, 10: 1818.

Morrison, D. K. 2012. Map kinase pathways. Cold Spring Harbor Perspectives in Biology, 4(11).

Nakagawa, J., Saio, M., Tamakawa, N., Suwa, T., Frey, A. B., Nonaka, K., et al. 2007. Tnf expressed by tumor-associated macrophages, but not microglia, can eliminate glioma. International Journal of Oncology, 30(4): 803-811.

Nerys, A., Braga-Dias, L. P., Pezzuto, P., Cotta-de-Almeida, V. and Tanuri, A. 2018. Comparison of the editing patterns and editing efficiencies of talen and crispr-cas9 when targeting the human ccr5 gene. Genetics and Molecular Biology, 41(1): 167-179.

Ng, S. R., Rideout, W. M., Akama-Garren, E. H., Bhutkar, A., Mercer, K. L., Schenkel, J. M., et al. 2020. Crispr-mediated modeling and functional validation of candidate tumor suppressor genes in small cell lung cancer. Proceedings of the National Academy of Sciences USA, 117(1): 513-521.

Nie, X. B., Liu, H. Y., Liu, L., Wang, Y. D. and Chen, W. D. 2020. Emerging roles of wnt ligands in human colorectal cancer. Frontiers in Oncology, 10.

Papageorgis, P., Lambert, A. W., Ozturk, S., Gao, F., Pan, H., Manne, U., et al. 2010. Smad signaling is required to maintain epigenetic silencing during breast cancer progression. Cancer Research, 70(3): 968-978.

Paul, M. D. and Hristova, K. 2019. The rtk interactome: Overview and perspective on rtk heterointeractions. Chemical Reviews, 119(9): 5881-5921.

Pauls, S. D., Lafarge, S. T., Landego, I., Zhang, T. and Marshall, A. J. 2012. The phosphoinositide 3-kinase signaling pathway in normal and malignant b cells: Activation mechanisms, regulation and impact on cellular functions. Frontiers in Immunology, 3: 224.

Peng, D., Fu, L. and Sun, G. 2016. Expression analysis of the tgf-beta/smad target genes in adenocarcinoma of esophagogastric junction. Open Medicine (Warsaw, Poland), 11(1): 83-86.

Plotnikov, A., Zehorai, E., Procaccia, S. and Seger, R. 2011. The mapk cascades: Signaling components, nuclear roles and mechanisms of nuclear translocation. Biochimica et Biophysica Acta, 1813(9): 1619-1633.

Pon, J. R. and Marra, M. A. 2015. Driver and passenger mutations in cancer. Annual Review of Pathology: Mechanisms of Disease, 10: 25-50.

Prattapong, P., Ngernsombat, C., Aimjongjun, S. and Janvilisri, T. 2020. CRISPR/Cas9-mediated double knockout of srpk1 and srpk2 in a nasopharyngeal carcinoma cell line. Cancer Rep-Us, 3(2).

Raghavendra, N. M., Pingili, D., Kadasi, S., Mettu, A. and Prasad, S. 2018. Dual or multitargeting inhibitors: The next generation anticancer agents. European Journal of Medicinal Chemistry, 143: 1277-1300.

Ramirez, D. 2016. Computational methods applied to rational drug design. The Open Medicinal Chemistry Journal, 10: 7-20.

Rath, P. C. and Aggarwal, B. B. 1999. Tnf-induced signaling in apoptosis. Journal of Clinical Immunology, 19(6): 350-364.

Regad, T. 2015. Targeting rtk signaling pathways in cancer. Cancers, 7(3): 1758-1784.

Sabel, M. S., Sondak, V. K. and Sussman, J. J. 2007. Surgical foundations: Essentials of surgical oncology. Philadelphia: Mosby Elsevier.

Sangwan, V. and Park, M. 2006. Receptor tyrosine kinases: Role in cancer progression. Current Oncology, 13(5): 191-193.

Schatoff, E. M., Leach, B. I. and Dow, L. E. 2017. Wnt signaling and colorectal cancer. Current Colorectal Cancer Reports, 13(2): 101-110.

Schirrmacher, V. 2019. From chemotherapy to biological therapy: A review of novel concepts to reduce the side effects of systemic cancer treatment (review). International Journal of Oncology, 54(2): 407-419.

Shalem, O., Sanjana, N. E., Hartenian, E., Shi, X., Scott, D. A., Mikkelson, T., et al. 2014. Genome-scale crispr-cas9 knockout screening in human cells. Science, 343(6166): 84-87.

Sledz, C. A. and Williams, B. R. 2005. Rna interference in biology and disease. Blood, 106(3): 787-794.

Smurnyy, Y., Cai, M., Wu, H., McWhinnie, E., Tallarico, J. A., Yang Y., et al. 2014. DNA sequencing and crispr-cas9 gene editing for target validation in mammalian cells. Nature Chemical Biology, 10(8): 623-U152.

Snigdha, K., Gangwani, K. S., Lapalikar, G. V., Singh, A. and Kango-Singh, M. 2019. Hippo signaling in cancer: Lessons from drosophila models. Frontiers in Cell and Developmental Biology, 7: 85.

Stadtmauer, E. A., Fraietta, J. A., Davis, M. M., Cohen, A. D., Weber, K. L., Lancaster, E., et al. 2020. Crispr-engineered t cells in patients with refractory cancer. Science, 367(6481): 1001-1015.

Swann, J. B., Vesely, M. D., Silva, A., Sharkey, J., Akira, S., Schreiber, R. D., et al. 2008. Demonstration of inflammation-induced cancer and cancer immunoediting during primary tumorigenesis. Proceedings of the National Academy of Sciences USA, 105(2): 652-656.

Takaku, K., Oshima, M., Miyoshi, H., Matsui, M., Seldin, M. F. and Taketo, M. M. 1998. Intestinal tumorigenesis in compound mutant mice of both dpc4 (smad4) and apc genes. Cell, 92(5): 645-656.

Tian, X., Gu, T., Patel, S., Bode, A. M., Lee, M. H. and Dong, Z. 2019. CRISPR-Cas9 – An evolving biological tool kit for cancer biology and oncology. Nature Partner Journals Precision Oncology, 3: 8.

Tohme, S., Simmons, R. L. and Tsung, A. 2017. Surgery for cancer: A trigger for metastases. Cancer Research, 77(7): 1548-1552.

Touil, Y., Igoudjil, W., Corvaisier, M., Dessein, A. F., Vandomme, J., Monte, D., et al. 2014. Colon cancer cells escape 5fu chemotherapy-induced cell death by entering stemness and quiescence associated with the c-yes/yap axis. Clinical Cancer Research, 20(4): 837-846.

Tsai, S. Q., Zheng, Z., Nguyen, N. T., Liebers, M., Topkar, V. V., Thapar, V., et al. 2015. Guide-seq enables genome-wide profiling of off-target cleavage by crispr-cas nucleases. Nature Biotechnology, 33(2): 187-197.

Varfolomeev, E. E. and Ashkenazi, A. 2004. Tumor necrosis factor: An apoptosis junkie? Cell, 116(4): 491-497.

Verstraete, K. and Savvides, S. N. 2012. Extracellular assembly and activation principles of oncogenic class iii receptor tyrosine kinases. Nature Reviews Cancer, 12(11): 753-766.

Wajant, H., Pfizenmaier, K. and Scheurich, P. 2003. Tumor necrosis factor signaling. Cell Death & Differentiation, 10(1): 45-65.

Wang, C., Xu, P., Zhang, L., Huang, J., Zhu, K. and Luo, C. 2018. Current strategies and applications for precision drug design. Frontiers in Pharmacology, 9: 787.

Wang, X. and Lin, Y. 2008. Tumor necrosis factor and cancer, buddies or foes? Acta Pharmacologica Sinica, 29(11): 1275-1288.

Waters, C. A., Strande, N. T., Wyatt, D. W., Pryor, J. M. and Ramsden, D. A. 2014. Nonhomologous end joining: A good solution for bad ends. DNA Repair, 17: 39-51.

Weisberg, E., Barrett, R., Liu, Q., Stone, R., Gray, N. and Griffin, J. D. 2009. Flt3 inhibition and mechanisms of drug resistance in mutant flt3-positive aml. Drug Resistance Updates, 12(3): 81-89.

Wintheiser, G. A. and Silberstein, P. 2020. Physiology, Tyrosine Kinase Receptors. Statpearls. Treasure Island (FL).

Wyld, L., Audisio, R. A. and Poston, G. J. 2015. The evolution of cancer surgery and future perspectives. Nature Reviews Clinical Oncology, 12(2): 115-124.

Xu, L., Wang, J., Liu, Y. L., Xie, L. F., Su, B., Mou, D. L., et al. 2019. Crispr-edited stem cells in a patient with hiv and acute lymphocytic leukemia. The New England Journal of Medicine, 381(13): 1240-1247.

Xu, W., Wei, Y. Y., Wu, S. S., Wang, Y., Wang, Z., Sun, Y., et al. 2015. Up-regulation of the hippo pathway effector taz renders lung adenocarcinoma cells harboring egfr-t790m mutation resistant to gefitinib. Cell and Bioscience, 5.

Xue, V. W., Wong, S. C. C. and Cho, W. C. S. 2020. Genome-wide crispr screens for the identification of therapeutic targets for cancer treatment. Expert Opinion on Therapeutic Targets, 24(11): 1147-1158.

Yamaoka, T., Kusumoto, S., Ando, K., Ohba, M. and Ohmori, T. 2018. Receptor tyrosine kinase-targeted cancer therapy. International Journal of Molecular Sciences, 19(11).

Yang, M., Zeng, C., Li, P. T., Qian, L. Y., Ding, B. N., Huang, L. H., et al. 2019. Impact of cxcr4 and cxcr7 knockout by CRISPR-Cas9 on the function of triple-negative breast cancer cells. OncoTargets and Therapy, 12: 3849-3858.

Yu, J. S. L. and Cui, W. 2016. Proliferation, survival and metabolism: The role of PI3K/AKT/mTOR signalling in pluripotency and cell fate determination. Development, 143(17): 3050-3060.

Zahreddine, H. and Borden, K. L. B. 2013. Mechanisms and insights into drug resistance in cancer. Frontiers in Pharmacology, 4.

Zeitouni, D., Pylayeva-Gupta, Y., Der, C. J. and Bryant, K. L. 2016. Kras mutant pancreatic cancer: No lone path to an effective treatment. Cancers, 8(4).

Zhan, T., Rindtorff, N. and Boutros, M. 2017. Wnt signaling in cancer. Oncogene. 36(11): 1461-1473.

Zhang, W. and Liu, H. T. 2002. Mapk signal pathways in the regulation of cell proliferation in mammalian cells. Cell Research, 12(1): 9-18.

Zhang, X., Yang, M., Shi, H., Hu, J., Wang, Y., Sun, Z., et al. 2017. Reduced e-cadherin facilitates renal cell carcinoma progression by wnt/beta-catenin signaling activation. Oncotarget, 8(12): 19566-19576.

Zhao, X., Mohaupt, M., Jiang, J., Liu, S., Li, B. and Qin, Z. 2007. Tumor necrosis factor receptor 2-mediated tumor suppression is nitric oxide dependent and involves angiostasis. Cancer Research, 67(9): 4443-4450.

Zhu, L. Y. and Chen, L. Q. 2019. Progress in research on paclitaxel and tumor immunotherapy. Cellular and Molecular Biology Letters, 24.

Zirn, B., Samans, B., Wittmann, S., Pietsch, T., Leuschner, I., Graf, N., et al. 2006. Target genes of the wnt/beta-catenin pathway in wilms tumors. Genes Chromosomes Cancer, 45(6): 565-574.

Zitvogel, L., Galluzzi, L., Mark and Kroemer, G. 2013. Mechanism of action of conventional and targeted anticancer therapies: Reinstating immunosurveillance. Immunity, 39(1): 74-88.

# The Advances of the CRISPR/Cas9 Technology for Correcting Genetic Disorders: Using Sarcoma as a Model System

**Pichaya Thanindratarn[1,2], Dylan C. Dean[1], Francis J. Hornicek[1] and Zhenfeng Duan[3]\***

[1] Department of Orthopedic Surgery, Sarcoma Biology Laboratory, David Geffen
School of Medicine, University of California, Los Angeles, CA, USA
[2] Department of Orthopedic Surgery, Chulabhorn Hospital, HRH Princess Chulabhorn
College of Medical Science, Chulabhorn Royal Academy, Bangkok, 10210, Thailand
[3] Department of Orthopedic Surgery, David Geffen School of Medicine at UCLA, 615
Charles E. Young. Dr. South, Los Angeles, CA 90095, USA

## Introduction

Sarcomas are rare mesenchymal tumors with heterogenous subtypes that exhibit a wide range of underlying molecular mechanisms and clinical behaviors (Thway et al. 2017). Surgery and chemotherapy are the mainstay of sarcoma treatment, which can significantly improve survival in early stages of some sarcomas (Harwood et al. 2015). However, the incidence of relapse in sarcoma is still high and the prognosis of chemoresistant and metastatic sarcoma remains poor (Harwood et al. 2015). Therefore, further research and therapeutic improvements are essential.

Sarcomas can be classified based on the genetic alterations involved in their development including oncogenic somatic mutations such as KIT or platelet-derived growth factor receptor-$\alpha$ (PDGFR-$\alpha$), mutations in gastrointestinal stromal tumor (GIST), DNA copy number alterations such as JUN gene amplification in dedifferentiated liposarcoma and myocardin gene amplification in leiomyosarcoma, and recurrent chromosomal translocations leading to oncogenic fusion genes such as EWS-FLI1 in Ewing sarcoma or SYT-SSX in synovial sarcoma (Taylor et al. 2011, Harwood et al. 2015, Movva et al. 2015). More commonly, pathogenesis is the result of complex chromosomal aberrations as in osteosarcoma and undifferentiated pleomorphic sarcoma (UPS) (Taylor et al. 2011, Harwood et al. 2015, Movva et al.

\* Corresponding author: zduan@mednet.ucla.edu

2015). In addition, a variety of protein kinases including tyrosine kinase receptors are overexpressed in sarcomas, both in translocation-associated sarcomas such as GIST, and karyotypically complex sarcomas such as osteosarcoma (Demicco et al. 2012, Zhang and Pollock 2014). These receptors along with their ligands represent some of the most attractive targets in sarcoma treatment.

While the development of adequate transgenic models has been elusive, sarcoma modeling via genetic engineering remains an important approach for sarcoma research. Conventional transgenic techniques mostly involve the time-consuming, sophisticated processes of germline manipulation and intensive animal cross-breeding that can model the complexity and multistep development of sarcoma mutations (Lu et al. 2015). RNA interference is another method for genetic engineering; however, it can usually provide only temporary and partial knockdown with some pervasive off-target effects (Agrawal, et al. 2003, Lu et al. 2015). Cyclization recombinase-*locus* of recombination in P1 (Cre-LoxP) consists of a single enzyme called Cre recombinase. Cre recognizes a 34-bp site on the P1 genome called LoxP and catalyzes reciprocal conservative DNA recombination between pairs of LoxP sites (Wilson and Kola 2001). The genetic engineering power of the Cre-LoxP system derives from the potential to generate more conditional mutants; however, conditional deletion of every gene in every cell type may not be genuinely achieved (Wilson and Kola 2001, Maizels 2013). Conditional deletions are only comparable to the promotors that regulate Cre expression, and transgenic Cre driver lines have exhibited some limitations such as expression outside the target tissue and inefficient cleavage leading to mosaicism (Wilson and Kola 2001, Maizels 2013).

Recently, clustered regularly interspaced short palindromic repeats (CRISPR)-associated Cas9 protein (CRISPR/Cas9), a new genome editing strategy, has been utilized in a variety of model organisms and cell types (Yin et al. 2019). Compared with RNA interference and Cre-LoxP, CRISPR/Cas9 is an exogenous genome-editing system that does not compete with the endogenous processes. CRISPR-Cas9 also functions at the DNA level of the targeting transcripts, which results in complete knockdown of a gene function. Moreover, CRISPR/Cas9 provides a larger targetable sequence space in which promotors of the gene may be targeted as well. Many advantages of CRISPR/Cas9 technology have been demonstrated compared with some incumbent genome editing technologies such as transcription activator-like effector nucleases (TALENs) and zinc finger nucleases (ZFNs) (Table 1) (Yin et al. 2019). In this chapter, we review development of the CRISPR/Cas9 system in genome editing and its potential applications in sarcoma research and treatment.

## Mechanisms of CRISPR/Cas9 in genome editing

CRISPR/Cas9 is an RNA-guided, targeted genome editing platform with remarkable potential in both basic research and clinical applications. The main components of the CRISPR/Cas9 system are a Cas endonuclease and a single guided RNA (sgRNA) or target-specific CRISPR RNA (crRNA). The Cas enzyme is comprised of two active catalytic domains: HNH and RuvC. The HNH domain is a single nuclease domain, while the RuvC domain contains three subdomains across the linear sequence. RuvC I locates near the N-terminal region of Cas9, whereas RuvC II and RuvC

**Table 1.** Comparison of the CRISPR/Cas9 system with ZFNs and TALENs

| | CRISPR/Cas9 | ZFNs | TALENs |
|---|---|---|---|
| Target DNA recognition | sgRNA | Zinc fingers (ZFs) | Repeat variable di-residues (RVDs) |
| DNA cleavage endonuclease | Cas9 | FokI endonuclease domain | FokI endonuclease domain |
| Construction of endonuclease | sgRNA | 3-4 ZF domains | 8-31 RVD repeats |
| Minimum number of DNA base being recognized | Single | Triple | Single |
| Size of recognized DNA sequences | 20 bp + NGG | 9-18 bp | (8-31 bp) × 2 |
| Restriction of target site | PAM sequences (NGG or NAG) | - | Binding site should start with a T base |
| Cytotoxicity | Low | High | Low |
| Advantages | Very high efficiency<br>Rapid construction<br>Easy delivery | High efficiency<br>Low off-target cleavage compared with CRISPR/Cas9 | High efficiency<br>Low off-target cleavage compared with CRISPR/Cas9 |
| Disadvantages | Possibly more frequent off-target cleavage | More difficult to assemble than CRISPR/Cas9<br>Remaining possibility of off-target cleavage | More difficult to assemble than CRISPR/Cas9<br>Remaining possibility of off-target cleavage |

**Abbreviations:** Base pairs (bp), Clustered regularly interspaced short palindromic repeats-associated Cas9 protein (CRISPR/Cas9), Protospacer adjacent motif (PAM), Repeat variable di-residues (RVDs), Single guided RNA (sgRNA), Transcription activator-like effector nuclease (TALEN), Zinc finger nuclease (ZFN).

III flank the HNH domain in the near middle of the protein. The HNH and RuvC nuclease domains are well-positioned for cleavage of the complementary and non-complementary strands of the target DNA, respectively.

The target sequence in the genome paired to the sgRNA sequence is immediately followed by the protospacer adjacent motif (PAM). The PAM sequence is located on the immediate 3' end of the sgRNA sequence, but is not part of the 20-nt guide sequence of the sgRNA (Figure 1). The CRISPR system uses Cas9, which forms complexes with the sgRNA, to cleave the DNA by 3-4 base pairs upstream of PAM and creates single Cas9-dependent double-strand breaks (DSBs) in a sequence-specific manner (Lino et al. 2018). The DSBs are then repaired either by non-homologous end joining (NHEJ)-mediated error-prone DNA repair or homologous directed repair (HDR)-mediated error-free DNA repair (Figure 2). The NHEJ-mediated repair can rapidly ligate the DSBs while also generating small insertion or deletion of specific sequences at target sites, which then disrupt and abolish the genomic element or its function. Likewise, HDR-mediated repair requires a homology-containing donor DNA sequence as a repair template (Figure 2).

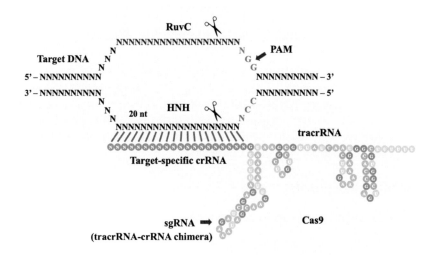

**Figure 1.** Overview of CRISPR/Cas9.

The core components of the CRISPR/Cas9 system are a Cas9 endonuclease and a single guide RNA (sgRNA). The Cas9 nuclease consists of two catalytic active domains: HNH and RuvC. The sgRNA binds to Cas9 and directs it to the *locus* of interest by a 20-nt guide sequence via base pairing to the genomic target. When Cas9 binds to the target DNA, the target sequence in the genomic DNA is paired to the sgRNA sequence and is immediately followed by the NGG sequence called the protospacer adjacent motif (PAM). The CRISPR/Cas9 system utilizes Cas9, which complexes with sgRNA to cleave DNA and generates double-strand breaks (DSBs) in a sequence-specific manner about 3-4 base pairs upstream of PAM.

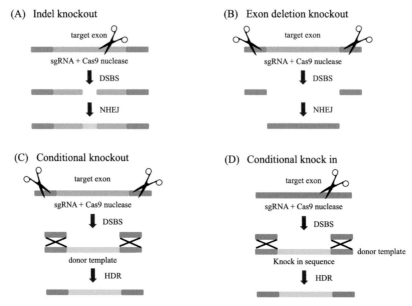

**Figure 2.** Genome editing technologies exploit endogenous DNA repair mechanisms.

Double-strand breaks (DSBs) induced by a nuclease at a specific site can be repaired either by non-homologous end joining (NHEJ) or homologous directed repair (HDR). Repair by NHEJ usually results in (A) the insertion or deletion of random base pairs, or (B) complete removal of the exon. HDR with a donor DNA template can be exploited to modify a gene by introducing precise nucleotide substitutions (C) or to achieve gene insertion (D).

# Potential applications of CRISPR/Cas9 in sarcoma

## Modeling pathogenesis in sarcoma with CRISPR/Cas9

Sarcoma pathogenesis is a multi-step pathway involving various genetic changes and epigenetic alterations. Genomic studies have identified a large number of genetic and epigenetic aberrations that occur in various types of human sarcomas (Sanchez-Rivera et al. 2014). Genome manipulating experiments of both normal and tumor cells are essential for better understanding of sarcoma tumorigenesis, and may inform the discovery of potential therapeutic targets. Several genetic engineering strategies have been applied to different sarcoma models such as Cre-LoxP, ZFNs, and TALENs; however, these techniques are generally labor-intensive, time consuming, and have relatively low efficiency for targeting a gene of interest (Gupta and Musunuru 2014). There is a great need for rapid, high efficiency, and less problematic means of genome editing. The CRISPR/Cas9 method has revolutionized the genetic engineering field and may overcome many limitations of previous approaches (Gupta and Musunuru 2014). Examples of recent sarcoma research modeling using the CRISPR/Cas9 system are summarized in Table 2.

**Table 2.** Modeling human sarcomas with the CRISPR/Cas9 system

| Type of sarcoma | Gene mutations | Significance | Reference |
|---|---|---|---|
| Sporadic/varied | P53 -/- | Mice deficient for P53 could be prone to sarcoma developing | Donehower et al. 1992 |
| Sporadic/varied | P53 mutation | Mice with P53 mutation were most prone to develop sarcoma | (Lang et al. 2004) |
| OS and PDSTS | RB and P53 deletion | P53 deletion in mesenchymal cells that give rise to osteoblasts is a powerful initiator of OS. RB deletion does not initiate sarcoma formation in mice, but it can accelerate sarcoma formation. | (Lin et al. 2009) |
| PDSTS | EWS-FLI1 or P53 deletion | In the P53 deletion, EWS-FLI1 accelerated the formation of sarcomas. | (Lin et al. 2009) |
| OS | TAX; P19ARF -/- | The loss of ARF selectively predisposed TAX transgenic mice to develop OS. | (Rauch et al. 2010) |
| OS | P53 and RB deletions | Osterix-Cre-mediated deletion of P53 and RB results in OS | (Ng et al. 2012) |
| OS, lymphosarcoma, leiomyosarcoma | RB and P107 deletions | RB haploinsufficiency coupled with P107 deletion possibly results in sarcoma formation | (Dannenberg et al. 2004) |
| Alveolar RMS | PAX3-FKHR transgene and P53 and RB deletion | Expression of the PAX3-FKHR transgene requires loss of P53 and RB for RMS formation | (Keller et al. 2004a) |
| RMS, various sarcoma | P53 and RB1 deletion with or without PTCH1 haploinsufficiency | Mutant P53, PTCH1, or RB1 in satellite cells gives rise to UPS; RB1 loss acts as an apparent modifier for sarcomas by inducing dedifferentiation. | (Rubin et al. 2011) |
| Pleomorphic RMS | Mutant KRAS expression and P53 deficiency | Cooperation of oncogenic KRAS and P53 deficiency results in the development of pleomorphic RMS in adult mice. | (Tsumura et al. 2006) |
| Pleomorphic RMS | Mutant KRAS expression and P53 mutation or loss | The combination of oncogenic KRAS and loss of P53 activity accelerates sarcoma formation. Mutation, but not a loss of single P53 allele, is sufficient for sarcoma development. | (Doyle et al. 2010) |

*(Contd.)*

**Table 2.** (*Contd.*)

| Type of sarcoma | Gene mutations | Significance | Reference |
|---|---|---|---|
| Synovial sarcoma | SYT-SSX fusion | Expression of SYT-SSX 2 fusion gene yields a highly penetrant and representative model of human synovial sarcoma. | (Haldar et al. 2007) |
| Liposarcoma | TLS-CHOP fusion with loss of P53 | Deletion of P53 cooperates in the formation of liposarcomas. | (Charytonowicz et al. 2012) |
| Neurofibromas and MPNSTs | NF1 deletion with or without P53 and P19ARF deletions | NF1 and P53 mutations lead to MPNSTs; INK4a and ARF deficiency cooperate with NF1 heterozygosity to yield MPNSTs. | (Joseph et al. 2008) |
| Uterine LMS | LMP2 deficient | LMP2 deficient mice spontaneously develop uterine LMS. | (Hayashi et al. 2011) |
| Uterine LMS | Overexpression of CRIPTO-1 | CR-1 plays an important role in the formation of uterine LMS. | (Strizzi et al. 2007) |
| Uterine LMS | Deletion of P53 and BRCA1 | Conditional deletion of P53 in mice results in development of uterine LMS and concurrent deletion of P53 and BRCA1 significantly accelerates the progression of tumors. | (Xing et al. 2009) |

**Abbreviations:** Leiomyosarcoma (LMS), Malignant peripheral nerve sheath tumor (MPNST), Osteosarcoma (OS), Poorly differentiated soft tissue sarcoma (PDSTS), Rhabdomyosarcoma (RMS).

## Genomic alterations using CRISPR/Cas9 in sarcomas

Previous genome sequencing studies have identified multiple driver gene mutations in various human sarcomas. CRISPR/Cas9 is capable of inducing *in vitro* and *in vivo* loss of function (LOF) and gain of function (GOF) mutations. Previous studies have successfully generated various malignant tumor models using the CRISPR/Cas9 system, including lung cancer, colorectal cancer, and hematologic malignancies (Heckl et al. 2014, Niu et al. 2014, Platt et al. 2014, Drost et al. 2015, Matano et al. 2015). Recently, CRISPR/Cas9 has been used to knockdown tumor suppressors cyclin-dependent kinase inhibitor 2A (CDKN2A), transformation-related protein 53 (TrP53), and phosphatase and tensin homolog (PTEN) to generate UPS and fusion-negative rhabdomyosarcoma mouse models (Albers et al. 2015).

Aberrations of the copy number and structure of chromosomes are frequently observed in sarcoma cells; however, to generate these abnormalities in a highly specific manner for functional research is somewhat difficult. CRISPR/Cas9 can facilitate complex manipulations of genomic structure, chromosomal rearrangements, oncogene amplifications, and tumor suppressor deletions (Lu et al. 2015). Gene fusions from specific translocations have proven to be essential in various types of sarcomas. Rhabdomyosarcoma (RMS) is a myogenic soft tissue sarcoma of childhood. The most common chromosomal translocation in RMS results in the expression of oncoprotein PAX3-FOXO1, which is important for cancer cell survival (Gryder et al. 2017). It is challenging to mimic this translocation in murine models since the PAX3 and FOXO1 genes in mice are in the opposite orientation on their respective chromosomes. To create a fusion-positive RMS mouse model, a translocation of PAX3-FOXO1 was engineered to invert FOXO1 orientation in murine Ewing sarcoma cells. Following this change, myoblasts and fibroblasts were isolated and cultured from these FOXO1$^{inv+/+}$ mice to virally deliver Cas9 and sgRNAs *in vitro* and produced a t(1;3) translocation, corresponding to the human t(2;13) translocation and PAX3-FOXO1 oncoprotein (Lagutina et al. 2015). This mouse model is a valuable tool for further investigation into molecular mechanisms underlying the initial stages of RMS implicated chromosomal translocations.

Ewing sarcoma is a rare and aggressive mesenchymal malignancy that occurs primarily in childhood. It is characterized by the t(11;22) (q24;q12) translocation, leading to fusion of a EWSR1-FLI1 oncogene (Torres et al. 2014). To create this chromosomal translocation using CRISPR/Cas9, four sgRNAs were designed to target intron 4 (F1, F2) of FLI1 and intron 7 (E1, E2) of EWSR1 (Torres et al. 2014). The t(11;22)/EWSR1-FLI1 chromosomal translocation was induced and characterized in HEK293 cells and in human primary mesenchymal stem cells (hMSCs) (Torres et al. 2014). Then, DNA DSBs in the target *loci* were generated by transfection of HEK293 cells with a plasmid expressing Cas9 and sgRNAs E1, E2, F1 and F2, as previously described (Torres et al. 2014). Thereafter, the group addressed whether the chromosomal translocation driven by CRISPR/Cas9 would be able to replicate the synthesis of the functional EWSR1-FLI1 fusion gene (Torres et al. 2014). The functionality of the fusion protein in the pool of hygromycin-selected t(11;22)-positive populations was then investigated. Upregulation of six well-known target genes of the EWSR1-FLI1 fusion protein, including EZH2, HMGA2, NKX2.2,

ID2, NR0B1, and SOX2, was confirmed by qRT-PCR. This indicated that this fusion protein was successfully produced after the induction of chromosomal translocation using CRISPR/Cas9 and mimicked activity of primary Ewing sarcoma cells (Torres et al. 2014).

### *In vivo* sarcoma models using CRISPR/Cas9

Genetically engineered mouse models (GEMM) of adult malignancies have been developed by CRISPR/Cas9 for the study of gene drivers and gene fusions such as EML4-ALK in lung cancers (Heckl et al. 2014, Maddalo et al. 2014, Platt et al. 2014, Sanchez-Rivera et al. 2014). Previous works developed the multiple lentiviral system (MuLE) that allowed multiple genetic alterations to be simultaneously introduced into mammalian cells (Albers et al. 2015). By transduction of single MuLE lentiviruses, investigators engineered tumors containing up to five different genetic alterations, identified genetic dependencies, conducted genetic interaction screens, and induced simultaneous CRISPR/Cas9-mediated knockout of three tumor-suppressor genes in primary cultured mouse cells. Intramuscular injection of MuLE viruses expressing oncogenic H-Ras$^{G12V}$ combined with knockdowns of CDKN2A, transformation-related protein 53 (Trp53), and phosphatase and tensin homolog (Pten) were able to generate undifferentiated pleomorphic sarcoma and fusion-negative RMS mouse models, illustrating MuLE could be applied for investigating the molecular mechanisms of human diseases (Albers et al. 2015).

To create a fusion-positive RMS mouse model, a translocation of PAX3-FOXO1 was engineered to invert FOXO1 orientation in murine Ewing sarcoma cells. Following this step, myoblasts and fibroblasts were then isolated and cultured from these FOXO1$^{inv+/+}$ mice to virally deliver Cas9 and sgRNAs *in vitro* and produced a t (1;3) translocation, correspondent to human t (2;13) translocation and PAX3-FOXO1 oncoprotein (Lagutina et al. 2015).

Recently, a combined Cre-loxP and CRISPR/Cas9 approach in GEMM production of pediatric and adult sarcomas uncovered the possibility of generating UPS and malignant peripheral nerve sheath tumors (MPNST) (Huang et al. 2017). UPS was generated in a Cre-loxP murine model conditionally expressing Kras and Cas9 following an intramuscular injection of the adenoviral Cre-recombinase and sgRNA for Trp53. In addition, intraneural injection of an adenoviral vector co-expressing Cas9 and single sgRNAs individually targeting neurofibromin 1 (NF1) and Trp53 developed MPNST in the sciatic nerve of wild-type mice (Huang et al. 2017). These sarcomas developed *in vivo* with a similar onset of their human counterparts, and with related characteristics of the sarcomas acquired with a Cre-loxP system. These CRISPR-based sarcoma models decrease both time and expenditure in research.

## Gene therapy with CRISPR/Cas9 in sarcoma

### Editing the genome for anti-sarcoma activity

The CRISPR/Cas9 system can also be used to investigate therapeutic agents and their resistance in sarcomas. Several studies have shown that CDK11 is an important kinase in various malignancies including in sarcoma cell growth and survival (Duan et al. 2012, Jia et al. 2014, Feng et al. 2015, Zhou et al. 2015). Recently, CRISPR/ Cas9 was used to determine the effect of targeting endogenous CDK11 at the DNA level in osteosarcoma cell lines (Feng et al. 2015). Migration and invasion activities were markedly reduced following CDK11 knockdown by CRISPR/Cas9, indicating that CDK11 may serve as a potential therapeutic target for osteosarcoma (Feng et al. 2015).

The major obstacle of osteosarcoma therapeutics is the development of multidrug resistance (MDR) (Susa et al. 2010). MDR1 gene, which encodes the membrane efflux pump P-glycoprotein (P-gp), plays a crucial role in drug resistance (Susa et al. 2010). MDR1 overexpression results in an active efflux of cytotoxic agents such as doxorubicin and paclitaxel from the cells, thus decreasing the intracellular drug concentrations to subtherapeutic levels (Susa et al. 2010). Similar to targeting of CDK11, CRISPR/Cas9 can knockout MDR1 in the drug-resistant osteosarcoma cell lines KHOSR2 and U2OSR2 and thus improve chemosensitivity (Liu et al. 2016).

## Genome-wide screens for anti-sarcoma drugs

While identifying drug resistance mutations is necessary for target confirmation, examination of actual drug-target interactions requires resistance mutation modification of wild-type models (Kasap et al. 2014, Wen et al. 2016). This has been burdensome and error-prone in mammalian cells, but CRISPR/Cas9 editing has recently streamlined validation experiments in different cancer models (Kasap et al. 2014, Wen et al. 2016).

Human exportin-1 (XPO1), also known as chromosome region maintenance 1 protein (CRM1), has been investigated as an anti-cancer target (Neggers et al. 2015). XPO1 overexpression in osteosarcoma correlates with poor prognosis and drug resistance (Yao et al. 2009). Selinexor (KPT-330), a specific XPO1 inhibitor, has recently shown high response rates in phase I trials of advanced solid tumor and hematologic malignancies (Yao et al. 2009, De Cesare et al. 2015, Kazim et al. 2015, Neggers et al. 2015). Mechanistically, selinexor suppressed the formation and function of XPO1 by binding to its cysteine 528 residue, resulting in accumulation of tumor suppressor proteins in the nucleus and thus cell cycle arrest and apoptosis (Yao et al. 2009, De Cesare et al. 2015, Kazim et al. 2015, Neggers et al. 2015). In this case, CRISPR/Cas9 was applied to introduce a single XPO1 C528S mutation in acute T cell leukemia Jurkat cells to validate the drug-target interaction, which confirmed that this mutation prevented Selinexor anti-cancer activity by blocking XPO1-Selinexor binding (Neggers et al. 2015).

Ispinesib, an inhibitor of kinetin spindle protein (KSP/HsEg5), has been studied in various clinical trials as an anti-cancer drug (Rath and Kozielski 2012). Ispinesib demonstrated a high level of *in vivo* anti-cancer activity against Ewing sarcoma (Houghton et al. 2007), which led to further clinical trials in sarcomas such as osteosarcoma and rhabdomyosarcoma (Souid et al. 2010). Previous sequencing and bioinformatic studies suggested that mutations in the Ala133 residue of KSP may induce ispinesib resistance (Kasap et al. 2014). CRISPR/Cas9 was applied to confirm this relation. Vectors harboring the Cas9 "nickase" and sgRNAs along with a template DNA bearing the mutation of interest were transfected into HeLa cells (Kasap et al. 2014). In the transfected cells, the Ala133 mutation was confirmed using Surveyor mutation-detection and Sanger sequencing of the genomic *locus*. The A133P substitution conferred more than a 150-fold resistance to ispinesib, a finding consistent with previous studies (Kasap et al. 2014).

Additionally, CRISPR/Cas9 systems can be used for genome-scale screening of drug-resistance mutations (Wen et al. 2016). The BRAF V600E mutation has been observed in a group of histiocytic tumors including Langerhans cell histiocytosis and histiocytic sarcoma (Chen et al. 2013, Idbaih et al. 2014, Di Liso et al. 2015). Vemurafenib, a specific BRAF inhibitor, showed therapeutic efficacy in some tumors with the BRAF V600E mutation (Kopetz et al. 2015). The genome-scale CRISPR/Cas9 knockout (GeCKO) library targeting 18,080 genes with 64,751 unique guide sequences was performed to identify which genes were involved in vemurafenib resistance (Shalem et al. 2014). Similarly, the Reference Sequence of NCBI (RefSeq), a library consisting of 70,290 guides targeting all human coding isoforms, was utilized for genomic screening of vemurafenib resistance (Konermann et al. 2015). The synergistic activation mediator (SAM), a protein complex modified for transcriptional activation of endogenous genes, was used in this study. The results revealed that the SAM system is potent, specific, and can facilitate genome-scale screening when combined with a compact pooled sgRNA library. The SAM screen properly confirmed the genes previously shown to yield drug resistance as well as new promising candidates by using individual sgRNA and complementary DNA overexpression (Konermann et al. 2015). Compared with RNA interference screens using a short hairpin RNA (shRNA) library, genome-scale screening using CRISPR/Cas9 demonstrated a higher reagent consistency and phenotypic effectiveness of individual sgRNAs (Shalem et al. 2014). Additionally, these studies highlight the significant advantages of CRISPR/Cas9 in genome-wide screens for anti-cancer drug development.

## Challenges and future directions of CRISPR/Cas9 in sarcoma

The CRISPR/Cas9 system has evolved as a powerful method for precise genome editing in mammalian cells, but limitations remain. It has the potential for generating off-target effects, although at a much smaller rate compared with other genome editing approaches (Cho et al. 2014, Tsai et al. 2015). Various strategies such as selecting the most unique target sequences have been implemented in order to reduce off-target effects (Cho et al. 2014). Specifically, ideal target sequences should differ from any other sites in the genome by at least two or three nucleotides in a 20-nt sequence. In addition, more powerful sgRNA, more potent Cas9 systems, and more efficient deliveries are in development (Cho et al. 2014, Tsai et al. 2015). The synergistic activation mediator (SAM) has been developed by engineering the sgRNA through appending a minimal hairpin aptamer to the tetraloop and stem loop 2 of sgRNA. SAM-mediated gene activation has more specificity with minimal off-target activity (Zhang et al. 2015). *Streptococcus pyogenes* Cas9-HF1 (SpCas9-HF1) was also demonstrated to be a high-fidelity CRISPR/Cas9 nuclease, and retains comparable on-target activities to wild-type SpCas9 without detectable off-target effects (Kleinstiver et al. 2016).

CRISPR/Cas9 systems continue to broaden their use cases for research and therapy in sarcoma. For instance, CRISPR interference (CRISPRi), the transcriptional suppression of target genes by Cas9 binding, can knockdown tumor-associated genes (Gilbert et al. 2014). Catalytically inactivated Cas9 (dead Cas or dCas9) can also serve as an RNA-guided DNA-binding domain, which can fuse to multiple effectors such as fluorescent proteins to obtain real-time imaging of the genomic *loci* of interest (Gilbert et al. 2014). If the effector is an epigenetic modifier, such as a DNA methyltransferase or histone acetyltransferase, it reaches the target sites directed by sgRNA and regulates epigenetic alteration, so called epi-genome editing (Lu et al. 2015). In sarcoma, genome level mutations and epigenetic aberrations synergistically drive sarcoma carcinogenesis. For these reasons, dCas9-mediated epigenome editing may prove a powerful tool for future epigenetic sarcoma research and treatment (Lu et al. 2015).

## Conclusion

Despite the short history of CRISPR-based genome editing, it has been incredibly transformative for oncology research as a whole and in emerging sarcoma works. CRISPR/Cas9 systems can generate unique cells and *in vivo* animal models, dramatically improve feasibility of new genetic modifications in the lab and future therapies.

## Acknowledgement

This work was supported, in part, by the Department of Orthopaedic Surgery at UCLA.

## References

Agrawal, N., Dasaradhi, P. V., Mohmmed, A., Malhotra, P., Bhatnagar, R. K. and Mukherjee, S. K. 2003. RNA interference: Biology, mechanism, and applications. Microbiology and Molecular Biology Reviews, 67(4): 657-685.

Albers, J., Danzer, C., Rechsteiner, M., Lehmann, H., Brandt, L. P., Hejhal, T. et al. 2015. A versatile modular vector system for rapid combinatorial mammalian genetics. The Journal of Clinical Investigation, 125(4): 1603-1619.

Charytonowicz, E., Terry, M., Coakley, K., Telis, L., Remotti, F., Cordon-Cardo, C. et al. 2012. PPARgamma agonists enhance ET-743-induced adipogenic differentiation in a transgenic mouse model of myxoid round cell liposarcoma. The Journal of Clinical Investigation, 122(3): 886-898.

Chen, W., Jaffe, R., Zhang, L., Hill, C., Block, A. M., Sait, S. et al. 2013. Langerhans cell sarcoma arising from chronic lymphocytic lymphoma/small lymphocytic leukemia: Lineage analysis and BRAF V600E mutation study. North American Journal of Medicine and Science, 5(6): 386-391.

Cho, S. W., Kim, S., Kim, Y., Kweon, J., Kim, H. S., Bae, S. et al. 2014. Analysis of off-target effects of CRISPR/Cas-derived RNA-guided endonucleases and nickases. Genome Research, 24(1): 132-141.

Dannenberg, J. H., Schuijff, L., Dekker, M., van der Valk, M. and te Riele, H. 2004. Tissue-specific tumor suppressor activity of retinoblastoma gene homologs p107 and p130. Genes & Development, 18(23): 2952-2962.

De Cesare, M., Cominetti, D., Doldi, V., Lopergolo, A., Deraco, M., Gandellini, P. et al. 2015. Anti-tumor activity of selective inhibitors of XPO1/CRM1-mediated nuclear export in diffuse malignant peritoneal mesothelioma: The role of survivin. Oncotarget, 6(15): 13119-13132.

Demicco, E. G., Maki, R. G., Lev, D. C. and Lazar, A. J. 2012. New therapeutic targets in soft tissue sarcoma. Advances in Anatomic Pathology, 19(3): 170-180.

Di Liso, E., Pennelli, N., Lodovichetti, G., Ghiotto, C., Dei Tos, A. P., Conte, P. et al. 2015. Braf mutation in interdigitating dendritic cell sarcoma: A case report and review of the literature. Cancer Biology & Therapy, 16(8): 1128-1135.

Donehower, L. A., Harvey, M., Slagle, B. L., McArthur, M. J., Montgomery Jr., C. A., Butel, J. S. et al. 1992. Mice deficient for p53 are developmentally normal but susceptible to spontaneous tumours. Nature, 356(6366): 215-221.

Doyle, B., Morton, J. P., Delaney, D. W., Ridgway, R. A., Wilkins, J. A. and Sansom, O. J. 2010. p53 mutation and loss have different effects on tumourigenesis in a novel mouse model of pleomorphic rhabdomyosarcoma. The Journal of Pathology, 222(2): 129-137.

Drost, J., van Jaarsveld, R. H., Ponsioen, B., Zimberlin, C., van Boxtel, R., Buijs, A., et al. 2015. Sequential cancer mutations in cultured human intestinal stem cells. Nature, 521(7550): 43-47.

Duan, Z., Zhang, J., Choy, E., Harmon, D., Liu, X., Nielsen, P., et al. 2012. Systematic kinome shRNA screening identifies CDK11 (PITSLRE) kinase expression is critical for osteosarcoma cell growth and proliferation. Clinical Cancer Research, 18(17): 4580-4588.

Feng, Y., Sassi, S., Shen, J. K., Yang, X., Gao, Y., Osaka, E., et al. 2015. Targeting CDK11 in osteosarcoma cells using the CRISPR-Cas9 system. Journal of Orthopaedic Research, 33(2): 199-207.

Gilbert, L. A., Horlbeck, M. A., Adamson, B., Villalta, J. E., Chen, Y., Whitehead, E. H., et al. 2014. Genome-Scale CRISPR-Mediated Control of Gene Repression and Activation. Cell, 159(3): 647-661.

Gryder, B. E., Yohe, M. E., Chou, H. C., Zhang, X., Marques, J., Wachtel, M., et al. 2017. PAX3-FOXO1 Establishes Myogenic Super Enhancers and Confers BET Bromodomain Vulnerability. Cancer Discovery, 7(8): 884-899.

Gupta, R. M. and Musunuru, K. 2014. Expanding the genetic editing tool kit: ZFNs, TALENs, and CRISPR-Cas9. Journal of Clinical Investigation, 124(10): 4154-4161.

Haldar, M., Hancock, J. D., Coffin, C. M., Lessnick, S. L. and Capecchi, M. R. 2007. A conditional mouse model of synovial sarcoma: Insights into a myogenic origin. Cancer Cell, 11(4): 375-388.

Harwood, J. L., Alexander, J. H., Mayerson, J. L. and Scharschmidt, T. J. 2015. Targeted chemotherapy in bone and soft-tissue sarcoma. Orthopedic Clinics of North America, 46(4): 587-608.

Hayashi, T., Horiuchi, A., Sano, K., Hiraoka, N., Kanai, Y., Shiozawa, T., et al. 2011. Molecular approach to uterine leiomyosarcoma: LMP2-deficient mice as an animal model of spontaneous uterine leiomyosarcoma. Sarcoma, 2011: 476498.

Heckl, D., Kowalczyk, M. S., Yudovich, D., Belizaire, R., Puram, R. V., McConkey, M. E., et al. 2014. Generation of mouse models of myeloid malignancy with combinatorial genetic lesions using CRISPR-Cas9 genome editing. Nature Biotechnology, 32(9): 941-946.

Houghton, P. J., Morton, C. L., Tucker, C., Payne, D., Favours, E., Cole, C. et al. 2007. The pediatric preclinical testing program: Description of models and early testing results. Pediatric Blood & Cancer, 49(7): 928-940.

Huang, J., Chen, M., Whitley, M. J., Kuo, H. C., Xu, E. S., Walens, A. et al. 2017. Generation and comparison of CRISPR-Cas9 and Cre-mediated genetically engineered mouse models of sarcoma. Nature Communications, 8: 15999.

Idbaih, A., Mokhtari, K., Emile, J. F., Galanaud, D., Belaid, H., de Bernard, S., et al. 2014. Dramatic response of a BRAF V600E-mutated primary CNS histiocytic sarcoma to vemurafenib. Neurology, 83(16): 1478-1480.

Jia, B., Choy, E., Cote, G., Harmon, D., Ye, S., Kan, Q. et al. 2014. Cyclin-dependent kinase 11 (CDK11) is crucial in the growth of liposarcoma cells. Cancer Letters, 342(1): 104-112.

Joseph, N. M., Mosher, J. T., Buchstaller, J., Snider, P., McKeever, P. E., Lim, M., et al. 2008. The loss of Nf1 transiently promotes self-renewal but not tumorigenesis by neural crest stem cells. Cancer Cell, 13(2): 129-140.

Kasap, C., Elemento, O. and Kapoor, T. M. 2014. DrugTargetSeqR: A genomics- and CRISPR-Cas9-based method to analyze drug targets. Nature Chemical Biology, 10(8): 626-628.

Kazim, S., Malafa, M. P., Coppola, D., Husain, K., Zibadi, S., Kashyap, T., et al. 2015. Selective nuclear export inhibitor KPT-330 enhances the antitumor activity of gemcitabine in human pancreatic cancer. Molecular Cancer Therapeutics, 14(7): 1570-1581.

Keller, C., Arenkiel, B. R., Coffin, C. M., El-Bardeesy, N., DePinho, R. A. and Capecchi, M. R. 2004a. Alveolar rhabdomyosarcomas in conditional Pax3:Fkhr mice: cooperativity of Ink4a/ARF and Trp53 loss of function. Genes & Development, 18(21): 2614-2626.

Keller, C., Hansen, M. S., Coffin, C. M. and Capecchi, M. R. 2004b. Pax3:Fkhr interferes with embryonic Pax3 and Pax7 function: Implications for alveolar rhabdomyosarcoma cell of origin. Genes & Development, 18(21): 2608-2613.

Kleinstiver, B. P., Pattanayak, V., Prew, M. S., Tsai, S. Q., Nguyen, N. T., Zheng, Z., et al. 2016. High-fidelity CRISPR-Cas9 nucleases with no detectable genome-wide off-target effects. Nature, 529(7587): 490-495.

Konermann, S., Brigham, M. D., Trevino, A. E., Joung, J., Abudayyeh, O. O., Barcena, C., et al. 2015. Genome-scale transcriptional activation by an engineered CRISPR-Cas9 complex. Nature, 517(7536): 583-588.

Kopetz, S., Desai, J., Chan, E., Hecht, J. R., O'Dwyer, P. J., Maru, D. et al. 2015. Phase II pilot study of vemurafenib in patients with metastatic BRAF-mutated colorectal cancer. Journal of Clinical Oncology, 33(34): 4032-4038.

Lagutina, I. V., Valentine, V., Picchione, F., Harwood, F., Valentine, M. B., Villarejo-Balcells, B., et al. 2015. Modeling of the human alveolar rhabdomyosarcoma Pax3-Foxo1 chromosome translocation in mouse myoblasts using CRISPR-Cas9 nuclease. PLoS Genet, 11(2): e1004951.

Lang, G. A., Iwakuma, T., Suh, Y. A., Liu, G., Rao, V. A., Parant, J. M., et al. 2004. Gain of function of a p53 hot spot mutation in a mouse model of Li-Fraumeni syndrome. Cell, 119(6): 861-872.

Lin, P. P., Pandey, M. K., Jin, F., Raymond, A. K., Akiyama, H. and Lozano, G. 2009. Targeted mutation of p53 and Rb in mesenchymal cells of the limb bud produces sarcomas in mice. Carcinogenesis, 30(10): 1789-1795.

Lino, C. A., Harper, J. C., Carney, J. P. and Timlin, J. A. 2018. Delivering CRISPR: A review of the challenges and approaches. Drug Delivery, 25(1): 1234-1257.

Liu, T., Li, Z., Zhang, Q., De Amorim Bernstein, K., Lozano-Calderon, S., Choy, E., et al. 2016. Targeting ABCB1 (MDR1) in multi-drug resistant osteosarcoma cells using the CRISPR-Cas9 system to reverse drug resistance. Oncotarget, 7(50): 83502-83513.

Lu, X. J., Qi, X., Zheng, D. H. and Ji, L. J. 2015. Modeling cancer processes with CRISPR-Cas9. Trends in Biotechnology, 33(6): 317-319.

Maddalo, D., Manchado, E., Concepcion, C. P., Bonetti, C., Vidigal, J. A., Han, Y. C. et al. 2014. In vivo engineering of oncogenic chromosomal rearrangements with the CRISPR/Cas9 system. Nature, 516(7531): 423-427.

Maizels, N. 2013. Genome engineering with Cre-loxP. The Journal of Immunology, 191(1): 5-6.

Matano, M., Date, S., Shimokawa, M., Takano, A., Fujii, M., Ohta, Y., et al. 2015. Modeling colorectal cancer using CRISPR-Cas9-mediated engineering of human intestinal organoids. Nature Medicine, 21(3): 256-262.

Movva, S., Wen, W., Chen, W., Millis, S. Z., Gatalica, Z., Reddy, S., et al. 2015. Multi-platform profiling of over 2000 sarcomas: Identification of biomarkers and novel therapeutic targets. Oncotarget, 6(14): 12234-12247.

Neggers, J. E., Vercruysse, T., Jacquemyn, M., Vanstreels, E., Baloglu, E., Shacham, S., et al. 2015. Identifying drug-target selectivity of small-molecule CRM1/XPO1 inhibitors by CRISPR/Cas9 genome editing. Chemistry & Biology, 22(1): 107-116.

Ng, A. J., Mutsaers, A. J., Baker, E. K. and Walkley, C. R. 2012. Genetically engineered mouse models and human osteosarcoma. Clinical Sarcoma Research, 2(1): 19.

Niu, Y., Shen, B., Cui, Y., Chen, Y., Wang, J., Wang, L., et al. 2014. Generation of gene-modified cynomolgus monkey via Cas9/RNA-mediated gene targeting in one-cell embryos. Cell, 156(4): 836-843.

Platt, R. J., Chen, S., Zhou, Y., Yim, M. J., Swiech, L., Kempton, H. R. 2014. CRISPR-Cas9 knockin mice for genome editing and cancer modeling. Cell, 159(2): 440-455.

Rath, O. and Kozielski, F. 2012. Kinesins and cancer. Nature Reviews Cancer, 12(8): 527-539.

Rauch, D. A., Hurchla, M. A., Harding, J. C., Deng, H., Shea, L. K., Eagleton, M. C., et al. 2010. The ARF tumor suppressor regulates bone remodeling and osteosarcoma development in mice. PLoS One, 5(12): e15755.

Rubin, B. P., Nishijo, K., Chen, H. I., Yi, X., Schuetze, D. P., Pal, R. et al. 2011. Evidence for an unanticipated relationship between undifferentiated pleomorphic sarcoma and embryonal rhabdomyosarcoma. Cancer Cell, 19(2): 177-191.

Sanchez-Rivera, F. J., Papagiannakopoulos, T., Romero, R., Tammela, T., Bauer, M. R., Bhutkar, A., et al. 2014. Rapid modelling of cooperating genetic events in cancer through somatic genome editing. Nature, 516(7531): 428-431.

Shalem, O., Sanjana, N. E., Hartenian, E., Shi, X., Scott, D. A., Mikkelson, T., et al. 2014. Genome-scale CRISPR-Cas9 knockout screening in human cells. Science, 343(6166): 84-87.

Souid, A. K., Dubowy, R. L., Ingle, A. M., Conlan, M. G., Sun, J., Blaney, S. M., et al. 2010. A pediatric phase I trial and pharmacokinetic study of ispinesib: Children's Oncology Group phase I consortium study. Pediatric Blood & Cancer, 55(7): 1323-1328.

Strizzi, L., Bianco, C., Hirota, M., Watanabe, K., Mancino, M., Hamada, S., et al. 2007. Development of leiomyosarcoma of the uterus in MMTV-CR-1 transgenic mice. The Journal of Pathology, 211(1): 36-44.

Susa, M., Iyer, A. K., Ryu, K., Choy, E., Hornicek, F. J., Mankin, H., et al. 2010. Inhibition of ABCB1 (MDR1) expression by an siRNA nanoparticulate delivery system to overcome drug resistance in osteosarcoma. PLoS One, 5(5): e10764.

Taylor, B. S., Barretina, J., Maki, R. G., Antonescu, C. R., Singer, S. and Ladanyi, M. 2011. Advances in sarcoma genomics and new therapeutic targets. Nature Reviews Cancer, 11(8): 541-557.

Thway, K., Noujaim, J., Jones, R. L. and Fisher, C. 2017. Advances in the pathology and molecular biology of sarcomas and the impact on treatment. Clinical Oncology (The Royal College of Radiologists), 29(8): 471-480.

Torres, R., Martin, M. C., Garcia, A. Cigudosa, J. C., Ramirez, J. C. and Rodriguez-Perales, S. 2014. Engineering human tumour-associated chromosomal translocations with the RNA-guided CRISPR-Cas9 system. Nature Communications, 5: 3964.

Tsai, S. Q., Zheng, Z., Nguyen, N. T., Liebers, M., Topkar, V. V., Thapar, V., et al. 2015. GUIDE-seq enables genome-wide profiling of off-target cleavage by CRISPR-Cas nucleases. Nature Biotechnology, 33(2): 187-197.

Tsumura, H., Yoshida, T., Saito, H., Imanaka-Yoshida, K. and Suzuki, N. 2006. Cooperation of oncogenic K-ras and p53 deficiency in pleomorphic rhabdomyosarcoma development in adult mice. Oncogene, 25(59): 7673-7679.

Wen, W. S., Yuan, Z. M., Ma, S. J., Xu, J. and Yuan, D. T. 2016. CRISPR-Cas9 systems: Versatile cancer modelling platforms and promising therapeutic strategies. International Journal of Cancer, 138(6): 1328-1336.

Wilson, T. J. and Kola, I. 2001. The LoxP/CRE system and genome modification. Methods in Molecular Biology, 158: 83-94.

Xing, D., Scangas, G., Nitta, M., He, L., Xu, X., Ioffe, Y. J., et al. 2009. A role for BRCA1 in uterine leiomyosarcoma. Cancer Research, 69(21): 8231-8235.

Yao, Y., Dong, Y., Lin, F., Zhao, H., Shen, Z., Chen, P., et al. 2009. The expression of CRM1 is associated with prognosis in human osteosarcoma. Oncology Reports, 21(1): 229-235.

Yin, H., Xue, W. and Anderson, D. G. 2019. CRISPR-Cas: A tool for cancer research and therapeutics. Nature Reviews Clinical Oncology, 16(5): 281-295.

Zhang, P. and Pollock, R. E. 2014. Epigenetic regulators: New therapeutic targets for soft tissue sarcoma. Cancer Cell & Microenvironment, 1(4).

Zhang, Y., Yin, C., Zhang, T., Li, F., Yang, W., Kaminski, R., et al. 2015. CRISPR/gRNA-directed synergistic activation mediator (SAM) induces specific, persistent and robust reactivation of the HIV-1 latent reservoirs. Scientific Reports, 5: 16277.

Zhou, Y., Han, C., Li, D., Yu, Z., Li, F., Li, F., et al. 2015. Cyclin-dependent kinase 11(p110) (CDK11(p110)) is crucial for human breast cancer cell proliferation and growth. Scientific Reports, 5: 10433.

# CRISPR/Cas9 Technology as a Strategy Against Viral Infections

**Huafeng Lin**[1,2‡], **Haizhen Wang**[3‡], **Jun He**[1]*, **Xiangwen Peng**[1‡], **Aimin Deng**[1],
**Lei Shi**[2], **Lei Ye**[2] **and Hanyue Gong**[2]

[1] Changsha Maternal and Child Health Care Hospital, Changsha 410007, PR China
[2] Institute of Food Safety and Nutrition, Jinan University, Guangzhou 510632, PR China
[3] Department of Life Science, Lv Liang University, Lvliang 033000, PR China

## Introduction

Currently, several infectious viruses, such as human immunodeficiency virus (HIV), hepatitis B virus (HBV), hepatitis C virus (HCV), human papillomavirus (HPV) and herpes simplex virus (HSV), are potentially threatening human health worldwide (Wahid et al. 2017). The inherent unique features of infectious viruses, specifically the formation of latent viral reservoirs in hosts, make strategies for the treatment of viral infections a significant challenging project (White et al. 2015). Moreover, many pathogenic human viruses are more likely to generate mutant strains for escape when they are threatened by drug intervention and even jump between different species spontaneously for survival (Parrish et al. 2008). Currently, antiviral treatments that attack viral proteins have shown promising results. Nevertheless, such antiviral drugs have faced setbacks and failures in some cases (Khan et al. 2018). The antiviral role of the CRISPR/Cas9 system initially in prokaryotes and its successful applications in mammalian cells (Cong et al. 2013, Jinek et al. 2012, Mali et al. 2013) makes it a potential candidate for combating human viruses. Moreover, CRISPR/Cas9 gene-edited technology has currently entered the field of biomedical engineering studies and gene therapy research and will eventually be used in the clinical setting.

Gene editing is a combinational process in which site-specific DNA cleavages are introduced by various nucleases, such as meganuclease, zinc-finger nucleases (ZFNs), transcription activator-like effector nucleases (TALENs) and CRISPR/Cas nuclease systems (Hryhorowicz et al. 2017, Khalil 2020), which subsequently trigger natural cellular pathways to repair DNA breaks. Canonically, for conducting site-specific genome manipulations, TALENs or ZFNs that function as pairs use

---

* Corresponding author: 19252702@qq.com
‡ These authors contributed equally to this work.

the catalytic domain of dimeric type IIS FokI endonuclease to function (Peng et al. 2014), while prototype CRISPR/Cas9 complexes employ two functional domains (RuvC-like domain and HNH domain) of Cas9 endonuclease to exert double-strand DNA cleavage activity (Batool et al. 2021). Cell DNA repair caused by DNA lesions is achieved either via the homology-directed repair (HDR) pathway with repair templates or the non-homologous end joining (NHEJ) pathway without repair templates (Fernández et al. 2017). To date, the CRISPR/Cas9 genome-editing system has been designed as a versatile instrument for targeted gene modifications in a broad range of animal species (Lin et al. 2019, Banan, 2020), the internal gut microbiota (Cresci et al. 2020, Hegde et al. 2019, Ramachandran and Bikard 2019) and their pathogenic viruses (Ashfaq and Khalid 2020), which causes changes in the relationships between the host and viruses. The success of genome editing of cultured human cells via the CRISPR/Cas9 system has instigated an important revolution for gene therapy in biomedical research since 2013 (Cho et al. 2013, Jinek et al. 2013, Mali et al. 2013). With increasing interest in gene-editing research, a new drug based on CRISPR/Cas9 gene-editing technology is entering the clinical era of treating human viral infection (Hirakawa et al. 2020).

In this chapter, we will provide an overview of the biological characteristics of the CRISPR/Cas machinery and its classification followed by a summary of the investigations that have employed CRISPR/Cas9 techniques for the prevention and control of several human infectious viruses. We will also discuss the potential limitations of the CRISPR/Cas9 system, including off-target effect, delivery challenge, viral escape issue, Cas9 cleavage activity, resistance to CRISPR/Cas9 and ethical concern.

# Biological characteristics of the CRISPR/Cas machinery and classification

In 1987, Japanese scholars initially discovered that CRISPR is a set of DNA repeat sequences of unidentified origin in the *Escherichia coli* genome (Ishino et al. 1987). The CRISPR/Cas modules in prokaryotes (approximately 40% of bacteria and 90% of archaea) provide these microorganisms with the ability to resist new invading genetic elements (e.g. plasmids, phages and viruses), constituting part of their adaptive immune system (Mohammadzadeh et al. 2020). Generally, the biological diversity of CRISPR/Cas systems depends on Cas-associated genes and their genomic evolution and variability, which deeply influence their classification (Makarova et al. 2015). According to the divergence of Cas protein homology, CRISPR/Cas systems are split into two distinct classes (Class 1 and Class 2) with six types (type I, II, III, IV, V and VI) and over 30 different subtypes (Escalona-Noguero et al. 2021, Yan et al. 2019). Among them, Class 1, which employs a multiprotein effector complex, is categorized into three types as follows: type I, type III and type IV. Class 2, which employs only a single effector protein for gene editing, comprises type II, type V, and type VI systems (Bayat et al. 2018, Makarova et al. 2015, Makarova et al. 2017, Zhang and Ye 2017). Typically, type II and type V Cas proteins are used for editing DNA, and type VI Cas nucleases (e.g. Cas13 nuclease) are used for programming RNA (Cao et al. 2020). In addition, type III CRISPR/Cas nucleases are capable of targeting and cleaving both the DNA and RNA sequences of invaders (Li et al. 2017b). The type II CRISPR/Cas system, commonly known as CRISPR/Cas9, is a binary complex formed by the Cas9 endonuclease, CRISPR-derived RNA (crRNA) and trans-activating RNA (tracrRNA) (Gebre et al. 2018, Saayman et al. 2015). For research convenience, crRNA and tracrRNA are usually designed to be chimeric single guide RNA (sgRNA) as an unit. Thus, this system has two simple requirements as follows: optimization of Cas9 endonuclease (Wright et al. 2015) and customization of the sgRNA (Cui et al. 2018). The CRISPR/Cas9 system has been extensively applied for RNA-guided genome-editing purposes to program the DNA sequences in the targeted genome (Jiang et al. 2015). The mechanism of action of this process is that the guide RNA molecule is recognized by the Cas9 protein in a sequence-independent manner (Nishimasu et al. 2014) via Watson-Crick base-pairing interactions to match the complementary strand of a specific DNA sequence with 17-20 nucleotides adjacent upstream to a short protospacer adjacent motif (PAM; e.g. 5'-NGG-3'). Then, the Cas9 RuvC-like domain cuts the noncomplementary DNA strand, and the Cas9 HNH domain simultaneously cleaves the complementary DNA strand (strand binding with gRNA) (Garneau et al. 2010, Jinek et al. 2012). The wild-type CRISPR/Cas9 system has various types of structural variants and Cas9 orthologs (Lin. et al. 2019, Ran et al. 2013). *Streptococcus pyogenes* Cas9 (*Sp*Cas9) and *Staphylococcus aureus* Cas9 (*Sa*Cas9) are the two most widely used types by investigators in biotechnology. Furthermore, Cas9 endonucleases from different bacterial species use different PAM sequences to recognize different *loci* (Lino et al. 2018). Basically, *Sp*Cas9 uses 'NGG' as its PAM site, whereas *Sa*Cas9 employs 'NNGRRT' as the PAM sequence (Xie et al. 2018).

Recently, the multifunctional features of the CRISPR/Cas9 machinery have enabled the expansion of the applications of this technology. The CRISPR/Cas9 system and its structural variants hold great promise for targeting viruses and viral reservoirs hidden in the human body.

## Applications of CRISPR/Cas9 to protect against infectious viruses

In general, host cells counteract viral infections by sensing viral nucleic acids or virus-encoded proteins, thereby triggering the expression of antiviral genes/proteins (Puschnik et al. 2017). However, sometimes the host immune system is tremendously damaged by a virus, resulting in the uncontrollability of the infection by host immunity. Consequently, the therapeutic challenge of targeting viral infections is how to capture these obligate parasites that rely on host metabolic machinery for proliferation. Theoretically, the CRISPR/Cas9 system can be employed not only to program viral DNA sequences but also to target and modify specific DNA sequences in eukaryotic host genomes (Soppe and Lebbink 2017). In addition, CRISPR/Cas9 with multiple sgRNAs has enabled Cas9 proteins to mutate several different functional genomic *loci* in parallel (Ota et al. 2014, Zhou et al. 2014). In summary, CRISPR/Cas systems, as a set of sophisticated genetic correction toolboxes, can be repurposed to achieve additional functions, including gene knockout, gene knockin, transcriptional activation, transcriptional inhibition, epigenetic modification, chromatin imaging and base editing (Barrangou and Dauda 2016, Cong et al. 2013, Hu et al. 2018, Miller et al. 2020, Sun and Wang 2020). Interestingly, a recent study has uncovered the existence of CRISPR/Cas-related systems involved in nondefense roles in the archaeal *Asgard superphylum* (Makarova et al. 2020). The use of CRISPR/Cas9 technology to remove infectious viruses from cells may potentially be available to any DNA virus or RNA virus with a DNA intermediate in its life cycle (Doudna and Charpentier 2014, Khalili et al. 2015, Mohammadzadeh et al. 2020). Therefore, the CRISPR/Cas9 methodologies raise high hopes for targeting viruses in different developmental phases of the viral life cycle and can potentially attain an effective therapeutic method for chronic infections caused by human viruses (Lee 2019). A series of studies on CRISPR/Cas9-based treatments of human immunodeficiency virus (HIV), hepatitis B virus (HBV), hepatitis C virus (HCV), human papilloma virus (HPV) and herpes simplex virus (HSV) is presented below to understand the current status of the CRISPR/Cas9-mediated antiviral strategy.

## HIV

According to the report of the United Nations Programme on AIDS (UNAIDS), more than 36.7 million people are infected with human immunodeficiency virus (HIV) around the world, and the number of new infections is greater than 5000 per day (Dash et al. 2019). There are three phases consisting of an early asymptomatic stage, an intermediate stage and a late stage in the progression of HIV infection to acquired immune deficiency syndrome (AIDS) (Nwankwo and Seker 2013). HIV comprises two main types: HIV-1 and HIV-2. HIV-1 is characterized by higher transmissibility and pathogenicity in humans than HIV-2 infection (Campbell-Yesufu and Gandhi 2011). Active HIV-1 replication *in vivo* causes severe CD4$^+$ T cell depletion, leading to AIDS (Doitsh et al. 2014, Vijayan et al. 2017). Great successes have been made by using antiretroviral therapy (ART) and highly active antiretroviral therapy (HAART) to prevent AIDS (Lu et al. 2018). However, these types of anti-HIV strategies, aimed at suppressing viral life cycles (Arribas and Eron 2013), are still unable to completely eradicate HIV due to the integration of viral genes in the host genome. With the advancement of gene-editing technologies, researchers have applied CRISPR/Cas9-based methods to address this elusive virus, unlocking many new possibilities for HIV prevention and cure (Dampier et al. 2014). Since the first two CRISPR/Cas9-based approaches for the prevention of HIV-1 infection were reported by Cho and Ebina in 2013 (Cho et al. 2013, Ebina et al. 2013), many studies using this technology to treat HIV-1/AIDS have been actively implemented in succession (Khalili et al. 2015). To date, there are two essential approaches that consist of targeting host genes and targeting viral genomes to protect against HIV-1 infection (Xiao et al. 2019). The most strategic editing targets for CRISPR/Cas9-based HIV eradication approaches include the C-C chemokine receptor 5 (CCR5) gene, C-C-C chemokine receptor 4 (CXCR4) gene, proviral DNA-encoding viral proteins and HIV 5' and 3' long terminal repeats (LTRs) (Bialek et al. 2016, Liu et al. 2017, Manjunath et al. 2013). In addition, the use of the CRISPR/Cas9 system to simultaneously target two coreceptor genes (CCR5 and CXCR4) has not yet been proven in relevant studies, although multiplexed CRISPR/Cas9 technology has gained certain achievements. Table 1 summarizes the recent studies for CRISPR/Cas9-based HIV-1/AIDS therapy (Binda et al. 2020, Cho et al. 2013, Ebina et al. 2013, Hou et al. 2015, Hu et al. 2014, Kaminski et al. 2016, Kaushik et al. 2019, Lebbink 2017, Liao et al. 2015, Liu et al. 2018, Wang et al. 2014, Ye et al. 2014, Zhu et al. 2015).

It has been more than 30 years since the discovery of HIV. However, there is still no effective anti-HIV vaccine available (Haynes 2015). In HIV patients who receive ART and who are older than 60 years old, latent viral reservoirs within resting memory CD4$^+$ T cells are still capable of causing damage (Siliciano et al. 2003). Thus, the objective of the eradication of HIV infections is effectively to purge dormant viral reservoirs. CRISPR/Cas9 gene-editing technology provides hope for the future use of personalized gene therapies in HIV-infected patients.

## HBV

Approximately 350-400 million people are chronic hepatitis B virus (HBV) carriers; therefore, hepatitis B is still an important health problem (Seo and Yano 2014, Trépo et al. 2014). Notably, hepatitis B causes 887,000 deaths annually according to the estimate of the World Health Organization (WHO) (Soriano et al. 2020). HBV, classified as one member of the family *Hepadnaviridae* (Locarnini et al. 2013), is a hepatotropic DNA virus that replicates by reverse transcription via a RNA intermediate in host hepatocytes. According to taxonomic classification, more than 8% of nucleotide mutations exist between any two of eight genotypes (A-H) of the HBV genome (Sunbul 2014). HBV leads to a relatively high incidence of liver cirrhosis and liver cancer in chronic HBV infection (El-Serag 2012, Lee 1997). Given that the chances of HBV-infected patients achieving a sustained viral response (SVR) and cure are low, novel and effective treatments against HBV are currently needed (Nassal 2015). The rapid advancement of CRISPR/Cas9 technology provides new opportunities for producing techniques aimed at preventing and treating HBV infectious diseases. Current antiviral treatments, including nucleoside analogs (NAs) and interferon-alpha (IFN-α), cannot eradicate chronic hepatitis B (CHB) owing to the persistence of covalently closed circular DNA (cccDNA) of HBV (Emery and Feld 2017). Gene therapy has become a promising methodology for the treatment of HBV infections, especially for effective targeting of cccDNA, and holds high promise for clinical applications, although it still requires overcoming some technical barriers (Bloom et al. 2018, Maepa et al. 2015). Preclinical studies on the suppression of HBV infection using the gene-editing platforms, ZFNs and TALENs, were first independently reported by two research groups (Bloom et al. 2013, Cradick et al. 2010). Additionally, similar studies on ZNF- and TALEN-based anti-HBV approaches have also made promising achievements (Dreyer et al. 2016, Weber et al. 2014). In 2014, Lin and colleagues first reported ground-breaking work employing the CRISPR/Cas9 system to counter HBV infection *in vitro* and *in vivo* (Lin et al. 2014). Since then, several studies have utilized CRISPR/Cas9 (or multiplex CRISPR/Cas9) to edit a single *locus* (which is usually in the conserved region of the HBV genome) to inhibit viral replication and production (Dong et al. 2015, Karimova et al. 2015, Kostyusheva et al. 2019a, Li et al. 2017a, Liu et al. 2015, Schiwon et al. 2018,

**Table 1.** Applications of CRISPR/Cas9 system in gene therapy of HIV infection

| Virus type | Target gene | Editing system | Number of gRNA | Cell/Animal model | Delivery methods | Reference |
|---|---|---|---|---|---|---|
| HIV | CCR5 | Cas9/gRNA | Single gRNA | HEK293T cells | Plasmid transfection | Cho et al. 2013 |
| HIV-1 | LTR | CRISPR/Cas9 | Two gRNAs | Jurkat cells, HeLa cells, T cells | Plasmid transfection | Ebina et al. 2013 |
| HIV-1 | LTR U3 region | Cas9/gRNA | Single gRNA, Multiple gRNAs | microglial, promonocytic, and T cells | Plasmid transfection | Hu et al. 2014 |
| HIV-1 | CCR5 | CRISPR/Cas9 | Multiple gRNAs | 293T cells, TZM.bl cells, CEMss-CCR5 cells | Lentiviral transduction | Wang et al. 2014 |
| HIV-1 | CCR5 (exon 4) | TALENs, CRISPR/Cas9 | Multiple gRNAs | iPSCs | The piggyBac transposon vectors, cotransfection | Ye et al. 2014 |
| HIV-1 | CXCR4 | CRISPR/Cas9 | Multiple gRNAs | Ghost-CXCR4 cells, Jurkat cells and primary human CD4+ T cells | Lentivirus-mediated delivery | Hou et al. 2015 |
| HIV-1 | LTR U3, T and R region | Multiplex CRISPR/Cas9 | Single gRNA, Multiple gRNAs | HEK293T cells, hPSCs | Plasmid transfection, lentiviral transduction | Liao et al. 2015 |
| HIV-1 | LTR, pol gene, and tat/rev | CRISPR/Cas9 | Ten gRNAs | Jurkat cell lines | Nucleo transfection | Zhu et al. 2015 |
| HIV-1 | LTR U3 region | saCas9/multiplex gRNAs | Multiple gRNAs | MEFs, transgenic mice, rats | Lentiviral delivery | Kaminski et al. 2016 |

*(Contd.)*

**Table 1.** (*Contd.*)

| Virus type | Target gene | Editing system | Number of gRNA | Cell/Animal model | Delivery methods | Reference |
|---|---|---|---|---|---|---|
| HIV-1 | LTR | CRISPR/Cas9 | Single gRNA, Multiple gRNAs | HEK293T cells, J.Lat FL cells, human T lymphoblast cells | Lentiviral transduction | Lebbink et al. 2017 |
| HIV-1 | CXCR4 | CRISPR/Cas9 | Multiple gRNAs | TZM-bl cells | Lipofectamine 2000 | Liu et al. 2018 |
| HIV-1 | LTR | CRISPR/Cas9 | Multiple gRNAs | Latent hmicroglia cells | Magnetic delivery | Kaushik et al. 2019 |
| HIV-1 | Proviral DNA | CRISPR/Cas9 | Two gRNAs | HEK 293T cells | Lentiviral transfection | Binda et al. 2020 |

Scott et al. 2017, Seeger and Sohn 2016, Zhu et al. 2016). To enhance the silencing effects for target genes, several research teams have developed applications based on the CRISPR/Cas9 method for simultaneous targeting of different *loci* (e.g. surface antigen region, X gene, reverse transcriptase (RT) gene and episomal cccDNA) from the HBV genome through cell culture methods or mouse models (Kennedy et al. 2015, Ramanan et al. 2015, Sakuma et al. 2016, Seeger and Sohn 2014, Wang et al. 2015, Zhen et al. 2015). Apart from the CRISPR/Cas9 system itself, several other studies involving the combination of CRISPR/Cas9 and other antiviral treatments (e.g. nucleoside molecules or nonnucleoside inhibitory systems) have also been developed for the purpose of extinguishing HBV genomes (Kostyusheva et al. 2019b, Soriano et al. 2020, Wang et al. 2017, Zheng et al. 2017).

Interestingly, a Cas9 variant, known as dead Cas9 (dCas9), has also been demonstrated to inhibit the replication of HBV without the need to dissect its genome (Kurihara et al. 2017). More importantly, a recent study has been conducted to investigate a potent inhibitor of the NHEJ DNA repair pathway, named 'NU7026', which impedes the degradation of cccDNA mediated by CRISPR/Cas9 cleavage (Kostyushev et al. 2019). This study provided a methodology for the verification of the activity and function of the CRISPR/Cas9 system in disabling the HBV genome. Recently, Yang and colleagues showed that the Cas9-associated base editing system holds promise to potentially cure CHB by permanent inactivation of integrated HBV DNA and cccDNA without generating double stand breaks (DSBs) in the genomic DNA of human cells (Yang et al. 2020).

The treatment of HBV-infected cells with CRISPR/Cas9 cannot be used to circumvent the problem of viral escape mutations similar to ZFN and TALEN therapies (Schinazi et al. 2018, Lee 2019). Furthermore, artificial models (either cell models or animal models) can simulate only persistent HBV infection in human hepatocytes, and they do not completely represent HBV infection *in vivo*. Taken together, additional studies are needed that use CRISPR/Cas9 alone or in conjunction with other therapies to inhibit virus replication *in vivo* and to effectively target integrated HBV genomes and eliminate HBV cccDNA formation in infected hepatocytes for functional HBV cure.

## HCV

Hepatitis C virus (HCV) infection is currently a major global health concern (Hanafiah et al. 2013). This condition leads to chronic liver diseases or liver cancer in approximately 170 million individuals annually worldwide (Lavanchy 2011). HCV is a single-stranded RNA virus whose genome consists of more than 9500 bp. The potential therapeutic utility of CRISPR/Cas technology for HCV RNA targeting has not been widely and systemically investigated, although the DNA-targeting capability of the CRISPR/Cas9 system has been extensively characterized (Moyo et al. 2018).

In 2015, Ren and colleagues identified three genes (CLDN1, OCLN and CD81) as essential for HCV cell-free entry and intercellular transmission using the NIrD (NS3-4A Inducible rtTA-mediated Dual-reporter) system and focused CRISPR/Cas9 library. These investigators also emphasized that this combined technology offers a robust and high-throughput method for identifying the key host factors of HCV infections (Ren et al. 2015). In the same year, Hopcraft and colleagues demonstrated that HCV replicates in the absence of miR-122 by using the CRISPR/Cas9 technique, which has important implications for understanding how the liver tropism of HCV is controlled and whether human cell types other than hepatocytes can also support infection. Furthermore, these results highlight potential problems with the use of miR-122-based drugs for the therapy of HCV infection (Hopcraft et al. 2015). In 2016, Yamauchi et al. used the CRISPR/Cas9 system to examine the role of STAT1 (signal transducer and activator of transcription 1) and STAT2 (signal transducer and activator of transcription 2) in the inhibition of HCV replication by IFN-α (interferon-alpha) and IFN-λ (interferon-lambda). The results revealed that IFN-α interrupts HCV replication solely by a STAT2-dependent pathway, while IFN-λ triggers ISG (IFN-stimulated gene) expression and restrains HCV replication exclusively via STAT1- and STAT2-dependent pathways (Yamauchi et al. 2016). By 2017, Senís and colleagues successfully employed TALEN and CRISPR/Cas9 nucleases to site-specifically integrate an anti-HCV shmiRNA (i.e. an shRNA embedded in a microRNA (miRNA) scaffold) into the liver-specific miR-122/*hcr locus* in hepatoma cells, and they acquired cellular clones that were hereditarily safeguarded against HCV infection. The final results showed that the expression of anti-HCV shmiRNA as well as miR-122 integrity and functionality were demonstrated in selected progeny cells (Senís et al. 2017). In 2018, Duan and coworkers designed an experiment using small interfering RNA (siRNA) (for overexpression) or the CRISPR/Cas9 system (for knockdown) to manipulate miR-130a (microRNA 130a) and its target genes. These researchers eventually concluded that miR-130a regulates HCV replication via its targeting of PKLR (pyruvate kinase in liver and red blood cells) and subsequent pyruvate production. These data provide new opinions regarding how miR-130a regulates steps of key metabolic enzymatic pathways, including steps involving PKLR and pyruvate, which are disrupted by HCV replication (Dun et al. 2018).

To date, CRISPR/Cas9 systems combined with other biotechnological methods are currently used to explore mechanisms for completely destroying HCV viral particles in hepatocytes. Significant progress will be made in the use of CRISPR/Cas technology for the treatment of HCV-associated diseases sooner or later.

## Human papilloma viruses HPV16 and HPV18

Human papillomaviruses (HPVs) are small double-stranded DNA viruses belonging to the *Papovaviridae* family with over 200 genetically distinct subtypes already identified (McBride 2017, Zhen and Li 2017). The HPV genome is approximately 8 kbp in length, and it encodes 9 or 10 ORFs and encompasses 8 early viral regulatory proteins (E1–E8) and two late capsid proteins (L1 and L2) (Ebrahimi et al. 2019). Because HPVs present epithelial tissue tropism (Harden and Munger 2017), sexual transmission (Ryndock and Meyers 2014) and oncogenic properties (Moens 2018), their important relations between individual infection and public health must be emphasized. Most HPV infections in humans are harmless and spontaneously clear (Crosbie et al. 2013). However, persistent high-risk type HPV (e.g. HPV-16 and HPV-18) infection is highly correlated with the development of cervical cancers in women (Gupta and Mania-Pramanik 2019). HPV is also responsible for other kinds of anogenital cancer, head and neck cancers and genital warts in men and women (Chen et al. 2018, Medeiros et al. 2020). Currently, there is no satisfactory clinical treatment for HPV infection because the virus circumvents host immune surveillance by means of 'self-dormancy', which makes it extremely difficult to remove its viral genome from an infected host cell in a latency state (Lee 2019). Based on the existing literature, tumor formation triggered by HPV-16 or HPV-18 has been mostly attributed to the HPV E6 and E7 oncoproteins, which are two prime intervention targets for anticancer therapy (Hoppe-Seyler et al. 2018, Pal and Kundu 2020). Theoretically, the HPV E6 and E7 genes suppress the p53 and retinoblastoma protein (pRB) cellular tumor suppressors, respectively (Kennedy and Cullen 2017). Therefore, overexpression of E6 or E7 induced by HPVs can cause malignant transformation of human cells with high probability through the activation of cellular oncogenes (e.g. *ras* or *fos*) (McLaughlin-Drubin and Munger 2009).

There is still an urgent need to develop novel effective therapies for HPV-associated carcinogenesis, although certain progress has been made in utilizing treatments for HPV. CRISPR/Cas9-based gene manipulation technology for HPV infection has been rapidly developed in recent years. Numerous articles have reported anti-HPV research and applications using the CRISPR/Cas9 system to disrupt the HPV genome (Cheng et al. 2018, Gao et al. 2020, Hsu et al. 2018, Hu et al. 2014, Inturi and Jemth 2020, Jubair et al. 2019, Kennedy et al. 2014, Lao et al. 2018, Liu et al. 2016, Yoshiba et al. 2019, Yu et al. 2014, 2017, Zhen et al. 2014, 2016, 2020) (Table 2). Based on these investigations, the CRISPR/Cas9 system, using E6/E7-specific gRNAs, has potential to act as an effective therapy for HPV-associated diseases in clinical settings.

## HSV

HSV-1 and HSV-2, two members of herpes simplex virus (HSV), are important human neurotropic pathogens that cause a wide variety of serious diseases, such as oral ulcers, genital ulcers and neonatal herpes, particularly in newborns and immunocompromised individuals (Aubert et al. 2020, De Silva Feelixge et al. 2018). Currently, nucleoside analog therapy inhibits HSV replication but does not cure latent viral infections or prevent virus reactivation from latency. Therefore, alternative strategies to combat HSV infection or move toward the elimination of dormant viruses from infected cells are needed (van Diemen and Lebbink 2017).

Several studies have described the use of meganucleases as a genome-editing strategy to inhibit HSV infections (Grosse et al. 2011, Aubert et al. 2014, 2016). The CRISPR/Cas9 system is generally believed to be a more efficacious and convenient technology for directed gene editing to counter HSV infection since the successful use of CRISPR/Cas9 in human cells in 2013 (Cong et al. 2013, Jinek et al. 2013). Bi and colleagues confirmed that anti-HSV-1 Cas9/gRNAs stimulate the formation of DSBs during virus replication in the HSV-1 genome where NHEJ subsequently introduces site-specific InDels (insertion/deletion) into the viral target *locus* (Bi et al. 2014). Turner and colleagues utilized the CRISPR/Cas9 genome-editing platform to measure the effect on HSV-1 production in which single- and double-knockout

**Table 2.** Examples of application of CRISPR/Cas9 system in the treatment of HPV

| Gene editing platforms | Virus types | Delivery patterns | gRNA targets | Cell or animal modes | References |
|---|---|---|---|---|---|
| CRISPR/Cas9 system | HPV-16 | Transfection | E7 | SiHa, Caski, C33A, and HEK293 cells | Hu et al. 2014 |
| CRISPR/Cas9 system | HPV-18 | LV transduction | E6, E7 | HeLa cells, SiHa cells, 293 T cells | Kennedy et al. 2014 |
| CRISPR/Cas9 system | HPV-16 | Plasmids transfection | E6 | SiHa and CaSki cells | Yu et al. 2014 |
| CRISPR/Cas9 system | HPV-16 | Plasmids and lipofectamine transfection | E6, E7 | SiHa and C33-A cells, BALB/C nude mice | Zhen et al. 2014 |
| CRISPR/Cas9 system | HPV-6, HPV-11 | Plasmids transfection | E7 | Human keratinocytes | Liu et al. 2016 |
| CRISPR/Cas9 system | HPV-16 | Plasmids and lipofectamine transfection | E6, E7 | Siha and C33-A cells | Zhen et al. 2016 |
| CRISPR/Cas9 system | HPV-18 | Plasmids transfection | E6, E7 | HeLa cells | Yu et al. 2017 |
| CRISPR/SaCas9 system | HPV-16 | AAV delivery | E6, E7 | 293 T cells | Hsu et al. 2018 |
| CRISPR/hCas9 system | HPV pseudotype virus | Plasmids transfection | E6 | SiHa cells, 293FT cells | Cheng et al. 2018 |
| CRISPR/SpCas9 system | HPV-18 | Micelle delivery, Lipofectamine | E7 | Hela cells | Lao et al. 2018 |
| WT Cas9, FokI-dCas9 | HPV-16, HPV-18 | liposomes | E6, E7 | Mouse model, CasKi cells, HeLa cells, HEK293T, Jurkat cells | Jubair et al. 2019 |
| CRISPR/Cas9 system | HPV-18 | Plasmids, AAV delivery | E6 | HeLa, HCS-2, and SKG-I cell lines | Yoshiba et al. 2019 |
| CRISPR/Cas9 system | HPV-16 | Plasmids transfection | E7 | SiHa cells, Hela cells, nude mice | Gao et al. 2020 |
| CRISPR/Cas9 system | HPV-18 | Plasmids transfection | E6, E7 | HeLa (CCL-2) cell lines | Inturi et al. 2020 |
| CRISPR/Cas9 system | HPV-16 | Lipofectamine delivery | E6, E7 | SiHa cell | Zhen et al. 2020 |

(KO) cell lines of TorA and TorB as well as their activators, LAP1 (lamina associated polypeptide 1) and LULL1 (luminal domain like LAP1), were generated. Their results supported the finding that LULL1 is the most important constituent of the Torsin system related to HSV production (Turner et al. 2015). Similarly, several studies have highlighted the technical specialties and applications for editing the HSV genome or constructing virus mutants by using the CRISPR/Cas9 system (Ebrahimi et al. 2020, Finnen and Banfield 2018, Karpov et al. 2019, Khodadad et al. 2020, Li et al. 2017c, Russell et al. 2015, Suenaga et al. 2014, Xu et al. 2016, 2017). Moreover, other studies have shown that the CRISPR/Cas9 system may have prophylactic and therapeutic potential in the treatment of HSV infections (Roehm et al. 2016, van Diemen et al. 2016). Interestingly, Cai and fellow researchers explored the therapeutic potential of *in vitro* transcribed gRNA (IVT-gRNA) by observing the antiviral activity of IVT-gRNA in HSV-1-infected Cas9$^+$ mice, which extended the applications of CRISPR by utilizing the immunostimulatory function of gRNAs (Cai et al. 2018). While using Epstein–Barr virus as a model, Wang and Quake in early 2014 employed the CRISPR/Cas9 system directly to target essential viral genome sequences and to develop a gene therapeutic strategy for herpesvirus (Wang and Quake 2014).

Taken together, although certain hurdles remain to be conquered, CRISPR/ Cas9-based antiviral strategies, either alone or in combination with other methods, have good prospects in curing latent HSV infections in clinical applications.

## Applications of CRISPR/Cas-based strategies on other human-associated viruses

CRISPR/Cas9 technology still has limitations in the treatment of certain human viruses, but other members of CRISPR/Cas systems have shown unexpected effects in the area of viral detection and diagnosis.

Recently, several research teams have used CRISPR/Cas toolkits (e.g. Cas13 orthologs) to target and detect a range of viruses, such as dengue (Myhrvold et al. 2018), influenza A (Abbott et al. 2020), lymphocyte choriomeningitis, vesicular stomatitis (Freije et al. 2019) and Zika (Myhrvold et al. 2018). Importantly, Cas12- and Cas13-based detection of SARS-CoV-2 has been reported (Broughton et al. 2020, Fozouni et al. 2021). The suitability and potential of CRISPR/Cas13 as a prophylactic or therapeutic approach for SARS-CoV-2 infections have also been elucidated (Abbott et al. 2020, Goudarzi et al. 2020).

# Potential challenges for the CRISPR/Cas9 strategy in human infectious viruses

The CRISPR/Cas9 platform holds considerable potential for therapeutic applications of human infectious viruses *in vivo* and *in vitro*. However, certain questions need to be addressed before its use in the clinic. The causes of these problems are multifaceted and can be grouped into several aspects as described below.

First, CRISPR/Cas9 off-target concerns are the major hurdle for antiviral therapeutic utilizations. Some researches have developed numerous methods, such as advanced versions or Cas9 nickases, to minimize off-target activity (Bellizzi et al. 2019) and cytotoxicities (Wang et al. 2016). As two related individual parts of the CRISPR/Cas9 system, the optimization of Cas9 proteins or upgradation of gRNAs can equally assist in reducing CRISPR/Cas9 off-target effects. Many investigations have focused on reducing the unwanted off-target activities of the CRISPR/Cas9 system (Eid and Mahfouz 2016) through the amelioration of Cas9 nucleases. Recently, two versions of Cas9 variants, termed "eSpCas9" (Slaymaker et al. 2016) and "SpCas9-HF1" (Kleinstiver et al. 2016), have significantly optimized the CRISPR/Cas9 genome-editing toolbox with their own higher specificity and exceptional precision. In addition, it is vital to design special and compatible gRNAs for CRISPR/Cas9. Currently, multiple website platforms are available for the optimal design of CRISPR gRNAs (e.g. https://www.genscript.com/gRNA-design-tool.html, and https://www.atum.bio/eCommerce/cas9/input). Moreover, there are other approaches whereby chemical modifications are incorporated into its structure to improve gRNA stability and activity in the cell. Additionally, most CRISPR/Cas spacers in bacteria naturally corresponding to foreign nucleic acids (Hille et al. 2018) are capable of precisely conferring adaptive bacterial immunity to defend against viruses. However, when this system works in animal cells, it usually neglects this 'implied condition' and thereby results in off-target effects, cytotoxicity and cellular resistance. For example, Kim and colleagues reported that the cytotoxicities caused by tailored CRISPR gRNAs (5'-ppp gRNAs) triggers RNA-mediated innate immune responses in human and murine cells (Kim et al. 2018).

Second, the delivery modality of CRISPR/Cas9 greatly influences its safety and therapeutic efficacy. To achieve successful gene manipulation using CRISPR/Cas9 reagents, delivery methods that maximize efficacy and minimize immune responses still need to be further developed (Luther et al. 2018, Schinazi et al. 2018). As a binary system, the CRISPR/Cas complex can be transported either together or separately. Specifically, Cas9 nucleases can be delivered to target cells in the format of DNA, mRNA or protein, and gRNAs can be transported in DNA or RNA form (Xu et al. 2019). Generally, the delivery patterns for the CRISPR/Cas9 system are categorized as viral vectors or nonviral vectors, both of which have their own unique advantages and disadvantages (Eoh and Gu 2019, Nelson and Gersbach 2016). Viral-based vectors commonly include adenovirus, adeno-associated virus (AAV), integration-deficient lentivirus and retrovirus (Tong et al. 2019). Additional viral carriers less frequently used for delivery include herpes simplex virus and poxvirus (Jin et al. 2014). As a complement, nonviral delivery modes have been rapidly developed for delivery purposes and are artificially divided into two classes: physical methods and chemical approaches (Li et al. 2018). Table 3 shows the types and applications of CRISPR/Cas9 delivery modalities.

Third, viral escape problems during gene therapy have raised many concerns among virus researchers. Many human viruses possess the capability to escape or inhibit the effect of pharmaceuticals (e.g. interferon) in their evolution. Specifically, in CRISPR/Cas9 interventions, viruses can escape the suppression via the acquisition of specific mutations at the target site that prevent gRNA binding without hindering viral replication (Binda et al. 2020). As a countermeasure, multiplex CRISPR/Cas9-based strategies have demonstrated to be more effective for the suppression of virus

**Table 3.** The summary of types and applications of CRISPR/Cas9 delivery

| Delivery modality | | Example | Form of Cas9 | *In vivo/In vitro* |
|---|---|---|---|---|
| Virus | | Lentivirus, Adenovirus, AAV, Baculovirus, Retrovirus, Herpes simplex virus | Cas9 plasmid | *In vivo* and *in vitro* |
| Bacterium | | *Salmonella* | Cas9 plasmid | *In vitro* |
| Chemical modes | Lipoid | Liposome, Lipofectamine, Exosome | Cas9 plasmid, Cas9 mRNA, Cas9 protein+sgRNA | *In vivo* and *in vitro* |
| | Nanoparticle | MSNs, Carbon nanotubes, Gold nanoparticles | Cas9 plasmid, Cas9 mRNA, Cas9 protein+sgRNA | *In vivo* and *in vitro* |
| | iTOP | iTOP | Cas9 protein+sgRNA, | *In vitro* |
| Physical modes | Electroporation | Electroporator | Cas9 plasmid, Cas9 mRNA, Cas9 protein+sgRNA | *In vitro* |
| | Injection | Intravenous injection, Microinjection, Hydrodynamic injection | Cas9 plasmid, Cas9 mRNA, Cas9 protein+sgRNA | *In vivo* and *in vitro* |
| | Gene gun | PDS-1000/He particle delivery system | Cas9 plasmid, Cas9 mRNA, Cas9 protein+sgRNA | *In vitro* |

MSNs: Mesoporous silica nanoparticles; iTOP: Induced transduction by osmocytosis and propanebetaine.

escape mutants and long-term maintenance of antiviral activity compared to the Cas9/sgRNA system (Lebbink 2017, Lee 2019, White et al. 2016).

Fourth, the problems associated with Cas9 cleavage activity and Cas9 immunogenicity need to be adequately considered before its translation into clinical practice. Cas9 proteins from different species have their different characteristics. Therefore, scientists need to select the appropriate nucleases according to the research requirements and study protocols. In addition, once the CRISPR/Cas9 system has been delivered into the target cell and activated, there are limited means to lower or shut off its activity (Pawluk et al. 2016), which may introduce new practical challenges and safety concerns to researchers. For example, excessive or prolonged Cas9 activity can exacerbate off-target effects. At present, several wild-type Cas9-specific "anti-CRISPRs (Acr)" provide biotechnological tools that can be used to adjust the activities of CRISPR/Cas9 for gene engineering (Rauch et al. 2017). Recently, more than 50 anti-CRISPR protein families have been characterized, providing various applications in genome engineering, such as posttranslational switches for the control of Cas9 or dCas9 activity (Wiegand et al. 2020).

## Ethical issues

CRISPR/Cas9 technology is rapidly developing, albeit still in its infancy stage in clinical use. More importantly the use of CRISPR/Cas systems for clinical applications are being confronted with ethical questions. First, the misuse of this novel technology will likely create certain ethical controversies. In the UK, scientists have gained license to edit human embryos with the use of CRISPR/Cas9 technology (Callaway 2016), which potentially shows the technical strengths of this system in prospective clinical applications. Nevertheless, CRISPR/Cas based clinical trials in the future must be efficiently supervised under newly established regulations (Shinwari et al. 2017) to boost CRISPR gene-editing technology to exactly serve humans. Second, the safety concerns induced by unwanted gene editing of CRISPR/Cas9 should be carefully assessed before it is clinically applied.

## Conclusion

The viruses mentioned above remain a global health threat, and continued efforts should be focused on developing an effective antiviral technique for the elimination of viral genomes from the host. CRISPR/Cas9 provides more potential opportunities for programmable gene-editing therapy and can become a powerful tool for antiviral medicine. However, lessons must be learned from traditional gene therapy, and scientists should maintain greater caution in moving forward with CRISPR systems to avoid adverse impacts and setbacks associated with the development of this new technology. Therefore, it is of utmost importance to identify and evaluate the adverse effects of antiviral CRISPR/Cas on host physiology and ensure that they do not damage the structural integrity and functionality of the host genome. Additionally, great efforts to develop CRISPR/Cas systems should be made to expand the toolbox, enabling a greater understanding of complex biological processes associated with hosts and viruses. With these related challenges resolved, CRISPR/Cas9 technology will be an attractive platform to extirpate numerous human viral infections.

## Fundings

This work was collectively supported by grants from National key research and development plan (2016YFD0500600), Guangdong provincial science and technology plan project (2017B020207004), Innovative province construction project—special topic on fighting against novel coronavirus pneumonia epidemic (2020SK3044), Youth fund project of Hunan natural science foundation (2019JJ50681) and Changsha biological resources sample bank establishment project (20200365).

## Conflict of interests

The authors declare no conflict of interest.

## References

Abbott, T. R., Dhamdhere, G., Liu, Y., Lin, X., Goudy, L. Zeng, L., et al. 2020. Development of CRISPR as an antiviral strategy to Combat SARS-CoV-2 and Influenza. Cell, 181: 865-876.

Arribas, J. R. and Eron, J. 2013. Advances in antiretroviral therapy. Current Opinion in HIV and AIDS, 8: 341-349.

Ashfaq, U. A. and Khalid, H. 2020. CRISPR/CAS9-mediated antiviral activity: A tool to combat viral infection. Critical Reviews in Eukaryotic Gene Expression, 30: 45-56.

Aubert, M., Boyle, N. M., Stone, D., Stensland, L., Huang, M.-L., Magaret, A. S., et al. 2014. In vitro inactivation of latent HSV by targeted mutagenesis using an HSV-specific homing endonuclease. Molecular Therapy Nucleic Acids, 3: e146.

Aubert, M., Madden, E. A., Loprieno, M., Feelixge, H. S. D., Stensland, L., Huang, M. L., et al. 2016. In vivo disruption of latent HSV by designer endonuclease therapy. JCI Insight, 1: e88468.

Aubert, M., Strongin, D. E., Roychoudhury, P., Loprieno, M. A., Haick, A. K., Klouser, L. M., et al. 2020. Gene editing and elimination of latent herpes simplex virus in vivo. Nature Communications, 11: 4148.

Banan, M. 2020. Recent advances in CRISPR/Cas9-mediated knock-ins in mammalian cells. Journal of Biotechnology, 308: 1-9.

Barrangou, R. and Doudna, J. A. 2016. Applications of CRISPR technologies in research and beyond. Nature Biotechnology, 34: 933-941.

Batool, A., Malik, F. and Andrabi, K. I. 2021. Expansion of the CRISPR/Cas genome-sculpting toolbox: Innovations, applications and challenges. Molecular Diagnosis & Therapy, 25: 41-57.

Bayat, H., Modarressi, M. H. and Rahimpour, A. 2018. The conspicuity of CRISPR-Cpf1 system as a significant breakthrough in genome editing. Current Microbiology, 75: 107-115.

Bellizzi, A., Ahye, N., Jalagadugula, G. and Wollebo, H. S. 2019. A broad application of CRISPR Cas9 in infectious diseases of central nervous system. Journal of Neuroimmune Pharmacology, 14: 578-594.

Bialek, J. K., Dunay, G. A., Voges, M., Schafer, C., Spohn, M., Stucka, R., et al. 2016. Targeted HIV-1 latency reversal using CRISPR/Cas9-derived transcriptional activator systems. PLoS One, 11: e0158294.

Binda, C. S., Klaver, B., Berkhout, B. and Das, A. T. 2020. CRISPR/Cas9 Dual-gRNA attack causes mutation, excision and inversion of the HIV-1 proviral DNA. Viruses, 12: 330.

Bi, Y., Sun, L., Gao, D., Ding, C., Li, Z., Li, Y., et al. 2014. High-efficiency targeted editing of large viral genomes by RNA-guided nucleases. PLoS Pathogens, 10: e1004090.

Bloom, K., Ely, A., Mussolino, C., Cathomen, T. and Arbuthnot, P. 2013. Inactivation of hepatitis B virus replication in cultured cells and in vivo with engineered transcription activator-like effector nucleases. Molecular Therapy: The Journal of the American Society of Gene Therapy, 21: 1889-1897.

Bloom, K., Maepa, M. B., Ely, A. and Arbuthnot, P. 2018. Gene therapy for chronic HBV – Can we eliminate cccDNA? Genes, 9: 207.

Broughton, J. P., Deng, X., Yu, G., Fasching, C. L., Servellita, V., Singh, J., et al. (2020). CRISPR/Cas12-based detection of SARS-CoV-2. Nature Biotechnology. 38: 870-874.

Callaway, E. 2016. UK scientists gain licence to edit genes in human embryos. Nature, 530: 18.

Cai, Y., Knudsen, A., Windross, S. J., Thomsen, M. K. and Paludan, S. R. 2018. Immunostimulatory guide RNAs mediate potent antiviral response. BioRxiv. 282558.

Campbell-Yesufu, O. T. and Gandhi, R. T. 2011. Update on human immunodeficiency virus (HIV)-2 infection. Clinical Infectious Diseases, 52: 780-787.

Cao, Y., Zhou, H., Zhou, X. and Li., F. 2020. Control of plant viruses by CRISPR/Cas system-mediated adaptive immunity. Frontiers in Microbiology, 11: 1-9.

Chen, S., Yu, X. and Guo, D. 2018. CRISPR-Cas targeting of host genes as an antiviral strategy. Viruses, 10: 40.

Cheng, Y. X., Chen, G. T., Yang, X., Wang, Y. Q. and Hong, L. 2018. Effects of HPV Pseudotype virus in cutting E6 gene selectively in SiHa cells. Current Medical Science, 38: 212-221.

Cho, S. W., Kim, S., Kim, J. M. and Kim, J. S. 2013. Targeted genome engineering in human cells with the Cas9 RNA-guided endonuclease. Nature Biotechnology, 31: 230-232.

Cong, L., Ran, F. A., Cox, D., Lin, S., Barretto, R., Habib, N., et al. 2013. Multiplex genome engineering using CRISPR/Cas systems. Science (New York, N.Y.), 339: 819-823.

Cradick, T. J., Keck, K., Bradshaw, S., Jamieson, A. C. and McCaffrey, A. P. 2010. Zinc-finger nucleases as a novel therapeutic strategy for targeting hepatitis B virus DNAs. Molecular therapy: The journal of the American Society of Gene Therapy, 18: 947-954.

Cresci, G. A. M., Lampe, J. W. and Gibson, G. 2020. Targeted approaches for in situ gut microbiome manipulation. Journal of Parenteral and Enteral Nutrition, 44: 581-588.

Crosbie, E. J., Einstein, M. H., Franceschi, S. and Kitchener, H. C. 2013. Human papillomavirus and cervical cancer. Lancet (London, England), 382: 889-899.

Cui, Y., Xu, J., Cheng, M., Liao, X. and Peng, S. 2018. Review of CRISPR/Cas9 sgRNA design tools. Interdisciplinary Sciences, Computational Life Sciences, 10: 455-465.

Dampier, W., Nonnemacher, M. R., Sullivan, N. T., Jacobson, J. M. and Wigdahl, B. 2014. HIV excision utilizing CRISPR/Cas9 technology: Attacking the proviral quasispecies in reservoirs to achieve a cure. MedCrave Online Journal of Immunology, 1: 00022.

Dash, P. K., Kaminski, R., Bella, R., Su, H., Mathews, S., Ahooyi, T. M., et al. 2019. Sequential LASER ART and CRISPR treatments eliminate HIV-1 in a subset of infected humanized mice. Nature Communications, 10: 2753.

De Silva Feelixge, H. S., Stone, D., Roychoudhury, P., Aubert, M. and Jerome, K. R. 2018. CRISPR/Cas9 and genome editing for viral disease – Is resistance futile? ACS Infectious Diseases, 4: 871-880.

Doitsh, G., Galloway, N. L., Geng, X., Yang, Z., Monroe, K. M., Zepeda, O. et al. 2014. Cell death by pyroptosis drives CD4 T-cell depletion in HIV-1 infection. Nature, 505: 509-514.

Dong, C., Qu, L., Wang, H., Wei, L., Dong, Y. and Xiong, S. 2015. Targeting hepatitis B virus cccDNA by CRISPR/Cas9 nuclease efficiently inhibits viral replication. Antiviral Research, 118: 110-117.

Doudna, J. A. and Charpentier, E. 2014. Genome editing. The new frontier of genome engineering with CRISPR-Cas9. Science, 346: 1258096.

Dreyer, T., Nicholson, S., Ely, A., Arbuthnot, P. and Bloom, K. 2016. Improved antiviral efficacy using TALEN-mediated homology directed recombination to introduce artificial primary miRNAs into DNA of hepatitis B virus. Biochemical and Biophysical Research Communications, 478: 1563-1568.

Duan, X., Li, S., Holmes, J.A., Tu, Z., Li, Y., Cai, D., et al. 2018. MicroRNA 130a regulates both hepatitis C virus and hepatitis b virus replication through a central metabolic pathway. Journal of Virology, 92.

Ebina, H., Misawa, N., Kanemura, Y. and Koyanagi, Y. 2013. Harnessing the CRISPR/Cas9 system to disrupt latent HIV-1 provirus. Scientific Reports, 3: 2510.

Ebrahimi, S., Teimoori, A., Khanbabaei, H. and Tabasi, M. 2019. Harnessing CRISPR/Cas9 system for manipulation of DNA virus genome. Reviews in Medical Virology, 29: e2009.

Ebrahimi S., Makvandi, M., Abbasi, S., Azadmanesh, K. and A. Teimoori. 2020. Developing oncolytic Herpes simplex virus type 1 through UL39 knockout by CRISPR-Cas9. Iranian Journal of Basic Medical Sciences, 23: 937-944.

Eid, A. and Mahfouz, M. M. 2016. Genome editing: The road of CRISPR/Cas9 from bench to clinic. Experimental and Molecular Medicine, 48: e265.

El-Serag, H. B. 2012. Epidemiology of viral hepatitis and hepatocellular carcinoma. Gastroenterology, 142: 1264-1273.

Emery, J. S. and Feld, J. J. 2017. Treatment of hepatitis B virus with combination therapy now and in the future. Best Practice & Research in Clinical Gastroenterology, 31: 347-355.

Eoh, J. and Gu, L. 2019. Biomaterials as vectors for the delivery of CRISPR-Cas9. Biomaterials Science, 7: 1240-1261.

Escalona-Noguero, C., López-Valls, M. and Sot, B. 2021. CRISPR/Cas technology as a promising weapon to combat viral infections. BioEssays, 2000315: 1-16.

Fernández, A., Josa, S. and Montoliu, L. 2017. A history of genome editing in mammals. Mammalian Genome, 28: 237-246.

Finnen, R. L. and Banfield, B. W. 2018. CRISPR/Cas9 mutagenesis of UL21 in multiple strains of Herpes simplex virus reveals differential requirements for pUL21 in viral replication. Viruses, 10: 258.

Fozouni, P., Son, S., Díaz de León Derby, M., Knott, G. J., Gray, C. N., D'Ambrosio, M. V., et al. 2021. Amplification-free detection of SARS-CoV-2 with CRISPR-Cas13a and mobile phone microscopy. Cell, 184: 323-333.

Freije, C. A., Myhrvold, C., Boehm, C. K., Lin, A. E., Welch, N. L., Carter, A. et al. 2019. Programmable inhibition and detection of RNA viruses using Cas13. Molecular Cell, 76: 826-837.

Gao, X., Jin, Z., Tan, X., Zhang, C., Zou, C., Zhang, W., et al. 2020. Hyperbranched poly (beta-amino ester) based polyplex nanopaticles for delivery of CRISPR/Cas9 system and treatment of HPV infection associated cervical cancer. Journal of Controlled Release, 321: 654-668.

Garneau, J. E., Dupuis, M. E., Villion, M., Romero, D. A., Barrangou, R., Boyaval, P., et al. 2010. The CRISPR/Cas bacterial immune system cleaves bacteriophage and plasmid DNA. Nature, 468: 67-71.

Gebre, M., Nomburg, J. L. and Gewurz, B. E. 2018. CRISPR-Cas9 genetic analysis of virus-host interactions. Viruses, 10.

Grosse, S., Huot, N., Mahiet, C., Arnould, S., Barradeau, S., Clerre, D. L., et al. 2011.

Meganuclease-mediated inhibition of HSV1 infection in cultured cells. Molecular Therapy, 19: 694-702.

Goudarzi, K. A., Nematollahi, M. H., Khanbabaei, H., Nave, H. H., Mirzaei, H. R., Pourghadamyari, H., et al. 2020. Targeted delivery of CRISPR/Cas13 as a promising therapeutic approach to treat SARS-CoV-2. Current Pharmaceutical Biotechnology, 21: 1.

Gupta, S. M. and Mania-Pramanik, J. 2019. Molecular mechanisms in progression of HPV-associated cervical carcinogenesis. Journal of Biomedical Science, 26: 28.

Hanafiah, K. M., Groeger, J., Flaxman, A. D. and Wiersma, S. T. 2013. Global epidemiology of hepatitis C virus infection: New estimates of age-specific antibody to HCV seroprevalence. Hepatology, 57: 1333-1342.

Harden, M. E. and Munger, K. 2017. Human papillomavirus molecular biology. Mutation Research-Reviews in Mutation Research, 772: 3-12.

Haynes, B. F. 2015. New approaches to HIV vaccine development. Current Opinion in Immunology, 35: 39-47.

Hegde, S., Nilyanimit, P., Kozlova, E., Anderson, E. R., Narra, H. P., Sahni, S. K., et al. 2019. CRISPR/Cas9-mediated gene deletion of the ompA gene in symbiotic Cedecea neteri impairs biofilm formation and reduces gut colonization of Aedes aegypti mosquitoes. PLoS Neglected Tropical Diseases, 13: e0007883.

Hille, F., Richter, H., Wong, S. P., Bratovic, M., Ressel, S. and Charpentier, E. 2018. The biology of CRISPR-Cas: Backward and forward. Cell, 172: 1239-1259.

Hirakawa, M. P., Krishnakumar, R., Timlin, J. A., Carney, J. P. and Butler, K. S. 2020. Gene editing and CRISPR in the clinic: Current and future perspectives. Bioscience Reports, 40: BSR20200127.

Hopcraft, S. E., Azarm, K. D., Israelow, B., Lévêque, N., Schwarz, M. C., Hsu, T. H., et al. 2015. Viral determinants of miR-122-independent hepatitis C virus replication. mSphere. 1: e00009-15.

Hoppe-Seyler, K., Bossler, F., Braun, J. A., Herrmann, A. L. and Hoppe-Seyler, F. 2018. The HPV E6/E7 Oncogenes: Key factors for viral carcinogenesis and therapeutic targets. Trends in Microbiology, 26: 158-168.

Hou, P., Chen, S., Wang, S., Yu, X., Chen, Y., Jiang, M. et al. 2015. Genome editing of CXCR4 by CRISPR/Cas9 confers cells resistant to HIV-1 infection. Scientific Reports, 5: 1-12.

Hryhorowicz, M., Lipiński, D., Zeyland, J. and Słomski, R. 2017. CRISPR/Cas9 Immune system as a tool for genome engineering. Archivum Immunologiae Et Therapiae Experimentalis, 65: 233-240.

Hsu, D. S., Kornepati, A. V., Glover, W., Kennedy, E. M. and Cullen, B. R. 2018. Targeting HPV16 DNA using CRISPR/Cas inhibits anal cancer growth. Future Virology, 13: 475-482.

Hu, J. H., Miller, S. M., Geurts, M. H., Tang, W., Chen, L., Sun, N. et al. 2018. Evolved Cas9 variants with broad PAM compatibility and high DNA specificity. Nature, 556: 57-63.

Hu, Z., Yu, L., Zhu, D., Ding, W., Wang, X., Zhang, C. et al. 2014. Disruption of HPV16-E7 by CRISPR/Cas system induces apoptosis and growth inhibition in HPV16 positive human cervical cancer cells. Biomed Research International, 1-9.

Inturi, R. and Jemth, P. 2020. CRISPR/Cas9-based inactivation of human papillomavirus oncogenes E6 and E7 induces senescence in cervical cancer cells. BioRxiv (in press).

Ishino, Y., Shinagawa, H., Makino, K., Amemura, M. and Nakata, A. 1987. Nucleotide sequence of the IAP gene, responsible for alkaline phosphatase isozyme conversion in Escherichia coli, and identification of the gene product. Journal of Bacteriology, 169: 5429-5433.

Jiang, F., Zhou, K., Ma, L., Gressel, S. and Doudna, J. A. 2015. Structural Biology: A Cas9-guide RNA complex preorganized for target DNA recognition. Science (New York, N.Y.), 348: 1477-1481.

Jin, L., Zeng, X., Liu, M., Deng, Y. and He, N. 2014. Current progress in gene delivery technology based on chemical methods and nano-carriers. Theranostics, 4: 240-255.

Jinek, M., Chylinski, K., Fonfara, I., Hauer, M., Doudna, J. A. and Charpentier, E. 2012. A programmable dual-RNA-guided DNA endonuclease in adaptive bacterial immunity. Science, 337: 816-821.

Jinek, M., East, A., Cheng, A., Lin, S., Ma, E. and Doudna, J. 2013. RNA-programmed genome editing in human cells. Elife, 2: e00471.

Jubair, L., Fallaha, S. and McMillan, N. A. J. 2019. Systemic Delivery of CRISPR/Cas9 targeting HPV oncogenes is effective at eliminating established tumors. Molecular Therapy, 27: 2091-2099.

Kaminski, R., Chen, Y., Fischer, T., Tedaldi, E., Napoli, A., Zhang, Y., et al. 2016. Elimination of HIV-1 genomes from human T-lymphoid cells by CRISPR/Cas9 gene editing. Scientific Reports, 6: 22555.

Karimova, M., Beschorner, N., Dammermann, W., Chemnitz, J., Indenbirken, D., Bockmann, J. H., et al. 2015. CRISPR/Cas9 nickase-mediated disruption of hepatitis B virus open reading frame S and X. Scientific Reports, 5: 13734.

Karpov, D. S., Karpov, V. L., Klimova, R. R., et al. 2019. A plasmid-expressed CRISPR/Cas9 system suppresses replication of HSV type I in a vero cell culture. Molecular Biology, 53: 70-78.

Kaushik, A., Yndart, A., Atluri, V., Tiwari, S., Tomitaka, A., Gupta, P., et al. 2019. Magnetically guided non-invasive CRISPR-Cas9/gRNA delivery across blood-brain barrier to eradicate latent HIV-1 infection. Scientific Reports, 9: 3928.

Kennedy, E. M., Kornepati, A. V., Goldstein, M., Bogerd, H. P., Poling, B. C., Whisnant, A. W., et al. 2014. Inactivation of the human papillomavirus E6 or E7 gene in cervical carcinoma cells by using a bacterial CRISPR/Cas RNA-guided endonuclease. Journal of Virology, 88: 11965-11972.

Kennedy, E. M., Bassit, L. C., Mueller, H., Kornepati, A. V. R., Bogerd, H. P., Nie, T. et al. 2015. Suppression of hepatitis B virus DNA accumulation in chronically infected cells using a bacterial CRISPR/Cas RNA-guided DNA endonuclease. Virology, 476: 196-205.

Kennedy, E. M. and Cullen, B. R. 2017. Gene editing: A new tool for viral disease. Annual Review of Medicine, 68: 401-411.

Khalil A. M. 2020. The genome editing revolution: Review. Journal, Genetic Engineering & Biotechnology, 18: 68.

Khalili, K., Kaminski, R., Gordon, J., Cosentino, L. and Hu, W. 2015. Genome editing strategies: Potential tools for eradicating HIV-1/AIDS. Journal of Neurovirology, 21: 310-321.

Khan, S., Mahmood, M. S., Rahman, S. U., Zafar, H., Habibullah, S., Khan, Z. et al. 2018. CRISPR/Cas9: The Jedi against the dark empire of diseases. Journal of Biomedical Science, 25: 29.

Khodadad, N., Fani, M., Jamehdor, S., Nahidsamiei, R., Makvandi, M., Kaboli, S., et al. 2020. A knockdown of the herpes simplex virus type-1 gene in all-in-one CRISPR vectors. Folia Histochemica et Cytobiologica, 58: 174-181.

Kim, S., Koo, T., Jee, H. G., Cho, H. Y., Lee, G., Lim, D. G. et al. 2018. CRISPR RNAs trigger innate immune responses in human cells. Genome Research, 28: 367-373. Advance online publication.

Kleinstiver, B. P., Pattanayak, V., Prew, M. S., Tsai, S. Q., Nguyen, N. T., Zheng, Z. et al. 2016. High-fidelity CRISPR-Cas9 nucleases with no detectable genome-wide off-target effects. Nature, 529: 490-495.

Kostyusheva, A. P., Brezgin, S. A., Zarifyan, D. N., Chistyakov, D. S., Gegechkory, V. I., Bayurova, E. O., et al. 2019a. Hepatitis B virus and site-specific nucleases: Effects of genetic modifications in CRISPR/Cas9 on antiviral activity. Russian Journal of Infection and Immunity, 9: 279-287.

Kostyusheva, A. P., Kostyushev, D. S., Brezgin, S. A., Zarifyan, D. N., Volchkova, E. V. and Chulanov, V. P. 2019b. Small molecular inhibitors of DNA double strand break repair pathways increase the anti-HBV activity of CRISPR/Cas9. Molecular Biology, 53: 274-285.

Kostyushev, D., Kostyusheva, A., Brezgin, S., Zarifyan, D., Utkina, A., Goptar, I., et al. 2019. Suppressing the NHEJ pathway by DNA-PKcs inhibitor NU7026 prevents degradation of HBV cccDNA cleaved by CRISPR/Cas9. Scientific Reports, 9: 1847.

Kurihara, T., Fukuhara, T., Ono, C., Yamamoto, S., Uemura, K., Okamoto, T., et al. 2017. Suppression of HBV replication by the expression of nickase- and nuclease dead-Cas9. Scientific Reports, 7: 6122.

Lao, Y. H., Li, M., Gao, M. A., Shao, D., Chi, C. W., Huang, D. et al. 2018. HPV oncogene manipulation using nonvirally delivered CRISPR/Cas9 or Natronobacterium gregoryi Argonaute. Advanced Science (Weinh), 5: 1700540.

Lavanchy, D. 2011. Evolving epidemiology of hepatitis C virus. Clinical Microbiology and Infection, 17: 107-115.

Lebbink, R. J., de Jong, D. C., Wolters, F., Kruse, E. M., van Ham, P. M., Wiertz, E. J., et al. 2017. A combinational CRISPR/Cas9 gene-editing approach can halt HIV replication and prevent viral escape. Scientific Reports, 7: 41968.

Lee, C. 2019. CRISPR/Cas9-based antiviral strategy: Current status and the potential challenge. Molecules, 24: 1349.

Lee, W. M. 1997. Hepatitis B virus infection. The New England Journal of Medicine, 337: 1733-1745.

Li, H., Sheng, C., Wang, S., Yang, L., Liang, Y., Huang, Y., et al. 2017a. Removal of integrated hepatitis B virus DNA using CRISPR-Cas9. Frontiers in Cellular and Infection Microbiology, 7: 91.

Li, Y., Zhang, Y., Lin, J., Pan, S., Han, W., Peng, N., et al. 2017b. Cmr1 enables efficient RNA and DNA interference of a III-B CRISPR-Cas system by binding to target RNA and crRNA. Nucleic Acids Research, 45: 11305-11314.

Li, Z., Bi, Y., Xiao, H., Sun, L., Ren, Y., Li, Y., et al. 2017c. CRISPR/Cas9 system-driven site-specific selection pressure on Herpes simplex virus genomes. Virus Research, 244: 286-295.

Li, L., Hu, S. and Chen, X. 2018. Non-viral delivery systems for CRISPR/Cas9-based genome editing: Challenges and opportunities. Biomaterials, 171: 207-218.

Liao, H. K., Gu, Y., Diaz, A., Marlett, J., Takahashi, Y., Li, M., et al. 2015. Use of the CRISPR/Cas9 system as an intracellular defense against HIV-1 infection in human cells. Nature Communications, 6: 6413.

Lin, S. R., Yang, H. C., Kuo, Y. T., Liu, C. J., Yang, T. Y., Sung, K. C., et al. 2014. The CRISPR/Cas9 system facilitates clearance of the intrahepatic HBV templates in vivo. Molecular Therapy - Nucleic Acids, 3: e186.

Lin, H., Deng, Q., Li, L. and Shi, L. 2019. Application and development of CRISPR/Cas9 technology in pig research. Gene Editing – Technologies and Applications. IntechOpen.

Lino, C. A., Harper, J. C., Carney, J. P. and Timlin, J. A. 2018. Delivering CRISPR: A review of the challenges and approaches. Drug Delivery, 25: 1234-1257.

Liu, X., Hao, R., Chen, S., Guo, D. and Chen, Y. 2015. Inhibition of hepatitis B virus by the CRISPR/Cas9 system via targeting the conserved regions of the viral genome. Journal of General Virology, 96: 2252-2261.

Liu, Y. C., Cai, Z. M. and Zhang, X. J. 2016. Reprogrammed CRISPR-Cas9 targeting the conserved regions of HPV6/11 E7 genes inhibits proliferation and induces apoptosis in E7-transformed keratinocytes. Asian Journal of Andrology, 18: 475-479.

Liu, Z., Chen, S., Jin, X., Wang, Q., Yang, K., Li, C. et al. 2017. Genome editing of the HIV co-receptors CCR5 and CXCR4 by CRISPR-Cas9 protects CD4(+) T cells from HIV-1 infection. Cell and Bioscience, 7: 47.

Liu, S., Wang, Q., Yu, X., Li, Y., Guo, Y., Liu, Z., et al. 2018. HIV-1 inhibition in cells with CXCR4 mutant genome created by CRISPR-Cas9 and piggyBac recombinant technologies. Scientific Reports, 8: 8573.

Locarnini, S., Littlejohn, M., Aziz, M. N. and Yuen, L. 2013. Possible origins and evolution of the hepatitis B virus (HBV). Seminars in Cancer Biology, 23: 561-575.

Lu, D. Y., Wu, H. Y., Yarla, N. S., Xu, B., Ding, J. and Lu, T. R. 2018. HAART in HIV/AIDS treatments: Future trends. Infectious Disorders Drug Targets, 18: 15-22.

Luther, D. C., Lee, Y. W., Nagaraj, H., Scaletti, F. and Rotello, V. M. 2018. Delivery approaches for CRISPR/Cas9 therapeutics in vivo: Advances and challenges. Expert Opinion on Drug Delivery, 15: 905-913.

Maepa, M. B., Roelofse, I., Ely, A. and Arbuthnot, P. 2015. Progress and prospects of anti-HBV gene therapy development. International Journal of Molecular Sciences, 16: 17589-17610.

Makarova, K. S., Wolf, Y. I., Alkhnbashi, O. S., Costa, F., Shah, S. A., Saunders, S. J., et al. 2015. An updated evolutionary classification of CRISPR/Cas systems. Nature Reviews Microbiology, 13: 722-736.

Makarova, K. S., Zhang, F. and Koonin, E. V. 2017. SnapShot: Class 1 CRISPR/Cas systems. Cell, 168: 946.e1.

Makarova, K. S., Wolf, Y. I., Shmakov, S. A., Liu, Y., Li, M. and Koonin, E. V. 2020. Unprecedented diversity of unique CRISPR-Cas-related systems and Cas1 homologs in Asgard archaea. The CRISPR Journal, 3: 156-163.

Mali, P., Yang, L., Esvelt, K. M., Aach, J., Guell, M., DiCarlo, J. E., et al. 2013. RNA-guided human genome engineering via Cas9. Science, 339: 823-826.

Manjunath, N., Yi, G., Dang, Y. and Shankar, P. 2013. Newer gene editing technologies toward HIV gene therapy. Viruses, 5: 2748-2766.

McBride, A. A. 2017. Oncogenic human papillomaviruses. Philosophical Transactions of the Royal Society of London, 372: 20160273.

McLaughlin-Drubin, M. E. and Munger, K. 2009. Oncogenic activities of human papillomaviruses. Virus Research, 143: 195-208.

Medeiros, R., Vaz, S., Rebelo, T. and Figueiredo-Dias, M. 2020. Prevention of human papillomavirus infection. Beyond cervical cancer: A brief review. Acta Medica Portuguesa, 33: 198-201.

Miller, S. M., Wang, T., Randolph, P. B., Arbab, M., Shen, M. W., Huang, T. P., et al. 2020. Continuous evolution of SpCas9 variants compatible with non-G PAMs. Nature Biotechnology, 38: 471-481.

Moens, U. 2018. Human Polyomaviruses and Papillomaviruses. International Journal of Molecular Sciences, 19: 2360.

Mohammadzadeh, I., Qujeq, D., Yousefi, T., Ferns, G. A., Maniati, M. and Vaghari-Tabari, M. 2020. CRISPR/Cas9 gene editing: A new therapeutic approach in the treatment of infection and autoimmunity. IUBMB Life, 72: 1603-1621.

Moyo, B., Bloom, K., Scott, T., Ely, A. and Arbuthnot, P. 2018. Advances with using CRISPR/Cas-mediated gene editing to treat infections with hepatitis B virus and hepatitis C virus. Virus Research, 244: 311-320.

Myhrvold, C., Freije, C. A., Gootenberg, J. S., Abudayyeh, O. O., Metsky, H. C., Durbin, A. F., et al. 2018. Field-deployable viral diagnostics using CRISPR-Cas13. Science (New York, N.Y.), 360: 444-448.

Nassal, M. 2015. HBV cccDNA: Viral persistence reservoir and key obstacle for a cure of chronic hepatitis B. Gut, 64: 1972-1984.

Nelson, C. E. and Gersbach, C. A. 2016. Engineering delivery vehicles for genome editing. Annual Review of Chemical and Biomolecular Engineering, 7: 637-662.

Nishimasu, H., Ran, F. A., Hsu, P. D., Konermann, S., Shehata, S. I., Dohmae, N., et al. 2014. Crystal structure of Cas9 in complex with guide RNA and target DNA. Cell, 156: 935-949.

Nwankwo, N. and Seker, H. 2013. HIV progression to AIDS: Bioinformatics approach to determining the mechanism of action. Current HIV Research, 11: 30-42.

Ota, S., Hisano, Y., Ikawa, Y. and Kawahara, A. 2014. Multiple genome modifications by the CRISPR/Cas9 system in zebrafish. Genes Cells, 19: 555-564.

Pal, A. and Kundu, R. 2020. Human papillomavirus E6 and E7: The cervical cancer hallmarks and targets for therapy. Frontiers in Microbiology, 10: 3116.

Parrish, C. R., Holmes, E. C., Morens, D. M., Park, E. C., Burke, D. S., Calisher, C. H., et al. 2008. Cross-species virus transmission and the emergence of new epidemic diseases. Microbiology and Molecular Biology Reviews, 72: 457-470.

Pawluk, A., Amrani, N., Zhang, Y., Garcia, B., Hidalgo-Reyes, Y., Lee, J. et al. 2016. Naturally occurring off-switches for CRISPR-Cas9. Cell, 167: 1829-1838.

Peng, Y., Clark, K. J., Campbell, J. M., Panetta, M. R., Guo, Y. and Ekker, S. C. 2014. Making designer mutants in model organisms. Development (Cambridge, England), 141: 4042-4054.

Puschnik, A. S., Majzoub, K., Ooi, Y. S. and Carette, J. E. 2017. A CRISPR toolbox to study virus-host interactions. Nature Reviews. Microbiology, 15: 351-364.

Ramachandran, G. and Bikard, D. 2019. Editing the microbiome the CRISPR way. Philosophical Transactions of the Royal Society of London. Series B, Biological Sciences, 374: 20180103.

Ramanan, V., Shlomai, A., Cox, D. B., Schwartz, R. E., Michailidis, E., Bhatta, A., et al. 2015. CRISPR/Cas9 cleavage of viral DNA efficiently suppresses hepatitis B virus. Scientific Reports, 5: 10833.

Ran, F. A., Hsu, P. D., Wright, J., Agarwala, V., Scott, D. A. and Zhang, F. 2013. Genome engineering using the CRISPR-Cas9 system. Nature Protocols, 8: 2281-2308.

Rauch, B. J., Silvis, M. R., Hultquist, J. F., Waters, C. S., McGregor, M. J., Krogan, N. J., et al. 2017. Inhibition of CRISPR-Cas9 with bacteriophage proteins. Cell, 168: 150-158.

Ren, Q., Li, C., Yuan, P., Cai, C., Zhang, L., Luo, G. G., et al. 2015. A dual-reporter system for real-time monitoring and high-throughput CRISPR/Cas9 library screening of the hepatitis C virus. Scientific Reports, 5: 8865.

Roehm, P. C., Shekarabi, M., Wollebo, H. S., Bellizzi, A., He, L., Salkind, J., et al. 2016. Inhibition of HSV-1 replication by gene editing strategy. Scientific Reports, 6: 23146.

Russell, T. A., Stefanovic, T. and Tscharke, D. C. 2015. Engineering herpes simplex viruses by infection-transfection methods including recombination site targeting by CRISPR/Cas9 nucleases. Journal of Virological Methods, 213: 18-25.

Ryndock, E. J. and Meyers, C. 2014. A risk for non-sexual transmission of human papillomavirus? Expert Review of Anti-Infective Therapy, 12: 1165-1170.

Saayman, S., Ali, S. A., Morris, K. V. and Weinberg, M. S. 2015. The therapeutic application of CRISPR/Cas9 technologies for HIV. Expert Opinion on Biological Therapy, 15: 819-830.

Sakuma, T., Masaki, K., Abe-Chayama, H., Mochida, K., Yamamoto, T. and Chayama, K. 2016. Highly multiplexed CRISPR-Cas9-nuclease and Cas9-nickase vectors for inactivation of hepatitis B virus. Genes Cells, 21: 1253-1262.

Schinazi, R. F., Ehteshami, M., Bassit, L. and Asselah, T. 2018. Towards HBV curative therapies. Liver International, 38: 102-114.

Schiwon, M., Ehrke-Schulz, E., Oswald, A., Bergmann, T., Michler, T., Protzer, U., et al. 2018. One-vector system for multiplexed CRISPR/Cas9 against hepatitis B virus cccDNA utilizing high-capacity adenoviral vectors. Molecular Therapy - Nucleic Acids, 12: 242-253.

Scott, T., Moyo, B., Nicholson, S., Maepa, M. B., Watashi, K., Ely, A., et al. 2017. ssAAVs containing cassettes encoding SaCas9 and guides targeting hepatitis B virus inactivate replication of the virus in cultured cells. Scientific Reports, 7: 7401.

Seeger, C. and Sohn, J. A. 2014. Targeting hepatitis B virus with CRISPR/Cas9. Molecular Therapy - Nucleic Acids, 3: e216.

Seeger, C. and Sohn, J. A. 2016. Complete spectrum of CRISPR/Cas9-induced mutations on HBV cccDNA. Molecular Therapy, 24: 1258-1266.

Senís, E., Mockenhaupt, S., Rupp, D., Bauer, T., Paramasivam, N., Knapp, B., et al. 2017. TALEN/CRISPR-mediated engineering of a promoterless anti-viral RNAi hairpin into an endogenous miRNA locus. Nucleic Acids Research, 45: e3.

Seo, Y. and Y. Yano. 2014. Short- and long-term outcome of interferon therapy for chronic hepatitis B infection. World Journal of Gastroenterology, 20: 13284-13292.

Shinwari, Z. K., Tanveer, F. and Khalil, A. T. 2017. Ethical issues regarding CRISPR mediated genome editing. Current Issues in Molecular Biology, 26: 103-110.

Siliciano, J. D., Kajdas, J., Finzi, D., Quinn, T. C., Chadwick, K., Margolick, J. B., et al. 2003. Long-term follow-up studies confirm the stability of the latent reservoir for HIV-1 in resting CD4+ T cells. Nature Medicine, 9: 727-728.

Slaymaker, I. M., Gao, L., Zetsche, B., Scott, D. A., Yan, W. X. and Zhang, F. 2016. Rationally engineered Cas9 nucleases with improved specificity. Science, 351: 84-88.

Soppe, J. A. and Lebbink, R. J. 2017. Antiviral goes viral: Harnessing CRISPR/Cas9 to combat viruses in humans. Trends in Microbiology, 25: 833-850.

Soriano, V., Barreiro, P., Cachay, E., Kottilil, S., Fernandez-Montero, J. V. and de Mendoza, C. 2020. Advances in hepatitis B therapeutics. Therapeutic Advances in Infectious Disease. 7: 2049936120965027.

Suenaga, T., Kohyama, M., Hirayasu, K. and Arase, H. 2014. Engineering large viral DNA genomes using the CRISPR-Cas9 system. Microbiology and Immunology, 58: 513-522.

Sun, W. and Wang, H. 2020. Recent advances of genome editing and related technologies in China. Gene Therapy, 27: 312-320.

Sunbul, M. 2014. Hepatitis B virus genotypes: Global distribution and clinical importance. World Journal of Gastroenterology, 20: 5427-5434.

Tong, S., Moyo, B., Lee, C. M., Leong, K. and Bao, G. 2019. Engineered materials for in vivo delivery of genome-editing machinery. Nature Reviews Materials, 4: 726-737.

Trépo, C., Chan, H. L. Y. and Lok, A. 2014. Hepatitis B virus infection. The Lancet, 384: 2053-2063.

Turner, E. M., Brown, R. S., Laudermilch, E., Tsai, P. L. and Schlieker, C. 2015. The Torsin activator LULL1 is required for efficient growth of herpes simplex virus 1. Journal of Virology, 89: 8444-8452.

van Diemen, F. R., Kruse, E. M., Hooykaas, M. J., Bruggeling, C. E., Schürch, A. C., van Ham, P. M., et al. 2016. CRISPR/Cas9-mediated genome editing of herpesviruses limits productive and latent infections. PLoS Pathogens, 12: e1005701.

van Diemen, F. R. and Lebbink, R. J., 2017. CRISPR/Cas9, a powerful tool to target human herpesviruses. Cellular Microbiology, 19: el005701.

Vijayan, K. K. V., Karthigeyan, K. P., Tripathi, S. P. and Hanna, L. E. 2017. Pathophysiology of CD4+ T-cell depletion in HIV-1 and HIV-2 infections. Frontiers in Immunology, 8: 580.

Wahid, B., Usman, S., Ali, A., Saleem, K., Rafique, S., Naz, Z., et al. 2017. Therapeutic strategies of clustered regularly interspaced palindromic repeats-cas systems for different viral infections. Viral Immunology, 30: 552-559.

Wang, J. and Quake, S. R. 2014. RNA-guided endonuclease provides a therapeutic strategy to cure latent herpesviridae infection. Proceedings of the National Academy of Sciences of the United States of America, 111: 13157-13162.

Wang, W., Ye, C., Liu, J., Zhang, D., Kimata, J. T. and Zhou, P. 2014. CCR5 gene disruption via lentiviral vectors expressing Cas9 and single guided RNA renders cells resistant to HIV-1 infection. PLoS One, 9: 1-26.

Wang, J., Xu, Z. W., Liu, S., Zhang, R. Y., Ding, S. L., Xie, X. M., et al. 2015. Dual gRNAs guided CRISPR/Cas9 system inhibits hepatitis B virus replication. World Journal of Gastroenterology, 21: 9554-9565.

Wang, Y., Wei, D., Zhu, X., Pan, J., Zhang, P., Huo, L., et al. 2016. A 'suicide' CRISPR-Cas9 system to promote gene deletion and restoration by electroporation in Cryptococcus neoformans. Scientific Reports, 6: 31145.

Wang, J., Chen, R., Zhang, R., Ding, S., Zhang, T., Yuan, Q. et al. 2017. The gRNA-miRNA-gRNA ternary cassette combining CRISPR/Cas9 with RNAi approach strongly inhibits hepatitis B virus Replication. Theranostics, 7: 3090-3105.

Weber, N. D., Stone, D., Sedlak, R. H., De Silva Feelixge, H. S., Roychoudhury, P., Schiffer, J. T., et al. 2014. AAV-mediated delivery of zinc finger nucleases targeting hepatitis B virus inhibits active replication. PLoS One, 9: e97579.

White, M. K., Hu, W. and Khalili, K. 2015. The CRISPR/Cas9 genome editing methodology as a weapon against human viruses. Discovery Medicine, 19: 255-262.

White, M. K., Hu, W. and Khalili, K. 2016. Gene editing approaches against viral infections and strategy to prevent occurrence of viral escape. PLoS Pathogens, 12: e1005953.

Wiegand, T., Karambelkar, S., Bondy-Denomy, J. and Wiedenheft, B. 2020. Structures and strategies of anti-CRISPR-mediated immune suppression. Annual Review of Microbiology, 74: 21-37.

Wright, A. V., Sternberg, S. H., Taylor, D. W., Staahl, B. T., Bardales, J. A., Kornfeld, J. E., et al. 2015. Rational design of a split-Cas9 enzyme complex. Proceedings of the National Academy of Sciences USA, 112: 2984-2989.

Xiao, Q., Guo, D. and Chen, S. 2019. Application of CRISPR/Cas9-based gene editing in HIV-1/AIDS therapy. Frontiers in Cellular and Infection Microbiology, 9: 69.

Xie, H., Tang, L., He, X., Liu, X. Zhou, C., Liu, J., et al. 2018. SaCas9 requires 5'-NNGRRT-3' PAM for sufficient cleavage and possesses higher cleavage activity than SpCas9 or FnCpf1 in human cells. Biotechnology Journal, 13: e1700561.

Xu, X., Fan, S., Zhou, J., Zhang, Y. and Li, Q. 2016. The mutated tegument protein UL7 attenuates the virulence of herpes simplex virus 1 by reducing the modulation of α-4 gene transcription. Virology Journal, 13: 152.

Xu, X., Guo, Y., Fan, S., Cui, P., Feng, M., Wang, L., et al. 2017. Attenuated phenotypes and analysis of a herpes simplex virus 1 strain with partial deletion of the UL7, UL41 and LAT genes. Virologica Sinica, 32: 404-414.

Xu, X., Wan, T., Xin, H., Li, D., Pan, H., Wu, J., et al. 2019. Delivery of CRISPR/Cas9 for therapeutic genome editing. The Journal of Gene Medicine, 21: e3107.

Yamauchi, S., Takeuchi, K., Chihara, K., Honjoh, C., Kato, Y., Yoshiki, H. et al. 2016. STAT1 is essential for the inhibition of hepatitis C virus replication by interferon-λ but not by interferon-α. Scientific Reports, 6: 38336.

Yan, F., Wang, W. and Zhang, J. 2019. CRISPR/Cas12 and Cas13: The lesser known siblings of CRISPR/Cas9. Cell Biology and Toxicology, 35(6): 489-492.

Yang, Y. C., Chen, Y. H., Kao, J. H., Ching, C., Liu, I. J., Wang, C. C., et al. 2020. Permanent inactivation of HBV genomes by CRISPR/Cas9-mediated non-cleavage base editing. Molecular Therapy Nucleic Acids, 20: 480-490.

Ye, L., Wang, J., Beyer, A. I., Teque, F., Cradick, T. J., Qi, Z., et al. 2014. Seamless modification of wild-type induced pluripotent stem cells to the natural CCR5Δ32 mutation confers resistance to HIV infection. Proceedings of the National Academy of Sciences of the United States of America, 111: 9591-9596.

Yoshiba, T., Saga, Y., Urabe, M., Uchibori, R., Matsubara, S., Fujiwara, H., et al. 2019. CRISPR/Cas9-mediated cervical cancer treatment targeting human papillomavirus E6. Oncology Letters, 17: 2197-2206.

Yu, L., Wang, X., Zhu, D., Ding, W., Wang, L., Zhang, C., et al. 2014. Disruption of human papillomavirus 16 E6 gene by clustered regularly interspaced short palindromic repeat/ Cas system in human cervical cancer cells. OncoTargets and Therapy, 8: 37-44.

Yu, L., Hu, Z., Gao, C., Feng, B. and Wang, H. 2017. Deletion of HPV18 E6 and E7 genes using dual sgRNA-directed CRISPR/Cas9 inhibits growth of cervical cancer cells. International Journal of Clinical & Experimental Medicine, 10: 9206-9213.

Zhang, Q. and Ye, Y. 2017. Not all predicted CRISPR/Cas systems are equal: Isolated Cas genes and classes of CRISPR like elements. BMC Bioinformatics, 18: 92.

Zhen, S., Hua, L., Takahashi, Y., Narita, S., Liu, Y. H. and Li, Y. 2014. In vitro and in vivo growth suppression of human papillomavirus 16-positive cervical cancer cells by CRISPR/ Cas9. Biochemical and Biophysical Research Communications, 450: 1422-1426.

Zhen, S., Hua, L., Liu, Y. H., Gao, L. C., Fu, J., Wan, D. Y., et al. 2015. Harnessing the clustered regularly interspaced short palindromic repeat (CRISPR)/CRISPR-associated Cas9 system to disrupt the hepatitis B virus. Gene Therapy, 22: 404-412.

Zhen, S., Lu, J. J., Wang, L. J., Sun, X. M., Zhang, J. Q., Li, X., et al. 2016. In vitro and in vivo synergistic therapeutic effect of cisplatin with human papillomavirus16 E6/E7 CRISPR/ Cas9 on cervical cancer cell line. Translational Oncology, 9: 498-504.

Zhen, S., Lu, J. J., Liu, Y. H., Chen, W. and Li, X. 2020. Synergistic antitumor effect on cervical cancer by rational combination of PD1 blockade and CRISPR-Cas9-mediated HPV knockout. Cancer Gene Therapy, 27: 168-178.

Zhen, S. and Li, X. 2017. Oncogenic human papillomavirus: Application of CRISPR/ Cas9 therapeutic strategies for cervical cancer. Cellular Physiology and Biochemistry: International Journal of Experimental Cellular Physiology, Biochemistry, and Pharmacology, 44: 2455-2466.

Zheng, Q., Bai, L., Zheng, S., Liu, M., Zhang, J., Wang, T., et al. 2017. Efficient inhibition of duck hepatitis B virus DNA by the CRISPR/Cas9 system. Molecular Medicine Reports, 16: 7199-7204.

Zhou, J., Shen, B., Zhang, W., Wang, J., Yang, J., Chen, L., et al. 2014. One-step generation of different immunodeficient mice with multiple gene modifications by CRISPR/Cas9 mediated genome engineering. International Journal of Biochemistry & Cell Biology, 46: 49-55.

Zhu, W., Lei, R., Duff, Y. L., Li, J., Guo, F., Wainberg, M. A., et al. 2015. The CRISPR/Cas9 system inactivates latent HIV-1 proviral DNA. Retrovirology, 12: 22.

Zhu, W., Xie, K., Xu, Y., Wang, L., Chen, K., Zhang, L., et al. 2016. CRISPR/Cas9 produces anti-hepatitis B virus effect in hepatoma cells and transgenic mouse. Virus Research, 217: 125-132.

# CRISPR/Cas9-based Genome and Epigenome Editing in Neuroscience Research

Nereo Kalebic[1]* and Wieland B. Huttner[2]

[1] Human Technopole, Via Cristina Belgioioso 171, 20157 Milan, Italy
[2] Max Planck Institute of Molecular Cell Biology and Genetics, Pfotenhauerstrasse 108, 01307 Dresden, Germany

## Genetic manipulation in neuroscience

Understanding how genes influence our behavior and how they underlie human neurological diseases has been one of the main research interests in neuroscience. From the early stages of brain development, genetic and epigenetic mechanisms enable production of a diversity of neuronal and glial cell types, their migration and maturation, and finally establishment of functional neuronal circuits. Moreover, various genetic and epigenetic mechanisms in the adult brain are responsible for a variety of fundamental brain functions in response to environmental stimuli. Manipulations of the genome and epigenome have thus had a profound impact on our understanding of brain development, function and related pathologies (Day 2019). The vast complexity of the human brain requires spatial and temporal precision in genetic manipulation. To experimentally and therapeutically manage the enormous cellular heterogeneity in mammalian brains, it is essential to be able to manipulate the genome with a cell-type level of specificity. It is further required to manipulate the genome with a high level of temporal precision, especially when dynamically regulated and developmental processes are under study (Sandoval et al. 2020).

Three major methods have been used to induce genetic manipulation in the brain: (i) transgenesis, (ii) RNAi and (iii) genome editing along with the related modifications. Transgenesis is based on the fundamental work on homologous recombination in embryonic stem cells, which allowed the generation of transgenic animals (Thomas and Capecchi 1987, Capecchi 2005). The spatial and temporal specificity is achieved by using various tissue-specific promoters, Cre-LoxP and Tet systems (Kos 2004). Although transgenesis remains the most comprehensive method

* Corresponding author: nereo.kalebic@fht.org

for genetic manipulation, it is very laborious and costly, with multiplexing being particularly difficult and time-consuming. RNAi (Fire et al. 1998) is much faster and cheaper, and it allows acute targeting at postnatal stages, but can suffer from severe off-target effects. Genome editing is discussed in detail below.

# Genome editing tools

Genome editing is a genetic manipulation in which genomic DNA is manipulated using engineered nucleases. Major genome editing tools are (i) zinc-finger nucleases (ZFNs), (ii) transcription activator-like effector nucleases (TALENs), and (iii) systems involving clustered regularly interspaced short palindromic repeats (CRISPR)/ CRISPR-associated (Cas) proteins. Whereas ZFNs and TALENs recognize genomic sequences through protein-DNA interactions, Cas proteins are guided to specific genomic sites by RNA molecules. Nucleases enable genome editing by performing a double-strand break of DNA at the target site. The cell attempts to repair the double-strand break through two possible pathways: non-homologous end joining (NHEJ) and homology-directed repair (HDR). NHEJ is rather error-prone, as the broken DNA ends are directly ligated, which can cause indels of various lengths that in turn can lead to a frameshift mutation and a disruption of gene expression. In contrast, HDR induces a precise recombination event using a DNA template that, if added into the cell, can introduce a specific mutation or insert a sequence into the target *locus* (Doudna and Charpentier 2014, Sander and Joung 2014).

## Zinc finger nucleases

ZFNs are composed of (i) a zinc-finger protein domain responsible for binding the DNA and (ii) the DNA-cleavage domain of the FokI endonuclease (Kim et al. 1996). The pioneering role of ZFNs among the genome editing tools enabled various improvements, which resulted in increased specificity and efficiency and in reduced off-target effects. Although ZFNs have already been employed in clinical trials (Jo et al. 2015), their use is decreasing, mainly because their synthesis is time-consuming and labor-intensive compared to the CRISPR/Cas9 system.

## TALENs

TALENs consist of bacterial DNA-binding proteins involved in transcriptional regulation fused to the DNA-cleavage domain of FokI (Boch et al. 2009, Moscou and Bogdanove 2009, Christian et al. 2010). TALENs are easier to generate than ZFNs, but they still require synthesis of a new protein, which also results in TALEN-mediated genome editing being more time-consuming and labor-intensive compared to the CRISPR/Cas9 system.

## CRISPR/Cas9

The CRISPR/Cas system is a genome editing tool engineered from the bacterial adaptive immune response, which bacteria use to defend themselves against phage infections. The first Cas enzyme repurposed for genome editing in eukaryotic cells was Cas9 from *Streptococcus pyogenes* (Jinek et al. 2012). Since then, Cas9 from other species and other Cas proteins targeting either DNA, such as Cas12a, or RNA, such as Cas13, have been used in genome editing (Shmakov et al. 2017). However, Cas9 from *S. pyogenes* still remains the most popular enzyme. The native CRISPR/Cas9 system in *S. pyogenes* involves two different RNA molecules (crRNA and tracrRNA), which to simplify matters were modified into a single guide RNA (gRNA) for genome editing purposes (Jinek et al. 2012). The protospacer adjacent motif (PAM) was found to be important for recognition of the target DNA, with the Cas9 endonuclease domain found to cleave the DNA and generate the double-strand break 3 nucleotides upstream of the PAM (Jinek et al. 2012). CRISPR/Cas9-mediated genome editing was for the first time applied to mammalian cells in 2013 (Cong et al. 2013, Jinek et al. 2013, Mali et al. 2013), which started a true genetic revolution (Doudna and Charpentier 2014).

The main difference between ZFNs and TALENs on the one side and CRISPR/Cas9 on the other is that the first two systems utilize proteins to bind to a specific DNA target, whereas the latter is guided to a specific DNA site by an RNA molecule. The obvious key advantage of the CRISPR/Cas9 system is that it does not require the design and generation of a new protein, but only of a new gRNA, which saves time, money and effort. This led to a massive development of various tools based on the CRISPR/Cas9 system, including the epigenome editors, as is described below.

# Modifications of the CRISPR/Cas9 system and epigenome editing

## Spatial and temporal specificity

Multiple modifications of the CRISPR/Cas9 system have been established since 2013. Importantly for neuroscience, spatial and temporal specificity has been achieved. To obtain such genome editing in an inducible manner, the CRISPR/Cas9 system was combined with various conventional systems used in mouse genetics, such as Cre-LoxP (Platt et al. 2014), Tet (Dow et al. 2015) and ERT–4-HT (Davis et al. 2015) as well as photoinducible systems (Nihongaki et al. 2015).

## Mutations of the Cas9 protein

Cas9 contains two nuclease domains: RuvC1 and HNH (Jinek et al. 2012). Mutating one of them generates a nickase, i.e. an enzyme that makes a single-strand break on DNA. The most significant mutation of the Cas9 protein was the generation of a catalytically dead Cas9 (dCas9), which has both of its nuclease domains mutated (Qi et al. 2013). dCas9 serves as a programmable DNA-binding module for fusion with various effector molecules which allow visualization of genomic *loci*, transcriptional control, epigenome editing, base editing, etc.

## Visualization of genomic *loci*

Fusion of dCas9 with fluorescent proteins such as GFP has been applied to visualize repetitive elements in telomeres and coding genes in living cells (Chen et al. 2013).

The same authors achieved visualization of non-repetitive genomic sequences by tiling an array of gRNAs along the target *locus* (Chen et al. 2013).

## Transcriptional control

The conventional ways of manipulating gene expression, notably cDNA overexpression and RNAi-mediated silencing, suffer from potential protein mis-localization and off-targeting, respectively. Fusion of dCas9 to various effector domains can help circumvent these shortcomings (Sandoval et al. 2020). Fusion of dCas9 with repression domains, such as KRAB, serves to achieve a reduction in transcription, known as CRISPR interference (CRISPRi), whereas a fusion of dCas9 with activation domains, such as VP64, is used to induce activation of transcription, known as CRISPR activation (CRISPRa) (Gilbert et al. 2013). In addition, gRNA can be modified and used as a scaffold molecule that is recognized by RNA-binding proteins, which in turn are fused to effector domains such as KRAB or VP64 (Zalatan et al. 2015). Finally, the mere targeting of dCas9 to the coding region of a gene can sterically block RNA polymerase binding, which can lead to a reduction in transcription (Gilbert et al. 2013, Qi et al. 2013).

## Epigenome editing

### DNA methylation

Fusion of dCas9 to catalytic domains of epigenetic enzymes enables manipulation of the epigenetic landscape. DNA methylation of CpG islands in promoter regions is associated with transcriptional repression (Gjaltema and Rots 2020). The first epigenome editing tools were engineered by fusion of ZF proteins or TALEs with DNA methylation enzymes (Xu and Bestor 1997, Maeder et al. 2013). Since dCas9 became available, its fusion with the DNA methyl transferase Dnmt3a and the methylcytosine dioxygenase Tet1 were reported, allowing both writing and erasing DNA methylation (Liu et al. 2016). Further advances allowed simultaneous targeting of dCas9 fusions with methyl transferases and the KRAB repressor to induce repressive transcriptional memory and epigenetic reprogramming (Amabile et al. 2016, Tarjan et al. 2019).

### Histone modifications

Various histone modifications are associated with gene expression. H3K4me1/3 and H3K27ac are associated with promoting gene expression, whereas H3K9me2/3 and H3K27me3 are associated with repressing gene expression. Fusion of dCas9 with catalytic domains of various histone modifiers allows editing of such histone epigenetic marks (Gjaltema and Rots 2020). To target histone acetylation, dCas9 was fused to histone acetyl-transferase P300 (Hilton et al. 2015) and deacetylase HDAC3 (Kwon et al. 2017). Editing histone methylation was achieved by dCas9 fusion with histone methyl-transferases EZH2 (Albert et al. 2017, O'Geen et al. 2017) or PRDM9 (Cano-Rodriguez et al. 2016) and histone demethylase LSD1 (Kearns et al. 2015).

*DNA looping*

Transcription is often regulated by the long-range interactions of proteins bound to the same DNA molecule as the transcription machinery through formation of large DNA loops. Several epigenome editing systems have been developed to modify transcription via manipulating DNA loops (Hao et al. 2017, Morgan et al. 2017), including a recently developed light-activated dynamic looping (LADL) system which allows inducible spatial co-localization of two genomic *loci* by application of blue light (Kim et al. 2019).

*Nuclear organization*

To investigate large-scale spatial genome organization and its role in cell function, Wang and colleagues developed a chemically inducible and reversible CRISPR-genome organization (CRISPR-GO) system that can control the spatial positioning of genomic *loci* relative to specific nuclear compartments (Wang et al. 2018b).

## RNA targeting

The Cas13 family of endonucleases were the first engineered RNA-targeting CRISPR systems, and they were shown to be more effective in silencing of gene expression than RNAi (Abudayyeh et al. 2017). Cas13d from *Ruminococcus flavefaciens* (CasRx) fused with a nuclear localization signal was found to be highly efficient in cleaving target RNA sequences, outperforming the effects of dCas9-KRAB (Konermann et al. 2018). A modified CasRx unable to complete the cleavage reaction (dCasRx) was targeted to RNA sequences that mediate splice selection and alter splice isoform ratios, which can be particularly important in some neurodegenerative diseases (see below) (Konermann et al. 2018).

## Base editing and prime editing

Base editing allows modifying the DNA or RNA sequence *in situ* without generating a double-strand break (Komor et al. 2016, Gaudelli et al. 2017). Instead, a fusion of dCas9 with various nucleobase deaminases enables direct mutation of single DNA and RNA bases, thereby allowing the efficient installation of point mutations in non-dividing cells. This can have important applications in the context of neurological diseases where single nucleotide polymorphisms were shown to play a paramount role (Rees and Liu 2018). Prime editors represent the newest technology allowing precise editing. They combine Cas9-nickase and an engineered reverse transcriptase programmed with a prime editing guide RNA (pegRNA) that both specifies the target site and encodes the desired edit (Anzalone et al. 2019).

## Delivery of CRISPR/Cas9 components to the brain

Cellular and molecular features of a mammalian brain can be studied both *ex vivo* and *in vivo*.

The *ex vivo* model systems, such as brain slices, are often used to study neuronal development or function. The first applications of CRISPR/Cas9 to hippocampal slice cultures were used to perform a knock-out of NMDA and AMPA receptor subunits via biolistic transfection (Incontro et al. 2014, Straub et al. 2014). To target neural progenitor cells in embryonic mouse neocortex, microinjection of a Cas9/gRNA complex was employed to perform a knock-out of GFP in a transgenic mouse line (Kalebic et al. 2016). Human fetal brain tissue was targeted by *ex vivo* electroporation to deliver CRISPR/Cas9 components (Kalebic et al. 2019). For *in vivo* genome editing in the brain, delivery of CRISPR/Cas9 components remains the most challenging step. In addition to the generation of transgenic animals, there are various methods, described in the following sections, that allow acute manipulation of the genome and epigenome in the developing or adult brain. These methods are particularly important in the context of potential therapy of human neurological diseases.

## Generation of genetically modified animals

Soon after its use in mammalian cells had been pioneered, the CRISPR/Cas9 system has been employed to generate genetically modified zebrafish, mice, rats and macaques (Hwang et al. 2013, Li et al. 2013a, Niu et al. 2014, Platt et al. 2014). Genome editing is used to generate both disruption of gene expression, i.e. a gene knock-out, and transgenic knock-in animal lines. Compared to the conventional generation of genetically modified animals, CRISPR/Cas9-based genome editing allows faster and easier production of animal lines and significantly facilitates multiplexing. Although generation of genetically modified animals remains the most thorough method for neuroscience research, the germline modifications associated with such generation can also produce undesirable developmental phenotypes, which may complicate the analysis at adult stages. Transgenic animals that carry a Cre-dependent Cas9 are helpful in this regard as they allow for the genome editing to take place in adults upon the delivery of Cre and gRNAs into the brain via viral vectors (Platt et al. 2014).

## Viral injection

Viral vectors have been commonly used to perform genetic manipulation via RNAi (Xia et al. 2002). They can be injected directly to the brain or can access it through the bloodstream (Murlidharan et al. 2014). Adeno-associated viruses (AAVs) are the most commonly used type of viruses in neuroscience research due to them being relatively safe and able to sustain long-term expression without genomic integration (Murlidharan et al. 2014). The greatest limitation of AAVs is their modest transgene capacity so that the large size of Cas9 from *S. pyogenes* poses great challenges (Platt et al. 2014, Swiech et al. 2015). To circumvent this, Cas9 from *S. pyogenes* and its gRNAs can be packaged in separate AAV vectors (Swiech et al. 2015), or smaller Cas9 proteins, such as the one coming from *S. aureus*, can be used (Ran et al. 2015). Retroviruses and lentiviruses enable stable integration of transgenes and are valuable tools when such integration is desired (Artegiani et al. 2012, Artegiani and Calegari 2013, Fricano-Kugler et al. 2016). However, the spread of retroviruses and lentiviruses is restricted, which enables targeting of only small areas of the brain.

## *In utero* electroporation

Since electroporation was shown to enable gene transfer into cells (Neumann et al. 1982), it has been applied to deliver DNA into various animals *in vivo*, including the rodent brain (Saito and Nakatsuji 2001, Tabata and Nakajima 2001). *In utero* electroporation comprises the injection of DNA into the lateral ventricle of the embryonic brain and the subsequent application of an electrical field (Walantus et al. 2007, Artegiani et al. 2012, Kalebic et al. 2020). The main limitations of the *in utero* electroporation method are the need for embryonic intervention and the targeting of neural progenitors rather than neurons. Thus, *in utero* electroporation is the method of choice for neurodevelopmental research, to study the biology of neural progenitors, neurogenesis, neural migration and maturation (Straub et al. 2014, Kalebic et al. 2016, Mikuni et al. 2016, Shinmyo et al. 2016, Albert et al. 2017).

## Polyethylenimine-mediated transfection

In addition to *in utero* electroporation, polyethylenimine-mediated transfection has been successfully used to perform a somatic gene transfer of CRISPR/Cas9 components to the mammalian brain during development (Zuckermann et al. 2015). This method has been used to induce medulloblastoma and glioblastoma in the mouse developing brain (Zuckermann et al. 2015).

## Delivery of Cas9/gRNA ribonucleoprotein complexes

The Cas9 protein, together with the appropriate gRNA, has initially been delivered to the mammalian brain mainly via an encoding DNA plasmid or mRNA. However, the delivery of nucleic acid-encoded Cas9 plus gRNA has several limitations. First, when delivered encoded on a plasmid, the expression of both Cas9 and gRNA will continue to occur until the plasmid is either diluted by cell division or silenced, which will increase the probability of off-target effects, as it has been shown that higher concentrations of Cas9 decrease specificity (Davis et al. 2015). Second, the size of a plasmid encoding Cas9 is substantial, which can present a problem when packaging it into viral particles or when electroporating it into neural progenitors *in utero*, as the plasmid size can affect the efficiency of electroporation (Lesueur et al. 2016). Third, when delivering the gRNAs and the Cas9 gene on the same plasmid, the number of gRNAs to be expressed is limited, which results in a reduced ability of multiplexing. Finally, the time needed for the Cas9 mRNA and the gRNA to be produced by transcription and for the Cas9 protein to be produced by translation take a substantial portion of time of the cell cycle, which during development is suboptimal since most cell fate decision occur during a single cell cycle (Kalebic et al. 2016). Delivering directly the Cas9 protein in a complex with gRNA can overcome all these limitations (Kalebic et al. 2016).

The Cas9 protein is slightly cationic at physiological pH, but the complex of Cas9 and gRNA becomes anionic. Because of its electrical charge, it can be directly electroporated into cells (Kim et al. 2014, Liang et al. 2015). Such complex can hence be delivered to the embryonic mammalian brain via *in utero* electroporation. This approach has been successfully established for the embryonic mouse brain

(Kalebic et al. 2016) and has subsequently been applied to disrupt expression of various genes in mouse and ferret neural progenitors *in vivo* (Tavano et al. 2018, Güven et al. 2020, Xing et al. 2020). Cas9/gRNA complexes were also injected into the spinal cord lumen of the axolotl and subsequently electroporated (Fei et al. 2016). Importantly, such delivery led to a greater efficiency of gene knock-out compared to the plasmid-based delivery (Fei et al. 2016). For delivery into adult animals, the Cas9/gRNA complex can also be integrated into cationic liposomes, vehicles commonly used to deliver molecules into cells. This system has been successfully applied to achieve CRISPR/Cas9-mediated genome editing in the mouse inner ear *in vivo*, resulting in efficient genetic modification in hair cells (Zuris et al. 2015). Finally, a direct injection of cell-penetrable Cas9/gRNA complexes was performed in the hippocampus, striatum and cortex of the adult mouse brain (Staahl et al. 2017). Engineered Cas9 containing multiple SV40 nuclear localization signals was found to exhibit enhanced cell penetrance, which facilitated the delivery into adult neurons *in vivo*.

## Nanocomplexes

Most recently, Cas9/gRNA complexes have been delivered to mouse brain via nanoparticles (Lee et al. 2018). In a system named CRISPR-Gold, gold nanoparticles conjugated to single-stranded DNA are used to recruit Cas9 ribonucleoproteins. The entire complex is coated with a cationic polymer that facilitates delivery across the plasma membrane (Foss and Wilson 2018, Lee et al. 2018). CRISPR nanocomplexes can also be generated by fusing Cas9 ribonucleoproteins to an amphiphilic R7L10 peptide (Park et al. 2019). This peptide facilitates the cellular entry and shows a sustained expression of Cas9 for over a week, which was reduced below detection levels after three weeks (Park et al. 2019). A magnetically guided non-invasive delivery of Cas9/gRNA bound to magneto-electric nanoparticles across the blood-brain barrier was recently achieved to inhibit a latent HIV-1 infection in microglia, suggesting a new powerful tool for genome editing-based personalized nano-medicine in the future (Kaushik et al. 2019).

## Spatiotemporal control of Cas9/gRNA complexes

Achieving spatiotemporal control of the delivery and/or activity of the Cas9/gRNA complexes is essential for therapeutic applications, and several responsive systems have recently been developed (Zhang et al. 2021). Photocaged gRNA that regulates the interaction between the Cas9/gRNA complexes and the genomic DNA has been applied to perform a UV light-controlled genome editing in zebrafish *in vivo* (Zhou et al. 2020). The upconversion nanoparticles that can convert a near-infrared light into local UV light for the cleavage of photosensitive molecules were combined with the near-infrared light-responsive nanocarrier of Cas9/gRNA complexes to achieve a light-controllable release of Cas9/gRNA (Pan et al. 2019). Ultrasound-responsive microbubbles were incorporated into lipid nanoparticles to enable a local delivery Cas9/gRNA complexes upon ultrasound activation (Ryu et al. 2020). This approach has been successfully applied to target genes involved in androgenic alopecia in dermal papilla cells of mouse hair follicles *in vivo* (Ryu et al. 2020). Moreover, various targeted delivery systems for Cas9/gRNA complexes have recently been developed (Zhang et al. 2021). For example, nanoparticles containing an RGD (arginine-glycine-aspartic acid) analogue, which has a high affinity to integrins and neuropilin-1, were used to deliver Cas9/gRNA complexes into mouse brain tumors *in vivo* (Chen et al. 2017).

# Cell-type specificity in genome and epigenome editing in the brain

All mammalian cells preferentially employ NHEJ over HDR. This should be considered when analyzing the effects of genome editing in cycling cells (e.g. neural progenitors) *versus* postmitotic cells (e.g. neurons, certain glial cells such as mature oligodendrocytes). It is generally thought that the HDR occurs in the S and G2 phases of the cell cycle, that is, it is confined to cycling cells (van Gent and van der Burg 2007, Pardo et al. 2009). NHEJ, in contrast, can occur irrespective of the cell cycle stage, which makes it the primary way to perform DNA repair in neurons. Also, inducing double-strand breaks in neurons can cause non-desired effect as such breaks appear to be part of normal neuronal physiology and seem to be linked to learning and memory (Suberbielle et al. 2013). Impaired repair of double-strand breaks, moreover, appears to be linked with pathological states, such as Alzheimer's disease (Suberbielle et al. 2013, Lee et al. 2016). This poses a problem when aiming at generating precise mutations in neurons, which is required in many neuropathological conditions (Heidenreich and Zhang 2016). Nevertheless, a recent report shows that HDR is operating in neurons of the adult brain and can be used to efficiently tag endogenous proteins (Nishiyama et al. 2017), which opens new routes to a broader application of genome editing in neurons.

## Targeting neural progenitors

Targeting neural progenitors during embryonic development is still the most comprehensive way of generating precise mutations in both these cells and their entire neuronal or glial progeny. The first applications of CRISPR/Cas9-mediated genome editing to neural progenitors were performed via *in utero* electroporation. Using this method, the transcription factor Tbr2, important for the generation of a subset of neural progenitors, was knocked-out in mouse embryonic neocortex (Kalebic et al. 2016), which led to a phenotype similar to that of the Tbr2 knock-out mice generated via homologous recombination in embryonic stem cells (ESCs) (Sessa et al. 2008). Grin1, the gene encoding the NMDA receptor subunit GluN1, was also knocked-out through *in utero* electroporation of neural progenitors, generating a mutation that was inherited to the neuronal progeny (Straub et al. 2014). The same approach has be used to knock out the transcriptional modifier Satb2, which is important for the function of callosally-projecting neurons (Shinmyo et al. 2016). The first knock-in in neurons performed via *in utero* electroporation of neural progenitors was the SLENDR system, which enabled detection of the subcellular localization of a protein via insertion of N- or C-terminal epitope tags (Mikuni et al. 2016). Even longer sequences, such as a GFP-encoding sequence, were successfully inserted into neuronal genomes by targeting the embryonic progenitors (Mikuni et al. 2016).

## Targeting neurons

Direct targeting of neurons, that is, without passing through the progenitor stage, was first achieved via viral injections. Using the AAV-mediated stereotactic delivery of the CRISPR/Cas9 components to the adult mouse brain, the transcriptional modifier Mecp2 was targeted, resulting in a phenotype similar to that observed in Mecp2 mutant mice and patients with Rett syndrome (Swiech et al. 2015). To efficiently package Cas9 into AAVs, a truncated mouse Mecp2 promoter and a minimal polyadenylation signal were used (Swiech et al. 2015). The first virus-independent approach consisted of a direct injection of cell-permeable Cas9 ribonucleoproteins, which was successfully applied to adult mouse brain by Doudna and colleagues (Staahl et al. 2017). The efficiency of genome editing was shown to increase in a dose-dependent manner with increasing doses of cell-permeable Cas9/gRNA complexes (Staahl et al. 2017). Finally, nanoparticle-mediated delivery systems like CRISPR-Gold can also be administered by direct intracranial injection, as shown when targeting the metabotropic glutamate receptor 5 (mGluR5) gene in the striatum (Lee et al. 2018).

The diversity of neuronal and neural progenitor types and subtypes (Pollen et al. 2014, Camp et al. 2015, Nowakowski et al. 2017) poses a significant challenge when aiming at precise genome editing. The most extensively used genetic tool for achieving site-specific recombination is the Cre-loxP system (Kos 2004). This system has very early been employed to perform CRISPR/Cas9-mediated site-specific genome editing in mouse (Platt et al. 2014).

Epigenome editing and dCas9-mediated manipulations of transcription *in vivo* are even more challenging due to the large size of effector domains that add to the already large size of dCas9. In addition, the magnitude of effect of the early systems was subtle, and only tools of the second and third generations showed more robust effects. An important approach allowing transcription activation *in vivo* was achieved by generation of the SPH (SunTag-p65-HSF1) platform, which was built by combining two second generation systems, SunTag and SAM (synergistic activation mediator) (Zhou et al. 2018). Finally, a Cre-dependent SPH transgenic mouse line enabled tissue specificity and circumvented delivery problems. This system was used to modulate neuronal function *in vivo* by directly converting astrocytes into neurons by activating Ascl1, Neurog2 and Neurod1 (Zhou et al. 2018). To achieve a site-specific recombination, a lentivirus-mediated approach can also be used (Artegiani and Calegari 2013). A lentivirus-based approach has recently been used to achieve a repression of expression of synaptotagmin I by a dCas9-KRAB fusion system in mice (Zheng et al. 2018). The system was modified to restrict the targeting to glutamatergic (CaMKIIa) or GABAergic (VGAT) neurons in the mouse dentate gyrus *in vivo*. The subsequent shifting of the inhibitory-excitatory ratio in the hippocampus resulted in alterations in spatial and contextual fear memory (Zheng et al. 2018).

# Genome and epigenome editing across model organisms

Neuroscience has massively benefited from the use of a whole range of model organisms, from the "specialist" species such as zebra finch, bat, and naked mole rat to the classical genetically tractable species such as worms, flies, zebrafish and mice (Laurent 2020). On the one hand, the vast complexity of the human brain, its cognitive functions and pathological conditions require model organisms that resemble human brain development and pathologies as much as possible, hence the need to use mammals including primates. On the other hand, the complexity of mammalian brains in general has induced the modeling of fundamental neurobiological processes in simpler organisms, mainly invertebrates and lower vertebrates. Transgenesis and RNAi have favored genetic manipulation in a small set of model organisms, but the genome editing revolution has recently contributed to the expansion of the list of species used in neurobiological research. The CRISPR/Cas9 system can now be applied to any species, as long as its genome has been sequenced, allowing scientists to generate better disease models and new drugs, to control disease vectors such as mosquitos, to generate improved pets and farm animals, and to resurrect extinct species (Reardon 2016).

## Non-mammalian species

So-called simple model organisms have been invaluable for various genetic screens using either chemical mutagenesis or RNAi, which have revealed lists of genes involved in fundamental neurobiological processes (Patton and Zon 2001, Jorgensen and Mango 2002, St Johnston 2002). The ZFN- and TALEN-based genome editing allowed for precise genetic modifications in those organisms (Wood et al. 2011), but only the revolution of CRISPR/Cas9-mediated genome editing had a profound impact on the generation of animal models. Rapid genetic screens performed by applying CRISPR/Cas9-based genome editing have already revealed various genes involved in fundamental functions of the nervous system (Shah et al. 2015). In a large-scale study, CRISPR/Cas9-based genome editing of synthetic target arrays for lineage tracing (GESTALT) was combined with single-cell RNA-seq to elucidate lineage relationships in the developing zebrafish (McKenna et al. 2016). In addition, CRISPR/Cas9 screens were performed in zebrafish to examine mutant phenotypes of schizophrenia-associated genes identified in human genome-wide association studies (Thyme et al. 2019).

## Rodents

Among mammals, mouse and rat have contributed the most to neuroscience research. Thanks to the availability of a much larger genetic toolbox, mouse became a more prominent model in neuroscience research than rat (Ellenbroek and Youn 2016). Indeed, mouse was one of the first species to which CRISPR/Cas9-mediated genome editing was applied for the generation of genetically modified animals (Wang et al. 2013, Yang et al. 2013). Of importance for neuroscience, the generation of Cas9-transgenic mice allowed a simplified CRISPR/Cas9-based genome editing *in vivo* since only an AAV was needed to deliver a gRNA to the target location in the adult animals (Platt et al. 2014). Mouse was also the first species where CRISPR/Cas9 components were delivered via *in utero* electroporation to target neural progenitors (Kalebic et al. 2016) or neurons (Straub et al. 2014, Mikuni et al. 2016, Shinmyo et al. 2016).

Although its transgenesis remained scarce, rat is an important animal model in behavioral neuroscience research. The genome editing revolution allowed an important advancement of rat transgenesis, with the generation of transgenic lines via germline mutations (Li et al. 2013a, b, Ma et al. 2014). Further advances were made with the generation of transgenic rats that can facilitate genomic modification in adult neurons via Cre-dependent Cas9 systems (Back et al. 2019).

## Carnivores

Among the larger mammals that have been used as models in neuroscience, the ferret takes a prominent place. Ferret has become a widely used animal model for studies of development and plasticity of sensory processing (Lohse et al. 2019) and of perinatal brain injury (Empie et al. 2015). Because of its expanded brain with a characteristic folding pattern, which mouse and rat lack, ferret has become an essential model in developmental neurobiology (Kawasaki 2017). Genetic manipulation in ferret can be achieved by (i) transgenesis (Johnson et al. 2018, Yu et al. 2019), (ii) postnatal electroporation or viral injection (Borrell et al. 2006, Borrell 2010), and (iii) *in utero* electroporation during embryonic development (Kawasaki et al. 2012, 2013, Kalebic et al. 2018, 2020). Considering its importance as a model organism with an expanded and folded brain, the first ferrets generated by genome editing were used to model microcephaly and lissencephaly, neurodevelopmental malformations that affect the size and folding of the brain, respectively. TALEN-mediated knock-out of *Aspm*, the gene most commonly associated with recessive microcephaly in humans, resulted in a reduction in brain size, cortical surface area and cortical thickness, similar to human patients (Johnson et al. 2018). CRISPR/Cas9-mediated knock-out in one-cell stage ferret embryos of *Dcx* (*doublecortin*), a gene implicated in neuronal migration, resulted in neocortical disorganization and lissencephaly (Kou et al. 2015). The CRISPR/Cas9 system has recently been used to perform an acute knock-in (Tsunekawa et al. 2016) and knock-out (Shinmyo et al. 2017, Güven et al. 2020, Xing et al. 2020) of genes in the embryonic ferret brain by *in utero* electroporation.

## Non-human primates

The most relevant *in vivo* models to study the human brain are undoubtedly non-human primates. Of particular interest are macaque and marmoset since tools have been established to generate genetically modified animals of these two species (Chan et al. 2001, Sasaki et al. 2009). The first macaques generated by genome editing (TALEN-mediated) modeled mutations in the neurologically important gene *MECP2* (Liu et al. 2014). In the same year, the first transgenic macaques were generated via CRISPR/Cas9-mediated gene transfer into one-cell embryos (Niu et al. 2014). CRISPR/Cas9-mediated knock-out and knock-in have also been performed in marmoset embryos (Kumita et al. 2019, Yoshimatsu et al. 2019). Genome editing in non-human primates has primarily been applied to modeling of various neurological diseases. CRISPR/Cas9-mediated knock-out of *SIRT6* in macaque delayed neuronal differentiation and resulted in postnatal death (Zhang et al. 2018). Deletion of *PINK1*, a gene involved in early onset of Parkinson's disease, recapitulated severe neurodegeneration phenotypes (Yang et al. 2019). Particularly interesting to model in non-human primates are neurodevelopmental disorders whose phenotypes cannot be fully recapitulated in rodents. Disruption of *SHANK3*, one of the causative genes of the autism spectrum disorders, led to abnormalities of neuronal connectivity and macaque behavior, which included repetitive behavior, impaired learning and social interaction (Tu et al. 2019, Zhou et al. 2019). To successfully model polygenic diseases in non-human primates, multiplexing base editing has recently been performed in macaque zygotes (Zhang et al. 2020). Thanks to such technological advancements and the fact that non-human primates can recapitulate very well many of the features of human neuronal circuit formation and brain regional identities, the research on genetically modified non-human primates has truly the potential to advance our knowledge and to facilitate development of treatments for neuropsychiatric disorders (Feng et al. 2020). In light of the presumptive future increase in the use of genetically modified non-human primates in neuroscience research, it is important to emphasize strong ethical considerations and the need for a responsible use of these animals, which should always involve a clear communication of the reasons for such research to the public (for a recent perspective on possibilities and limitations of genetically modified non-human primates in neuroscience, see (Feng et al. 2020)).

## Genome and epigenome editing and neuropathological conditions

Neurological and psychiatric diseases, including neurodevelopmental disorders and neurodegeneration, constitute a major societal burden, and the development of new effective therapies is critically required. CRISPR/Cas9-mediated genome and epigenome editing may contribute to preventing or curing various diseases by inducing genetic or epigenetic changes at endogenous *loci* (Doudna 2020, Duarte and Deglon 2020). Current challenges in developing such effective therapies stem from the facts that many neurological diseases are strikingly polygenic, can involve complex interactions between genes and environment, and often show specificity for certain neuronal types, circuits or brain regions (Sullivan et al. 2012, Russo and Nestler 2013, Lee et al. 2016, Hyman 2018).

### Induced pluripotent stem cells

CRISPR/Cas9-mediated genome editing in human induced pluripotent stem cells (IPSCs) is a promising research avenue for studying such complex polygenic neurological disorders (Heidenreich and Zhang 2016). IPSCs, being cycling cells, are amenable to HDR, which enables various genome editing approaches. IPSCs can be differentiated into neurons and used to examine processes that cannot be studied *in vivo*, such as human neuronal development. IPSCs are also valuable for studying regeneration of lost neurons and modeling of neuronal function in the central nervous system (Amamoto and Arlotta 2014). CRISPR/Cas9-mediated genome editing can be performed simultaneously with the *in vitro* generation of neurons to target genes associated with neurological disorders (Rubio et al. 2016). Finally, patient-derived IPSCs can be used to rescue genomic mutations that cause the disease, allowing scientists to obtain a model identical to patients' cells except for the corrected DNA sequence (Lee et al. 2016).

For example, some phenotypes of Alzheimer's disease, the main cause of dementia whose hallmark is the presence of scattered extracellular senile plaques due to the accumulation of amyloid-β (Aβ), are successfully modeled *in vitro*. Neurons derived (often via IPSCs) from cells taken from patients with familial or sporadic Alzheimer's disease show typical cellular and molecular phenotypes of the disease (Israel et al. 2012). Similar phenotypes were observed in IPSC-derived neurons expressing ApoE4, the major genetic risk factor for Alzheimer's disease (Wang et al. 2018a, Kampmann 2020). Genome editing-mediated conversion of ApoE4 to ApoE3 rescued these phenotypes, whereas the introduction of ApoE4 into the neurons recapitulated them (Wang et al. 2018a). IPSC-derived neurons were utilized to test potential treatments by small molecules, which led to amelioration of the phenotypes (Wang et al. 2018a).

Another neurodegenerative disease that has been extensively modeled in IPSCs is Parkinson's disease, which is caused by loss of dopaminergic neurons and is associated with progressive motor dysfunction (Safari et al. 2020). Genetic linkage studies have implicated *α-synuclein* (*SNCA*) as a key gene for the pathogenesis of Parkinson's disease. Jaenisch and colleagues performed CRISPR/Cas9-based genome editing in IPSCs, exploiting the genome-wide epigenetic information, to identify a common risk variant in a distal enhancer that controls *SNCA* expression, thus functionally connecting genetic variation with disease-relevant phenotypes (Soldner et al. 2016).

## Cerebral organoids

IPSC-derived three-dimensional cerebral organoids and the recently developed assembloids hold great potential to investigate complex human genetic states and to model aspects of human brain development and pathology (Quadrato and Arlotta 2017). Of importance for translational research, cerebral organoids can be of human origin, containing human genomic material, and can be derived from patient cells. They are fairly easy to manipulate and cost-effective. However, they cannot fully replace a living organism as they still cannot replicate the *in vivo* microenvironment, brain-like architecture and blood-derived signals in addition to remaining primarily a developmental model whose application to neurodegenerative diseases is still challenging (Quadrato and Arlotta 2017, Panoutsopoulos 2020). The CRISPR/Cas9 technology can be applied to cerebral organoids in two ways. First, a stable IPSC or ESC line that contains the desired genome editing event can be established, and all the subsequently generated organoid cells will contain that modification (Iefremova et al. 2017, Fiddes et al. 2018). Second, CRISPR/Cas9 components can be acutely introduced into grown organoids, targeting only a subset of cells (Bian et al. 2018).

## Animal models

Considering the enormous importance of animal models for studying human neurological disorders, it is not surprising that many disease models have already been generated by using CRISPR/Cas9-based genome editing. In addition to generating animal models, CRISPR/Cas9-based genome editing has been employed to functionally examine specific genes and mutations in already existing animal (typically mouse) models. For example, genome editing was applied to Huntington's disease, which is an inherited neurodegenerative disorder caused by an expansion of the CAG trinucleotide in the *HTT* gene (Duarte and Deglon 2020). CRISPR/Cas9-based genome editing of the disease-causing allele was performed via AAV or lentiviral delivery to the transgenic mouse models, and it achieved an efficient correction of the mutant protein (Merienne et al. 2017, Monteys et al. 2017, Yang et al. 2017). Similarly, *in vivo* genome editing was employed to rescue the phenotypes in the mouse models of amyotrophic lateral sclerosis (Gaj et al. 2017), fragile X syndrome (Lee et al. 2018), *MECP2* duplication syndrome (Yu et al. 2020), and others (Duarte and Deglon 2020).

Rodents are a suboptimal model system to study those neurological disorders that require more complex brain features and a longer lifespan of the organism (Safari et al. 2020). As discussed above, non-human primates have been used to study various neurodevelopmental disorders (Tu et al. 2019, Zhou et al. 2019) and Parkinson's disease (Yang et al. 2019). In addition, Guangxi Bama minipigs are emerging as a valuable model system for studying Parkinson's disease. CRISPR/Cas9-based multiplexing that targets *DJ-1*, *Parkin* and *PINK1*, three important genes involved in the pathogenesis of Parkinson's disease, was employed to generate transgenic minipigs (Wang et al. 2016). Furthermore, animals carrying disease-causing mutations in *SCNA* have also been recently generated (Zhu et al. 2018).

## CRISPR screens

The CRISPR/Cas9 technology in combination with the IPSC technology has been used to perform various screens to identify genetic factors associated with diseases and to evaluate the functional consequences of the risk genes and *loci* that were previously identified through human genetic studies (Kampmann 2020). For example, a large-scale CRISPR screen in IPSC-derived neurons identified neuron-specific essential genes as well as genes required for proper neuronal morphology (Tian et al. 2019). Moreover, CRISPR screens have been performed in cerebral organoids, where, for example, a CRISPR-lineage tracing at cellular resolution in heterogenous tissue (CRISPR-LICHT) has recently been employed to test 173 microcephaly candidate genes (Esk et al. 2020). Finally, CRISPR screens have also been performed *in vivo* in mouse to identify genetic factors that regulate glioblastoma pathogenesis (Chow et al. 2017), or in combination with single-cell RNA-seq (termed Perturb-Seq) to functionally evaluate 35 risk genes implicated in the autism spectrum disorder/neurodevelopmental delay (ASD/ND) (Jin et al. 2020).

# Future directions

The CRISPR/Cas9-based genome and epigenome editing will almost certainly continue to be one of the most important technologies to examine complex genetic and epigenetic underpinnings of brain development, function and pathologies. This will likely occur concomitant with further technological advances in generating improved genome and epigenome editing tools. Of particular relevance for the purpose of neuroscience research will be future progress towards greater cell-type specificity and inducibility of the various editing tools, improving the precision of editing and targeting, and advancing multiplexing techniques. Of importance for future therapies, delivery of the CRISPR/Cas9 components to the adult brain remains the most challenging step. Viral vectors are increasingly being associated with safety concerns, in particular in the central nervous system, since a prolonged exposure to the virus and exogenous proteins can lead to an adverse immune response and neurotoxicity (Colella et al. 2018). Hence, CRISPR nanocomplexes currently constitute one of the most promising routes towards injectable ribonucleoprotein therapies in neurological disorders (Lee et al. 2018). Further translational research will likely aim at generating better animal models of human diseases and improving the *in vitro* model systems. Since CRISPR/Cas9-based genome and epigenome editing can be applied to virtually any species, it is probable that emerging model organisms will be increasingly humanized to model specific human neurological traits and pathologies. IPSCs are a promising route for CRISPR screens and identification of cellular and molecular mechanisms implicated in neurological diseases. However, IPSC-derived neurons resemble young ones, whereas older neurons would be a better model for many neurological diseases. In addition to the lack of the tissue context and the ability to model neuronal circuits, IPSCs cannot be used to generate all the disease-relevant cell types (Kampmann 2020). Some, but not all, of these shortcomings can be overcome by cerebral organoids and assembloids, suggesting that these three-dimensional cultures are a likely future route for modeling human neurological disorders *in vitro*. Combining such systems with model organisms via xenografting is a particularly important approach to study the functional consequences of human genomic changes in an *in vivo* context.

# References

Abudayyeh, O. O., Gootenberg, J. S., Essletzbichler, P., Han, S., Joung, J., Belanto, J. J., et al. 2017. Rna targeting with crispr-cas13. Nature, 550: 280-284.

Albert, M., Kalebic, N., Florio, M., Lakshmanaperumal, N., Haffner, C., Brandl, H., et al. 2017. Epigenome profiling and editing of neocortical progenitor cells during development. The EMBO Journal, 36: 2642-2658.

Amabile, A., Migliara, A., Capasso, P., Biffi, M., Cittaro, D., Naldini, L., et al. 2016. Inheritable silencing of endogenous genes by hit-and-run targeted epigenetic editing. Cell, 167: 219-232 e214.

Amamoto, R. and Arlotta, P. 2014. Development-inspired reprogramming of the mammalian central nervous system. Science, 343: 1239882.

Anzalone, A. V., Randolph, P. B., Davis, J. R., Sousa, A. A., Koblan, L. W., Levy, J. M., et al. 2019. Search-and-replace genome editing without double-strand breaks or donor DNA. Nature, 576: 149-157.

Artegiani, B., Lange, C. and Calegari, F. 2012. Expansion of embryonic and adult neural stem cells by in utero electroporation or viral stereotaxic injection. Journal of Visualized Experiments. DOI: 10.3791/4093

Artegiani, B. and Calegari, F. 2013. Lentiviruses allow widespread and conditional manipulation of gene expression in the developing mouse brain. Development, 140: 2818-2822.

Back, S., Necarsulmer, J., Whitaker, L. R., Coke, L. M., Koivula, P., Heathward, et al. 2019. Neuron-specific genome modification in the adult rat brain using CRISPR-Cas9 transgenic rats. Neuron, 102: 105-119 e108.

Bian, S., Repic, M., Guo, Z., Kavirayani, A., Burkard, T., Bagley, J. A., et al. 2018. Genetically engineered cerebral organoids model brain tumor formation. Nature Methods, 15: 631-639.

Boch, J., Scholze, H., Schornack, S., Landgraf, A., Hahn, S., Kay, S., et al. 2009. Breaking the code of DNA binding specificity of tal-type III effectors. Science, 326: 1509-1512.

Borrell, V. 2010. In vivo gene delivery to the postnatal ferret cerebral cortex by DNA electroporation. Journal of Neuroscience Methods, 186: 186-195.

Borrell, V., Kaspar, B. K., Gage, F. H. and Callaway, E. M. 2006. In vivo evidence for radial migration of neurons by long-distance somal translocation in the developing ferret visual cortex. Cereb Cortex, 16: 1571-1583.

Camp, J. G., Badsha, F., Florio, M., Kanton, S., Gerber, T., Wilsch-Brauninger, M., et al. 2015. Human cerebral organoids recapitulate gene expression programs of fetal neocortex development. Proceedings of the National Academy of Sciences USA, 112: 15672-15677.

Cano-Rodriguez, D., Gjaltema, R. A., Jilderda, L. J., Jellema, P., Dokter-Fokkens, J., Ruiters, M. H., et al. 2016. Writing of h3k4me3 overcomes epigenetic silencing in a sustained but context-dependent manner. Nature Communications, 7: 12284.

Capecchi, M. R. 2005. Gene targeting in mice: Functional analysis of the mammalian genome for the twenty-first century. Nature Reviews Genetics, 6: 507-512.

Chan, A. W., Chong, K. Y., Martinovich, C., Simerly, C. and Schatten, G. 2001. Transgenic monkeys produced by retroviral gene transfer into mature oocytes. Science, 291: 309-312.

Chen, B., Gilbert, L. A., Cimini, B. A., Schnitzbauer, J., Zhang, W., Li, G. W., et al. 2013. Dynamic imaging of genomic loci in living human cells by an optimized CRISPR/Cas system. Cell, 155: 1479-1491.

Chen, Z., Liu, F., Chen, Y., Liu, J., Wang, X., Chen, A. T., et al. 2017. Targeted delivery of CRISPR/Cas9-mediated cancer gene therapy via liposome-templated hydrogel nanoparticles. Advanced Functional Materials, 27.

Chow, R. D., Guzman, C. D., Wang, G., Schmidt, F., Youngblood, M. W., Ye, L., et al. 2017. Aav-mediated direct in vivo CRISPR screen identifies functional suppressors in glioblastoma. Nature Neuroscience, 20: 1329-1341.

Christian, M., Cermak, T., Doyle, E. L., Schmidt, C., Zhang, F., Hummel, A., et al. 2010. Targeting DNA double-strand breaks with tal effector nucleases. Genetics, 186: 757-761.

Colella, P., Ronzitti, G. and Mingozzi, F. 2018. Emerging issues in AAV-mediated in vivo gene therapy. Molecular Therapy—Methods & Clinical Development, 8: 87-104.

Cong, L., Ran, F. A., Cox, D., Lin, S. L., Barretto, R., Habib, et al. 2013. Multiplex genome engineering using CRISPR/cas systems. Science, 339: 819-823.

Davis, K.M., Pattanayak, V., Thompson, D. B., Zuris, J. A. and Liu, D. R. 2015. Small molecule-triggered cas9 protein with improved genome-editing specificity. Nature Chemical Biology, 11: 316-318.

Day, J. J. 2019. Genetic and epigenetic editing in nervous system. Dialogues in Clinical Neuroscience, 21: 359-368.

Doudna, J. A. 2020. The promise and challenge of therapeutic genome editing. Nature, 578: 229-236.

Doudna, J. A. and Charpentier, E. 2014. Genome editing. The new frontier of genome engineering with CRISPR-Cas9. Science, 346: 1258096.

Dow, L. E., Fisher, J., O'Rourke, K. P., Muley, A., Kastenhuber, E. R., Livshits, G., et al. 2015. Inducible in vivo genome editing with CRISPR-Cas9. Nature Biotechnology, 33: 390-394.

Duarte, F. and Deglon, N. 2020. Genome editing for cns disorders. Frontiers in Neuroscience, 14: 579062.

Ellenbroek, B. and Youn, J. 2016. Rodent models in neuroscience research: Is it a rat race? Disease Models & Mechanisms, 9: 1079-1087.

Empie, K., Rangarajan, V. and Juul, S. E. 2015. Is the ferret a suitable species for studying perinatal brain injury? International Journal of Developmental Neuroscience, 45: 2-10.

Esk, C., Lindenhofer, D., Haendeler, S., Wester, R. A., Pflug, F., Schroeder, B., et al. 2020. A human tissue screen identifies a regulator of ER secretion as a brain-size determinant. Science, 370: 935-941.

Fei, J. F., Knapp, D., Schuez, M., Murawala, P., Zou, Y., Pal Singh, S., et al. 2016. Tissue-and time-directed electroporation of cas9 protein-grna complexes in vivo yields efficient multigene knockout for studying gene function in regeneration. NPJ Regenerative Medicine, 1: 16002.

Feng, G., Jensen, F. E., Greely, H. T., Okano, H., Treue, S., Roberts, A. C., et al. 2020. Opportunities and limitations of genetically modified nonhuman primate models for neuroscience research. Proceedings of the National Academy of Sciences USA, 117: 24022-24031.

Fiddes, I. T., Lodewijk, G. A., Mooring, M., Bosworth, C. M., Ewing, A. D., Mantalas, G. L., et al. 2018. Human-specific notch2nl genes affect notch signaling and cortical neurogenesis. Cell, 173: 1356-1369.

Fire, A., Xu, S., Montgomery, M. K., Kostas, S. A., Driver, S. E. and Mello, C. C. 1998. Potent and specific genetic interference by double-stranded RNA in caenorhabditis elegans. Nature, 391: 806-811.

Foss, D. V. and Wilson, R. C. 2018. Emerging strategies for genome editing in the brain. Trends in Molecular Medicine, 24: 822-824.

Fricano-Kugler, C. J., Williams, M. R., Salinaro, J. R., Li, M. and Luikart, B. 2016. Designing, packaging, and delivery of high titer crispr retro and lentiviruses via stereotaxic injection. Journal of Visualized Experiments. DOI: 10.3791/53783

Gaj, T., Ojala, D. S., Ekman, F. K., Byrne, L. C., Limsirichai, P. and Schaffer, D. V. 2017. In vivo genome editing improves motor function and extends survival in a mouse model of als. Science Advances, 3: eaar3952.

Gaudelli, N. M., Komor, A. C., Rees, H. A., Packer, M. S., Badran, A. H., Bryson, D. I. and Liu, D. R. 2017. Programmable base editing of a*t to g*c in genomic DNA without DNA cleavage. Nature, 551: 464-471.

Gilbert, L. A., Larson, M. H., Morsut, L., Liu, Z., Brar, G. A., Torres, S. E., et al. 2013. Crispr-mediated modular RNA-guided regulation of transcription in eukaryotes. Cell, 154: 442-451.

Gjaltema, R. A. F. and Rots, M. G. 2020. Advances of epigenetic editing. Current Opinion in Chemical Biology, 57: 75-81.

Güven, A., Kalebic, N., Long, K.R., Florio, M., Vaid, S., Brandl, H., et al. 2020. Extracellular matrix-inducing sox9 promotes both basal progenitor proliferation and gliogenesis in developing neocortex. eLife 9: e49808.

Hao, N., Shearwin, K. E. and Dodd, I. B. 2017. Programmable DNA looping using engineered bivalent dcas9 complexes. Nature Communications, 8: 1628.

Heidenreich, M. and Zhang, F. 2016. Applications of CRISPR-Cas systems in neuroscience. Nature Reviews Neuroscience, 17: 36-44.

Hilton, I. B., D'Ippolito, A. M., Vockley, C. M., Thakore, P. I., Crawford, G. E., Reddy, T. E., et al. 2015. Epigenome editing by a CRISPR-Cas9-based acetyltransferase activates genes from promoters and enhancers. Nature Biotechnology, 33: 510-517.

Hwang, W. Y., Fu, Y., Reyon, D., Maeder, M. L., Tsai, S. Q., Sander, J. D., et al. 2013. Efficient genome editing in zebrafish using a CRISPR-Cas system. Nature Biotechnology, 31: 227-229.

Hyman, S. E. 2018. The daunting polygenicity of mental illness: Making a new map. Philosophical Transactions of the Royal Society B: Biological Sciences, 373.

Iefremova, V., Manikakis, G., Krefft, O., Jabali, A., Weynans, K., Wilkens, R., et al. 2017. An organoid-based model of cortical development identifies non-cell-autonomous defects in WNT signaling contributing to miller-dieker syndrome. Cell Reports, 19: 50-59.

Incontro, S., Asensio, C. S., Edwards, R. H. and Nicoll, R. A. 2014. Efficient, complete deletion of synaptic proteins using CRISPR. Neuron, 83: 1051-1057.

Israel, M. A., Yuan, S. H., Bardy, C., Reyna, S. M., Mu, Y., Herrera, C., et al. 2012. Probing sporadic and familial alzheimer's disease using induced pluripotent stem cells. Nature, 482: 216-220.

Jin, X., Simmons, S. K., Guo, A., Shetty, A. S., Ko, M., Nguyen, L., et al. 2020. In vivo perturb-seq reveals neuronal and glial abnormalities associated with autism risk genes. Science, 370.

Jinek, M., Chylinski, K., Fonfara, I., Hauer, M., Doudna, J. A. and Charpentier, E. 2012. A programmable dual-rna-guided DNA endonuclease in adaptive bacterial immunity. Science, 337: 816-821.

Jinek, M., East, A., Cheng, A., Lin, S., Ma, E. and Doudna, J. 2013. Rna-programmed genome editing in human cells. eLife 2: e00471.

Jo, Y. I., Kim, H. and Ramakrishna, S. 2015. Recent developments and clinical studies utilizing engineered zinc finger nuclease technology. Cellular and Molecular Life Sciences, 72: 3819-3830.

Johnson, M. B., Sun, X., Kodani, A., Borges-Monroy, R., Girskis, K. M., Ryu, S. C., et al. 2018. Aspm knockout ferret reveals an evolutionary mechanism governing cerebral cortical size. Nature, 556: 370-375.

Jorgensen, E. M. and Mango, S. E. 2002. The art and design of genetic screens: Caenorhabditis elegans. Nature Reviews Genetics, 3: 356-369.

Kalebic, N., Taverna, E., Tavano, S., Wong, F. K., Suchold, D., Winkler, S., et al. 2016. CRISPR/Cas9-induced disruption of gene expression in mouse embryonic brain and single neural stem cells in vivo. EMBO Reports, 17: 338-348.

Kalebic, N., Gilardi, C., Albert, M., Namba, T., Long, K. R., Kostic, M., et al. 2018. Human-specific arhgap11b induces hallmarks of neocortical expansion in developing ferret neocortex. eLife 7: e41241.

Kalebic, N., Gilardi, C., Stepien, B., Wilsch-Brauninger, M., Long, K. R., Namba, T., et al. 2019. Neocortical expansion due to increased proliferation of basal progenitors is linked to changes in their morphology. Cell Stem Cell, 24: 535-550.

Kalebic, N., Langen, B., Helppi, J., Kawasaki, H. and Huttner, W. B. 2020. In vivo targeting of neural progenitor cells in ferret neocortex by in utero electroporation. Journal of Visualized Experiments, 159. DOI: 10.3791/61171.

Kampmann, M. 2020. CRISPR-based functional genomics for neurological disease. Nature Reviews Neurology, 16: 465-480.

Kaushik, A., Yndart, A., Atluri, V., Tiwari, S., Tomitaka, A., Gupta, P., et al. 2019. Magnetically guided non-invasive CRISPR-Cas9/grna delivery across blood-brain barrier to eradicate latent hiv-1 infection. Scientific Reports, 9: 3928.

Kawasaki, H. 2017. Molecular investigations of development and diseases of the brain of higher mammals using the ferret. Proceedings of the Japan Academy, Series B, Physical and Biological Sciences, 93: 259-269.

Kawasaki, H., Iwai, L. and Tanno, K. 2012. Rapid and efficient genetic manipulation of gyrencephalic carnivores using in utero electroporation. Molecular Brain, 5: 24.

Kawasaki, H., Toda, T. and Tanno, K. 2013. In vivo genetic manipulation of cortical progenitors in gyrencephalic carnivores using in utero electroporation. Biology Open, 2: 95-100.

Kearns, N. A., Pham, H., Tabak, B., Genga, R. M., Silverstein, N. J., Garber, M., et al. 2015. Functional annotation of native enhancers with a Cas9-histone demethylase fusion. Nature Methods, 12: 401-403.

Kim, J. H., Rege, M., Valeri, J., Dunagin, M. C., Metzger, A., Titus, K. R., et al. 2019. Ladl: Light-activated dynamic looping for endogenous gene expression control. Nature Methods 16: 633-639.

Kim, S., Kim, D., Cho, S. W., Kim, J. and Kim, J. S. 2014. Highly efficient RNA-guided genome editing in human cells via delivery of purified Cas9 ribonucleoproteins. Genome Research, 24: 1012-1019.

Kim, Y. G., Cha, J. and Chandrasegaran, S. 1996. Hybrid restriction enzymes: Zinc finger fusions to fok i cleavage domain. Proceedings of the National Academy of Sciences USA, 93: 1156-1160.

Komor, A. C., Kim, Y. B., Packer, M. S., Zuris, J. A. and Liu, D. R. 2016. Programmable editing of a target base in genomic DNA without double-stranded DNA cleavage. Nature, 533: 420-424.

Konermann, S., Lotfy, P., Brideau, N. J., Oki, J., Shokhirev, M. N. and Hsu, P. D. 2018. Transcriptome engineering with RNA-targeting type VI-d CRISPR effectors. Cell, 173: 665-676 e614.

Kos, C. H. 2004. Cre/loxp system for generating tissue-specific knockout mouse models. Nutrition Reviews, 62: 243-246.

Kou, Z., Wu, Q., Kou, X., Yin, C., Wang, H., Zuo, Z., et al. 2015. CRISPR/Cas9-mediated genome engineering of the ferret. Cell Research, 25: 1372-1375.

Kumita, W., Sato, K., Suzuki, Y., Kurotaki, Y., Harada, T., Zhou, Y., et al. 2019. Efficient generation of knock-in/knock-out marmoset embryo via CRISPR/Cas9 gene editing. Scientific Reports 9: 12719.

Kwon, D. Y., Zhao, Y. T., Lamonica, J. M. and Zhou, Z. 2017. Locus-specific histone deacetylation using a synthetic CRISPR-Cas9-based hdac. Nature Communications, 8: 15315.

Laurent, G. 2020. On the value of model diversity in neuroscience. Nature Reviews Neuroscience, 21: 395-396.

Lee, B., Lee, K., Panda, S., Gonzales-Rojas, R., Chong, A., Bugay, V., et al. 2018. Nanoparticle delivery of CRISPR into the brain rescues a mouse model of fragile x syndrome from exaggerated repetitive behaviours. Nature Biomedical Engineering, 2: 497-507.

Lee, H. B., Sundberg, B. N., Sigafoos, A. N. and Clark, K. J. 2016. Genome engineering with tale and crispr systems in neuroscience. Frontiers in Genetics, 7: 47.

Lesueur, L. L., Mir, L. M. and Andre, F. M. 2016. Overcoming the specific toxicity of large plasmids electrotransfer in primary cells in vitro. Molecular Therapy - Nucleic Acids 5: e291.

Li, D., Qiu, Z., Shao, Y., Chen, Y., Guan, Y., Liu, M., et al. 2013a. Heritable gene targeting in the mouse and rat using a CRISPR-Cas system. Nature Biotechnology, 31: 681-683.

Li, W., Teng, F., Li, T. and Zhou, Q. 2013b. Simultaneous generation and germline transmission of multiple gene mutations in rat using CRISPR-Cas systems. Nature Biotechnology, 31: 684-686.

Liang, X. Q., Potter, J., Kumar, S., Zou, Y. F., Quintanilla, R., Sridharan, M., et al. 2015. Rapid and highly efficient mammalian cell engineering via Cas9 protein transfection. Journal of Biotechnology, 208: 44-53.

Liu, H., Chen, Y., Niu, Y., Zhang, K., Kang, Y., Ge, W., et al. 2014. Talen-mediated gene mutagenesis in rhesus and cynomolgus monkeys. Cell Stem Cell, 14: 323-328.

Liu, X. S., Wu, H., Ji, X., Stelzer, Y., Wu, X., Czauderna, S., et al. 2016. Editing DNA methylation in the mammalian genome. Cell, 167: 233-247 e217.

Lohse, M., Bajo, V. M. and King, A. J. 2019. Development, organization and plasticity of auditory circuits: Lessons from a cherished colleague. European Journal of Neuroscience, 49: 990-1004.

Ma, Y., Shen, B., Zhang, X., Lu, Y., Chen, W., Ma, J., et al. 2014. Heritable multiplex genetic engineering in rats using CRISPR/Cas9. PloS One 9: e89413.

Maeder, M. L., Angstman, J. F., Richardson, M. E., Linder, S. J., Cascio, V. M., Tsai, S. Q., et al. 2013. Targeted DNA demethylation and activation of endogenous genes using programmable tale-tet1 fusion proteins. Nature Biotechnology, 31: 1137-1142.

Mali, P., Yang, L. H., Esvelt, K. M., Aach, J., Guell, M., DiCarlo, J. E., et al. 2013. Rna-guided human genome engineering via Cas9. Science, 339: 823-826.

McKenna, A., Findlay, G. M., Gagnon, J. A., Horwitz, M. S., Schier, A. F. and Shendure, J. 2016. Whole-organism lineage tracing by combinatorial and cumulative genome editing. Science, 353: aaf7907.

Merienne, N., Vachey, G., de Longprez, L., Meunier, C., Zimmer, V., Perriard, G., et al. 2017. The self-inactivating kamicas9 system for the editing of cns disease genes. Cell Reports, 20: 2980-2991.

Mikuni, T., Nishiyama, J., Sun, Y., Kamasawa, N. and Yasuda, R. 2016. High-throughput, high-resolution mapping of protein localization in mammalian brain by in vivo genome editing. Cell, 165: 1803-1817.

Monteys, A. M., Ebanks, S. A., Keiser, M. S. and Davidson, B. L. 2017. CRISPR/Cas9 editing of the mutant huntingtin allele in vitro and in vivo. Molecular Therapy, 25: 12-23.

Morgan, S. L., Mariano, N. C., Bermudez, A., Arruda, N. L., Wu, F., Luo, Y., et al. 2017. Manipulation of nuclear architecture through CRISPR-mediated chromosomal looping. Nature Communications, 8: 15993.

Moscou, M. J. and Bogdanove, A. J. 2009. A simple cipher governs DNA recognition by tal effectors. Science, 326: 1501.

Murlidharan, G., Samulski, R. J. and Asokan, A. 2014. Biology of adeno-associated viral vectors in the central nervous system. Frontiers in Molecular Neuroscience, 7: 76.

Neumann, E., Schaefer-Ridder, M., Wang, Y. and Hofschneider, P. H. 1982. Gene transfer into mouse lyoma cells by electroporation in high electric fields. The EMBO Journal, 1: 841-845.

Nihongaki, Y., Kawano, F., Nakajima, T. and Sato, M. 2015. Photoactivatable CRISPR-Cas9 for optogenetic genome editing. Nature Biotechnology, 33: 755-760.

Nishiyama, J., Mikuni, T. and Yasuda, R. 2017. Virus-mediated genome editing via homology-directed repair in mitotic and postmitotic cells in mammalian brain. Neuron, 96: 755-768 e755.

Niu, Y. Y., Shen, B., Cui, Y. Q., Chen, Y. C., Wang, J. Y., Wang, L., et al. 2014. Generation of gene-modified cynomolgus monkey via Cas9/rna-mediated gene targeting in one-cell embryos. Cell, 156: 836-843.

Nowakowski, T. J., Bhaduri, A., Pollen, A. A., Alvarado, B., Mostajo-Radji, M. A., Di Lullo, E., et al. 2017. Spatiotemporal gene expression trajectories reveal developmental hierarchies of the human cortex. Science, 358: 1318-1323.

O'Geen, H., Ren, C., Nicolet, C. M., Perez, A. A., Halmai, J., Le, V. M., et al. 2017. Dcas9-based epigenome editing suggests acquisition of histone methylation is not sufficient for target gene repression. Nucleic Acids Research, 45: 9901-9916.

Pan, Y., Yang, J., Luan, X., Liu, X., Li, X., Yang, J., et al. 2019. Near-infrared upconversion-activated CRISPR-Cas9 system: A remote-controlled gene editing platform. Science Advances, 5: eaav7199.

Panoutsopoulos, A. A. 2020. Organoids, assembloids, and novel biotechnology: Steps forward in developmental and disease-related neuroscience. Neuroscientist: 1073858420960112.

Pardo, B., Gomez-Gonzalez, B. and Aguilera, A. 2009. DNA repair in mammalian cells: DNA double-strand break repair: How to fix a broken relationship. Cellular and Molecular Life Sciences, 66: 1039-1056.

Park, H., Oh, J., Shim, G., Cho, B., Chang, Y., Kim, S., et al. 2019. In vivo neuronal gene editing via CRISPR-Cas9 amphiphilic nanocomplexes alleviates deficits in mouse models of alzheimer''s disease. Nature Neuroscience, 22: 524-528.

Patton, E. E. and Zon, L. I. 2001. The art and design of genetic screens: Zebrafish. Nature Reviews Genetics, 2: 956-966.

Platt, R. J., Chen, S., Zhou, Y., Yim, M. J., Swiech, L., Kempton, H. R., et al. 2014. CRISPR-Cas9 knockin mice for genome editing and cancer modeling. Cell, 159: 440-455.

Pollen, A. A., Nowakowski, T. J., Shuga, J., Wang, X., Leyrat, A. A., Lui, J. H., et al. 2014. Low-coverage single-cell mrna sequencing reveals cellular heterogeneity and activated signaling pathways in developing cerebral cortex. Nature Biotechnology, 32: 1053-1058.

Qi, L. S., Larson, M. H., Gilbert, L. A., Doudna, J. A., Weissman, J. S., Arkin, A. P., et al. 2013. Repurposing crispr as an rna-guided platform for sequence-specific control of gene expression. Cell, 152: 1173-1183.

Quadrato, G. and Arlotta, P. 2017. Present and future of modeling human brain development in 3d organoids. Current Opinion in Cell Biology, 49: 47-52.

Ran, F. A., Cong, L., Yan, W. X., Scott, D. A., Gootenberg, J. S., Kriz, A. J., et al. 2015. In vivo genome editing using staphylococcus aureus Cas9. Nature, 520: 186-191.

Reardon, S. 2016. Welcome to the CRISPR zoo. Nature, 531: 160-163.

Rees, H. A. and Liu, D. R. 2018. Base editing: Precision chemistry on the genome and transcriptome of living cells. Nature Reviews Genetics, 19: 770-788.

Rubio, A., Luoni, M., Giannelli, S. G., Radice, I., Iannielli, A., Cancellieri, C., et al. 2016. Rapid and efficient CRISPR/Cas9 gene inactivation in human neurons during human pluripotent stem cell differentiation and direct reprogramming. Scientific Reports, 6: 37540.

Russo, S. J. and Nestler, E. J. 2013. The brain reward circuitry in mood disorders. Nature Reviews Neuroscience, 14: 609-625.

Ryu, J. Y., Won, E. J., Lee, H. A. R., Kim, J. H., Hui, E., Kim, H. P., et al. 2020. Ultrasound-activated particles as CRISPR/Cas9 delivery system for androgenic alopecia therapy. Biomaterials, 232: 119736.

Safari, F., Hatam, G., Behbahani, A. B., Rezaei, V., Barekati-Mowahed, M., Petramfar, P., et al. 2020. CRISPR system: A high-throughput toolbox for research and treatment of parkinson's disease. Cellular and Molecular Neurobiology, 40: 477-493.

Saito, T. and Nakatsuji, N. 2001. Efficient gene transfer into the embryonic mouse brain using in vivo electroporation. Developmental Biology, 240: 237-246.

Sander, J. D. and Joung, J. K. 2014. CRISPR-Cas systems for editing, regulating and targeting genomes. Nature Biotechnology, 32: 347-355.

Sandoval, A., Jr., Elahi, H. and Ploski, J. E. 2020. Genetically engineering the nervous system with CRISPR-Cas. eNeuro 7.

Sasaki, E., Suemizu, H., Shimada, A., Hanazawa, K., Oiwa, R., Kamioka, M., et al. 2009. Generation of transgenic non-human primates with germline transmission. Nature, 459: 523-527.

Sessa, A., Mao, C. A., Hadjantonakis, A. K., Klein, W. H. and Broccoli, V. 2008. Tbr2 directs conversion of radial glia into basal precursors and guides neuronal amplification by indirect neurogenesis in the developing neocortex. Neuron, 60: 56-69.

Shah, A. N., Davey, C. F., Whitebirch, A. C., Miller, A. C. and Moens, C. B. 2015. Rapid reverse genetic screening using CRISPR in zebrafish. Nature Methods, 12: 535-540.

Shinmyo, Y., Tanaka, S., Tsunoda, S., Hosomichi, K., Tajima, A. and Kawasaki, H. 2016. CRISPR/Cas9-mediated gene knockout in the mouse brain using in utero electroporation. Scientific Reports, 6: 20611.

Shinmyo, Y., Terashita, Y., Dinh Duong, T. A., Horiike, T., Kawasumi, M., Hosomichi, K., et al. 2017. Folding of the cerebral cortex requires cdk5 in upper-layer neurons in gyrencephalic mammals. Cell Reports, 20: 2131-2143.

Shmakov, S., Smargon, A., Scott, D., Cox, D., Pyzocha, N., Yan, W., et al. 2017. Diversity and evolution of class 2 CRISPR-Cas systems. Nature Reviews Microbiology, 15: 169-182.

Soldner, F., Stelzer, Y., Shivalila, C. S., Abraham, B. J., Latourelle, J. C., Barrasa, M. I., et al. 2016. Parkinson-associated risk variant in distal enhancer of alpha-synuclein modulates target gene expression. Nature, 533: 95-99.

St Johnston, D. 2002. The art and design of genetic screens: Drosophila melanogaster. Nature Reviews Genetics, 3: 176-188.

Staahl, B. T., Benekareddy, M., Coulon-Bainier, C., Banfal, A. A., Floor, S. N., Sabo, J. K., et al. 2017. Efficient genome editing in the mouse brain by local delivery of engineered Cas9 ribonucleoprotein complexes. Nature Biotechnology, 35: 431-434.

Straub, C., Granger, A. J., Saulnier, J. L. and Sabatini, B. L. 2014. Crispr/Cas9-mediated gene knock-down in post-mitotic neurons. PloS One, 9: e105584.

Suberbielle, E., Sanchez, P. E., Kravitz, A. V., Wang, X., Ho, K., Eilertson, K., et al. 2013. Physiologic brain activity causes DNA double-strand breaks in neurons, with exacerbation by amyloid-beta. Nature Neuroscience, 16: 613-621.

Sullivan, P. F., Daly, M. J. and O'Donovan, M. 2012. Genetic architectures of psychiatric disorders: The emerging picture and its implications. Nature Reviews Genetics, 13: 537-551.

Swiech, L., Heidenreich, M., Banerjee, A., Habib, N., Li, Y., Trombetta, J., et al. 2015. In vivo interrogation of gene function in the mammalian brain using CRISPR-Cas9. Nature Biotechnology, 33: 102-106.

Tabata, H. and Nakajima, K. 2001. Efficient in utero gene transfer system to the developing mouse brain using electroporation: Visualization of neuronal migration in the developing cortex. Neuroscience, 103: 865-872.

Tarjan, D. R., Flavahan, W. A. and Bernstein, B. E. 2019. Epigenome editing strategies for the functional annotation of ctcf insulators. Nature Communications, 10: 4258.

Tavano, S., Taverna, E., Kalebic, N., Haffner, C., Namba, T., Dahl, A., et al. 2018. Insm1 induces neural progenitor delamination in developing neocortex via downregulation of the adherens junction belt-specific protein plekha7. Neuron, 97: 1299-1314.

Thomas, K. R. and Capecchi, M. R. 1987. Site-directed mutagenesis by gene targeting in mouse embryo-derived stem cells. Cell, 51: 503-512.

Thyme, S. B., Pieper, L. M., Li, E. H., Pandey, S., Wang, Y., Morris, N. S., et al. 2019. Phenotypic landscape of schizophrenia-associated genes defines candidates and their shared functions. Cell, 177: 478-491 e420.

Tian, R., Gachechiladze, M. A., Ludwig, C. H., Laurie, M. T., Hong, J. Y., Nathaniel, D., et al. 2019. CRISPR interference-based platform for multimodal genetic screens in human ipsc-derived neurons. Neuron, 104: 239-255 e212.

Tsunekawa, Y., Terhune, R. K., Fujita, I., Shitamukai, A., Suetsugu, T. and Matsuzaki, F. 2016. Developing a de novo targeted knock-in method based on in utero electroporation into the mammalian brain. Development, 143: 3216-3222.

Tu, Z., Zhao, H., Li, B., Yan, S., Wang, L., Tang, Y., et al. 2019. CRISPR/Cas9-mediated disruption of shank3 in monkey leads to drug-treatable autism-like symptoms. Human Molecular Genetics, 28: 561-571.

van Gent, D. C. and van der Burg, M. 2007. Non-homologous end-joining, a sticky affair. Oncogene, 26: 7731-7740.

Walantus, W., Castaneda, D., Elias, L. and Kriegstein, A. 2007. In utero intraventricular injection and electroporation of e15 mouse embryos. Journal of Visualized Experiments, 239. DOI: 10.3791/239.

Wang, C., Najm, R., Xu, Q., Jeong, D. E., Walker, D., Balestra, M. E., et al. 2018a. Gain of toxic apolipoprotein e4 effects in human ipsc-derived neurons is ameliorated by a small-molecule structure corrector. Nature Medicine, 24: 647-657.

Wang, H., Xu, X., Nguyen, C. M., Liu, Y., Gao, Y., Lin, X., et al. 2018b. CRISPR-mediated programmable 3d genome positioning and nuclear organization. Cell, 175: 1405-1417 e1414.

Wang, H., Yang, H., Shivalila, C. S., Dawlaty, M. M., Cheng, A. W., Zhang, F., et al. 2013. One-step generation of mice carrying mutations in multiple genes by CRISPR/Cas-mediated genome engineering. Cell, 153: 910-918.

Wang, X., Cao, C., Huang, J., Yao, J., Hai, T., Zheng, Q., et al. 2016. One-step generation of triple gene-targeted pigs using CRISPR/Cas9 system. Scientific Reports, 6: 20620.

Wood, A. J., Lo, T. W., Zeitler, B., Pickle, C. S., Ralston, E. J., Lee, et al. 2011. Targeted genome editing across species using zfns and talens. Science, 333: 307.

Xia, H., Mao, Q., Paulson, H. L. and Davidson, B. L. 2002. Sirna-mediated gene silencing in vitro and in vivo. Nature Biotechnology, 20: 1006-1010.

Xing, L., Kalebic, N., Namba, T., Vaid, S., Wimberger, P. and Huttner, W. B. 2020. Serotonin receptor 2a activation promotes evolutionarily relevant basal progenitor proliferation in the developing neocortex. Neuron, 108: 1113-1129 e1116.

Xu, G. L. and Bestor, T. H. 1997. Cytosine methylation targetted to pre-determined sequences. Nature Genetics, 17: 376-378.

Yang, H., Wang, H., Shivalila, C. S., Cheng, A. W., Shi, L. and Jaenisch, R. 2013. One-step generation of mice carrying reporter and conditional alleles by CRISPR/Cas-mediated genome engineering. Cell, 154: 1370-1379.

Yang, S., Chang, R., Yang, H., Zhao, T., Hong, Y., Kong, H. E., et al. 2017. CRISPR/Cas9-mediated gene editing ameliorates neurotoxicity in mouse model of huntington's disease. Journal of Clinical Investigation, 127: 2719-2724.

Yang, W., Liu, Y., Tu, Z., Xiao, C., Yan, S., Ma, X., et al. 2019. CRISPR/Cas9-mediated pink1 deletion leads to neurodegeneration in rhesus monkeys. Cell Research, 29: 334-336.

Yoshimatsu, S., Okahara, J., Sone, T., Takeda, Y., Nakamura, M., Sasaki, E., et al. 2019. Robust and efficient knock-in in embryonic stem cells and early-stage embryos of the common marmoset using the CRISPR-Cas9 system. Scientific Reports, 9: 1528.

Yu, B., Yuan, B., Dai, J. K., Cheng, T. L., Xia, S. N., He, L. J., et al. 2020. Reversal of social recognition deficit in adult mice with mecp2 duplication via normalization of mecp2 in the medial prefrontal cortex. Neuroscience Bulletin, 36: 570-584.

Yu, M., Sun, X., Tyler, S. R., Liang, B., Swatek, A. M., Lynch, T. J., et al. 2019. Highly efficient transgenesis in ferrets using CRISPR/Cas9-mediated homology-independent insertion at the rosa26 locus. Scientific Reports, 9: 1971.

Zalatan, J. G., Lee, M. E., Almeida, R., Gilbert, L. A., Whitehead, E. H., La Russa, M., et al. 2015. Engineering complex synthetic transcriptional programs with CRISPR RNA scaffolds. Cell, 160: 339-350.

Zhang, S., Shen, J., Li, D. and Cheng, Y. 2021. Strategies in the delivery of Cas9 ribonucleoprotein for CRISPR/Cas9 genome editing. Theranostics, 11: 614-648.

Zhang, W., Aida, T., Del Rosario, R. C. H., Wilde, J. J., Ding, C., Zhang, X., et al. 2020. Multiplex precise base editing in cynomolgus monkeys. Nature Communications, 11: 2325.

Zhang, W., Wan, H., Feng, G., Qu, J., Wang, J., Jing, Y., et al. 2018. Sirt6 deficiency results in developmental retardation in cynomolgus monkeys. Nature, 560: 661-665.

Zheng, Y., Shen, W., Zhang, J., Yang, B., Liu, Y. N., Qi, H., et al. 2018. CRISPR interference-based specific and efficient gene inactivation in the brain. Nature Neuroscience, 21: 447-454.

Zhou, H., Liu, J., Zhou, C., Gao, N., Rao, Z., Li, H., et al. 2018. In vivo simultaneous transcriptional activation of multiple genes in the brain using CRISPR-dcas9-activator transgenic mice. Nature Neuroscience, 21: 440-446.

Zhou, W., Brown, W., Bardhan, A., Delaney, M., Ilk, A. S., Rauen, R. R., et al. 2020. Spatiotemporal control of CRISPR/Cas9 function in cells and zebrafish using light-activated guide rna. Angewandte Chemie International Edition in English, 59: 8998-9003.

Zhou, Y., Sharma, J., Ke, Q., Landman, R., Yuan, J., Chen, H., et al. 2019. Atypical behaviour and connectivity in shank3-mutant macaques. Nature, 570: 326-331.

Zhu, X. X., Zhong, Y. Z., Ge, Y. W., Lu, K. H. and Lu, S. S. 2018. CRISPR/Cas9-mediated generation of guangxi bama minipigs harboring three mutations in alpha-synuclein causing parkinson's disease. Scientific Reports, 8: 12420.

Zuckermann, M., Hovestadt, V., Knobbe-Thomsen, C. B., Zapatka, M., Northcott, P. A., Schramm, K., et al. 2015. Somatic CRISPR/Cas9-mediated tumour suppressor disruption enables versatile brain tumour modelling. Nature Communications, 6: 7391.

Zuris, J. A., Thompson, D. B., Shu, Y., Guilinger, J. P., Bessen, J. L., Hu, J. H., et al. 2015. Cationic lipid-mediated delivery of proteins enables efficient protein-based genome editing in vitro and in vivo. Nature Biotechnology, 33: 73-80.

# Non-viral Alternatives to Delivery of CRISPR-Cas Gene Editing Components

**Kate Senger and Benjamin Haley***

Department of Molecular Biology, Genentech, Inc., 1 DNA Way,
South San Francisco, CA 94080, USA

## Introduction and general considerations

While already a mainstay in basic research laboratories, an ever-evolving suite of molecular technologies, fueled by the application of CRISPR-Cas for eukaryotic genome engineering, has increased the potential for success of gene and cell therapies in the clinic. However, to achieve a desired gene editing outcome, each element of the CRISPR-Cas system must traverse a gauntlet of biological membranes, organelles, and immunological defenses in order to reach the relevant compartment within a target cell type, be it through *in vitro*, *ex vivo*, or *in vivo* delivery. The composition of a CRISPR-Cas delivery system must therefore account for the unique parameters of each cell type, the duration of time needed to produce the gene edit, and the distinct barriers expected to impede entry of the editing molecules. Options for delivery are plentiful when working with immortalized cells *in vitro*. As one moves into primary cell/*ex vivo* applications or animal models, however, high efficiencies become harder to achieve, requiring more specialized agents for reagent stabilization, tissue selectivity, and/or reduced immunogenicity.

Cells in culture are, in theory, the least complicated to modify. The most challenging barriers to be surmounted are the plasma membrane, the acidic endosomal compartment, and the nuclear membrane. These membranes can be breached using a needle, as in the case of microinjection, or by using electrical pulses, which create transient pores. Shear forces can temporarily disrupt cell membranes as well. Encasing CRISPR agents in lipid-like particles, which become transferred to acidic endosomes upon cellular uptake, requires an escape mechanism for their contents to reach the cytosol. All of the above must occur while avoiding intra- or extra-

* Corresponding author: benjamih@gene.com

cellular nucleases, proteases, and immune sensors. Chemical modifications may be added to DNA oligonucleotides, mRNAs, and Cas-guiding RNAs to help confront these limitations. The editing reagents themselves may exhibit cellular toxicity, as is expected with exogenous DNA or RNA, or genotoxicity, induced by unintended DNA damage from Cas-nucleases.

When *in vivo* delivery is desired, all of the above factors still apply and are compounded by new considerations: circumvention of natural barriers like skin, epithelium, endothelium, or the blood brain barrier. The capillaries within certain tissues contain pores, or "fenestrae", between the endothelial cells that comprise them. Fenestrated tissues, such as the liver, are therefore more accessible to gene editing reagents with systemic injection. The size and shape of the injected particles can affect the rate of clearance by the body, and can change over time due to agglomeration. Achieving tissue or cell-type specificity is another challenge, which can be hindered if the deliverables bind non-specifically to endogenous proteins, or to unwanted cell types. Tissue selectivity may also be achieved by altering the charge of particles encompassing the gene editing elements. Direct injection into the cell or tissue of interest may bypass some of these limitations. Regardless of the delivery method, triggering the adaptive immune response is a risk to be avoided, and, unlike *in vitro* or *ex vivo* applications, may be affected by existing or *de novo* adaptive immune responses. Finally, clearance mechanisms by the liver, kidney, and resident phagocytes can limit the durability of gene editing effector function.

This chapter surveys several established and emerging non-viral methods for CRISPR-Cas delivery that have shown success in cultured cells and animal models. These methods have been broadly categorized as physical, nanoparticle-enabled, or protein transduction-mediated.

# Non-viral delivery of CRISPR-Cas gene editing reagents

## Gene editing modalities

Central to any gene modification process is the selection of an appropriate tool set used to generate a desired editing outcome. Aside from varied precision, genomic targeting space, and efficiency, all CRISPR-Cas-based systems, particularly those derived from the more commonly used Type-II CRISPR lineage, require a minimum of two components: a Cas protein and a guide RNA (gRNA) (Chylinski et al. 2014). Some of these, like the Cas nucleases or base editors, are designed to induce permanent genetic changes (Anzalone et al. 2020). On the other hand, Cas-based transcriptional activators and suppressors, or Cas-related technology designed to edit RNA, are likely to have transient effects in most cells when delivered through non-viral, non-integrating means (Engreitz et al. 2019). For creating discrete single nucleotide polymorphism (SNP) edits or when insertion of a replacement or reporter gene is preferred, a third modality, a DNA homology donor cassette, is added along with the Cas protein and gRNA (Nelson and Gersbach 2015). Because these tools often combine RNA, DNA, and protein reagents into a single mixture, a host of delivery approaches may need to be evaluated in order to achieve the highest rates of editing. The molecular characteristics for the three principal categories of CRISPR-Cas gene editing modalities are highlighted below:

## Cas proteins

The primary effector of CRISPR-Cas-mediated gene editing is any one of several Cas proteins. Of those currently in use, Cas9 derived from *Streptococcus pyogenes* (and to a lesser extent *Staphylococcus aureus)* has been the most broadly applied and adapted for either nuclease-based gene perturbation, base editing, or as a transcription modifier (Doudna 2020). More recently, alternative Cas enzymes such as Cas12a or Cas13 have been engineered for DNA and RNA editing, respectively (Engreitz et al. 2019). For *in vitro* and *ex vivo* applications, recombinant Cas9 complexed with an appropriate synthetic or *in vitro*-transcribed gRNA (hereafter termed Cas9-ribonucleoprotein or RNP) has proven to be effective for gene knockout or homology-directed repair (Kim et al. 2014). Alternatively, Cas expression plasmids, which may co-express gRNAs, have found wide-spread application, although the slower kinetics of plasmid turnover relative to Cas9-RNP may elicit both innate immune responses (to the plasmid DNA) and, potentially, more off-target effects though lingering Cas expression (Hsu et al. 2013). Engineered variants of Cas9 have been designed to reduce off-target DNA recognition or cutting, but this may be at the expense of on-target activity for some sites (Schmid-Burgk et al. 2020). Lastly, Cas expression from *in vitro*-transcribed mRNA has been applied largely for *in vivo* gene editing (Kowalski et al. 2019). Cas proteins encoded by mRNA may be particularly effective for *in vivo* purposes owing to the development of several nanoparticle-based technologies. These technologies enable (1) the expression of Cas proteins that may be difficult to produce or purify as recombinant material, (2) tissue-selective delivery, (3) dampened innate immune sensing through chemical modification of the mRNA nucleotides (Warren et al. 2010), and (4) a reduced likelihood of antibody-mediated neutralization of circulating *S. pyogenes* or *S. aureus* Cas9 protein (Charlesworth et al. 2019).

## Cas guide RNAs

Each Cas protein has distinct gRNA scaffold and gene targeting requirements (Moon et al. 2019). In turn, this will influence the genomic sequences that can be modified with each Cas variant. It will also impact experimental design, which could be limited by the size and stability of the individual gRNA molecules. These latter two elements should be weighed heavily in the context of CRISPR-Cas delivery *in vitro* or *in vivo*.

Regardless of the Cas variant used, the gRNA is comprised of primarily single-stranded RNA molecules with exposed 5′ and 3′ termini. This, in turn, leaves the gRNAs susceptible to RNase degradation. The two-part nature of some guides, for example those natively used by the Type-II Cas proteins (i.e. Cas9), requires those RNA strands to hybridize with one another in order for Cas to function. This technical incumbrance has been solved by linking the two strands into a now-termed single-guide RNA (sgRNA) (Jinek et al. 2012). sgRNAs produced through chemical synthesis affords the opportunity to introduce nucleotide modifications that can both stabilize the gRNAs against RNase attack while also imparting improved function

(Hendel et al. 2015). Production of sgRNAs through *in vitro* transcription is possible, but their use is not recommended due to the potential for activating deleterious innate immune pathways (Kim et al. 2018, Wienert et al. 2018). Both the stability and size of the sgRNAs is of minimal concern when this component is delivered by way of plasmid DNA or viral vectors (Wang et al. 2020).

*DNA homology donors*

The least trivial application of CRISPR-Cas9 for gene editing is precision genome repair via homologous recombination (HR). This requires (near) simultaneous delivery of the Cas protein and gRNA as well as a homologous donor DNA containing the edit of choice into the nucleus of a target cell. The homology donor can take on any number of forms, depending on target site accessibility, the size or complexity of the intended genome modification, and host cell type sensitivity to foreign DNA.

For introducing or removing SNPs and other small gene edits (<10 bp), single-strand oligodeoxynucleotide (ssODN) donor DNAs, up to 200 nt in length, have been used successfully *in vitro* and *in vivo* (Chen et al. 2011b), and these may contain subtle modifications to improve stability (Renaud et al. 2016). Long (>200 bp) single-stranded DNA (ssDNA) has also been used *in vitro*, but producing these molecules at scales needed for *in vivo* editing brings significant challenges (Miura et al. 2015).

For larger genome edits, such as full gene insertion, plasmid-based or PCR-generated donor DNA is often employed (Roth et al. 2018). In this context, the DNA to be inserted is generally flanked by at least 300 bp of homologous DNA to achieve reasonable levels of efficiency. Despite the broad utility of this approach, delivery of circular or linearized DNA may induce innate immune signaling and/or the plasmid can integrate randomly into the genome. In circumstances where plasmid DNA is too toxic or delivery is inefficacious, the donor DNA can be introduced by way of adeno-associated-virus (AAV) mediated infection (Wang et al. 2020). AAV is a single-stranded DNA virus that can be engineered to contain a homology donor cassette. Various studies have found that combining AAV-based donor DNAs with recombinant Cas9-RNPs can enhance the gene editing frequency above the combination of Cas9-RNP and plasmid-based homology cassettes (Gaj et al. 2017).

## Delivery mechanisms

Setting aside the many roadblocks that affect the survival of editing reagents in the human body when introduced systemically (e.g. stability in the bloodstream, metabolism by the liver, excretion by the kidneys), any effective CRISPR-Cas delivery technology must be able to bypass the plasma membrane. Several strides have been made in circumventing this barrier and allowing these molecules to cross over. This section categorizes delivery methods as being either physical, nanoparticle-enabled, or reliant on transduction of a protein component. Physical methods include injection, electroporation, and application of shear forces to cells through hydrodynamic injection or cell "squeezing" techniques. These methods are likely to work with the broadest range of CRISPR-Cas modalities, since they are constrained less by the size and charge of the deliverables. Nanoparticle-enabled methods cover an evolving variety of formulations, many of which are lipid-based, and some of which are built upon a metallic core. Protein-only transduction covers cell-penetrating peptides, and a method that combines sodium chloride with a small molecule to induce macropinocytosis. A summary of the advantages and disadvantages of each method is shown in Table 1.

## Physical delivery methods

**Hydrodynamic injection:** Hydrodynamic injection is one of the easiest methods for administering plasmid-based expression vectors to hepatocytes in mice, rats, and other animals (Herweijer and Wolff 2007). An early example of this process was demonstrated by the preparation of a 1 mL solution carrying 100 µg of a reporter plasmid followed by its delivery within a short time window (30 seconds) into the hepatic portal vein of a mouse (Budker et al. 1996). The reporter gene was expressed by hepatocytes throughout the liver. This was an improvement over direct plasmid

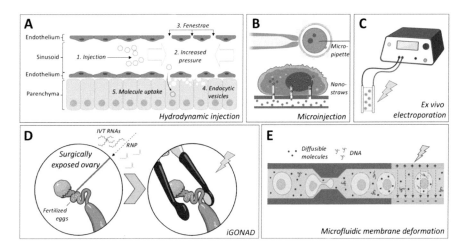

**Figure 1.** Physical methods of CRISPR-Cas delivery. A) Mechanism of pDNA uptake by tissues after hydrodynamic injection. A large (e.g. 10% of body weight) volume of solution containing pDNA (open circles) is rapidly (within 5-7 sec) injected into the tail vein or into the local vasculature of the tissue of interest (1). The resulting pressure (2). Enhances the fenestration between endothelial cells of the capillaries (3). Membrane pores, and/or large endocytic vesicles appear in the parenchyma (4). Through which pDNA molecules gain entry (5). B) Microinjection techniques use thin capillaries or needles to rupture cell membranes and deposit gene editing reagents directly inside cells. C) Cell lines and purified primary cell populations can be edited in a high-throughput fashion by applying electrical pulses in cuvettes, which opens transient pores in cell membranes. D) Tissues may be electroporated *in situ* after local injection of gene editing reagents. In the iGONAD technique, an ovary bearing recently fertilized eggs is surgically exposed. Gene editing reagents are injected into the oviduct, and then an electrical pulse is delivered using a specialized set of electrode-tipped tweezers. The ovary is returned to its original location and the incision sutured. E) Pushing cells at high speed through openings smaller than the cell diameter can create temporary membrane pores, through which molecules from the surrounding environment can diffuse. Some molecules efficiently diffuse into the cells, while others, such as DNA, require a subsequent electrical pulse for efficient delivery.

**Table 1:** Pros and cons of common delivery approaches

| Category | Delivery platform | Example use | CRISPR-Cas modality | Advantages | Drawbacks |
|---|---|---|---|---|---|
| **Physical** | Hydrodynamic injection | *In vivo* | pDNA | • Ease of execution in animal models | • Limited tissue specificity |
| | Zygote microinjection | *Ex vivo* | pDNA, mRNA, RNP | • Greatly speeds creation of genetically engineered animal models | • Low throughput<br>• Specialized skillset and equipment required<br>• Genetic mosaicism |
| | Electroporation | *In vivo, ex vivo, in vitro, in ovo* | pDNA, mRNA, RNP | • High throughput<br>• Ease of execution | • Impacts to cell viability and transcriptional profiles<br>• Some cell types more amenable to treatment than others |
| | Microfluidic membrane deformation | *Ex vivo, in vitro* | RNP | • High cell viability<br>• Minimal perturbation of cell transcriptome | • Uptake depends on size and diffusion rates of deliverables |
| **Nanoparticle-enabled** | Lipid nanoparticles (LNPs) | *In vivo, ex vivo, in vitro* | pDNA, mRNA, RNP | • Tunable composition<br>• Successful track record in the clinic | • Corona formation<br>• Challenges achieving tissue specificity<br>• Endosomal escape |
| **Protein transduction** | Cell-penetrating peptides (CPPs) | *In vivo, ex vivo, in vitro* | pDNA, RNP | • Small in size (<30 aa)<br>• Low toxicity at appropriate doses | • Challenges achieving tissue specificity<br>• Potentially immunogenic |
| | Induced transduction by osmocytosis and propanebetaine (iTOP) | *Ex vivo, in vitro* | RNP | • Transduces native proteins<br>• Allows control over protein levels delivered | • Negatively affects cell proliferation if not optimized |

injection into the liver parenchyma, which resulted in plasmid uptake only by cells proximal to the injection site (Hickman et al. 1994). Soon after, it was discovered that the plasmid solution could be injected via the tail vein resulting in efficient uptake and reporter gene expression by the liver (Liu et al. 1999, Zhang et al. 1999). Convenient access and high efficiency delivery has made the tail vein a preferred route of administration through a method often termed hydrodynamic tail vein (HTV) injection.

The parameters of HTV injection are straightforward. The molecule to be injected is first suspended in a sterile buffer. The injection volume is generally 10% of the animal's weight (2 mL volume for a 20g mouse), and the entire volume must be injected within roughly 5-7 seconds for maximal efficacy. The injected fluid temporarily enlarges the cardiac ventricles and pressurizes the inferior vena cava (Crespo et al. 2005, Zhang et al. 2004). This causes the solution to back-fill the hepatic vein (Crespo et al. 2005), making the liver visibly swell and blanch (Zhang et al. 2004, 1999). The pressure opens fenestrae of the liver microvasculature, leading to the formation of membrane pores (Zhang et al. 2004) and/or vesicles within hepatocytes approaching the diameter of nuclei (~10 μm) in size (Budker et al. 2006, Crespo et al. 2005), which are thought to comprise the mechanism of cellular plasmid DNA (pDNA) uptake (Wolff and Budker 2005) (Figure 1A). The aftereffects of HTV injection include transient liver damage, as evidenced by a temporary increase in serum alanine aminotransferase (ALT) post-injection (Budker et al. 2006, Liu et al. 1999, Rossmanith et al. 2002, Zhang et al. 1999), and increased hepatocyte mitosis (Budker et al. 2006). Despite this, multiple injections, spaced two weeks apart, have been well tolerated (Crespo et al. 2005). The efficiency of hepatocyte uptake of pDNA is typically under 50%, and ranges from 5% to 70% (Wolff and Budker 2005).

Many research groups have used HTV injection to introduce pDNA-encoded Cas9 and gRNAs to the liver in order to create or disrupt genetic diseases. In one such example, HTV-mediated delivery of CRISPR-Cas9 expression plasmids was used to disable tumor suppressor genes, with the goal of speeding up the creation of cancer models (Xue et al. 2014). This enabled simultaneous targeting of *Pten* and *Trp53*, achieving 4% and 6.4% indel frequencies of each gene in the liver, respectively. Three months post-injection, all injected mice developed liver tumors, illustrating the ease with which cancer-causing mutations can be introduced by this method. Separately, HTV injection has been employed to create a mouse model of fibrolamellar hepatocellular carcinoma (FL-HCC), a rare and highly fatal liver cancer (Engelholm et al. 2017). Here, two sites in the genome were targeted by Cas9 and unique gRNAs to induce a large (~400 kb) deletion of mouse chromosome 8, leaving behind a fusion of genes *Dnajb1* and *Prkaca* – a feature also found in FL-HCC patients, though not known to be causative for the disease. Of the injected mice, 80% developed oncogenic lesions with features similar to human FL-HCC, establishing a new model system for the study of this rare cancer.

CRISPR-Cas9-generated knockouts of certain genes can engender cells with a growth advantage, leading to the spreading of a genotype within the liver. This has been utilized by one group as a way to create a somatic liver knockout, or "SLiK" undefined. In this study, the authors start with mice lacking the fumarylacetoacetate hydrolase (*Fah*) gene, which is essential for tyrosine catabolism. Disabling the *Hpd* gene by HTV injection of plasmid-encoded Cas9/gRNAs confers resistance to the hepatotoxicity caused by loss of *Fah*, and gives those cells a growth advantage over cells with intact *Hpd*. The advantage is so large that the *Hpd*-edited hepatocytes almost completely replace the liver in as little as two months. HTV injection can also be used to deliver larger Cas9-fusion variants, for example, those expressing Cas9-base editor enzymes. A splice site mutation in the *Fah* gene and the resulting metabolic disorder was corrected in ~9.5% of recipients using a plasmid-encoded adenine base editor-Cas9 fusion/gRNA delivered by HTV injection in *Fah^{mut/mut}* mice (Song et al. 2020). This treatment partially restored normal *Fah* splicing and rescued weight loss in the animals.

The major limitation of HTV injection is its reduced effectiveness outside of hepatocytes. For example, various cell types in and around the liver, such as endothelial cells, Kupffer cells, and bile duct epithelium display modest, if any, plasmid uptake (Lewis and Wolff 2007). It is possible to deliver genes to extrahepatic tissues with HTV injection, but this occurs at a much-reduced efficiency (100-1000 fold) relative to liver (Liu et al. 1999). Better results may be obtained by injecting solutions of pDNA directly into the blood vessels that supply the organ of interest. For example, injecting the renal vein with a reporter plasmid can direct high levels of gene expression in the kidney (Kameda et al. 2004). Hydrodynamic injection of an anti-angiogenic gene expression vector into the carotid artery of rats bearing brain tumors led to vector expression by 10-30% of tumor cells, followed by tumor volume reduction and prolonged animal survival (Barnett et al. 2004).

In humans, the systemic injection of large volumes of fluid is not considered realistic or safe. Instead, balloon catheters may be directed to the organ of interest and inflated to create localized pressure prior to gene delivery (Kamimura et al. 2015). Yet another variation on the technique, hydrodynamic limb vein (HLV) injection, works through pDNA administration into a limb vein while simultaneously applying a tourniquet to block blood flow to or from the limb, resulting in reporter expression by skeletal muscle with up to 45% efficiency after a single injection, and sometimes up to 80% following multiple injections (Budker et al. 1998, Hagstrom et al. 2004). This method shows promise for clinical applications such as treating muscular dystrophy patients (Guen et al. 2020). While hydrodynamic injection is most commonly used with pDNA, it also has clear utility for delivering a range of other substances including dyes, siRNAs, peptides, proteins, antibodies, phage, polymers, and polystyrene microspheres (Kobayashi et al. 2001, 2004, Sebestyén et al. 2006, Zhang et al. 2004), suggesting it should be possible to deliver other CRISPR-Cas formats (preformed RNP, RNP-HDR template fusions, mRNA, nanoparticle) using this method.

**Microinjection:** Genetically-modified mice are essential tools for the study of developmental processes and disease. Until recently, the generation of engineered mouse models was a laborious and time-intensive process (Capecchi 2005). This included targeting the gene of interest in embryonic stem (ES) cells by homologous recombination, injecting the modified ES cells into a recipient blastocyst, generating chimeric mice, and mating their offspring. Following this procedure, it could take up to a year to have a working mouse model. The production of mice bearing multiple mutations was even more laborious, as intercrossing between single mutant mice was required. Now, CRISPR-Cas gene editing tools can be directly injected into fertilized mouse eggs, producing single edits, multiple simultaneous edits, and large (multi-kilobase) knock-ins. This is done in some labs by microinjection, a technique that uses a fine glass micropipette to inject materials into fertilized oocytes (Figure 1B).

Early studies aimed at creating "one-step" gene knock-out mice microinjected *in vitro* transcribed RNAs encoding Cas9 and gRNAs into fertilized eggs (Shen et al. 2013, Wang et al. 2013, Yang et al. 2013). These studies showed it was possible to disrupt one or more genes with high efficiency during early development, as well as produce knock-in, albeit with lower and varied efficiencies, through co-delivery of ssODNs or circular DNA templates. The genome modifications catalyzed by CRISPR-Cas9 do not always occur in a consistent and uniform fashion in each cell of the developing embryos, however, resulting in genetically mosaic animals: an outcome influenced by a wide variety of parameters. If injecting *in vitro* transcribed, Cas9-expressing mRNA, for example, there is a delay caused by protein translation, which in turn affects the timing of genome modification as the early embryonic cells divide. Cas9 catalytic activity, gRNA specificity, the presence of a repair template, any errors in the repair template, the stage at which the injection is performed, and the rapidity of development of the species of interest can all impact the rate of mosaicism. Compound, non-uniform insertion and deletion (i.e. indel) mutations near the Cas9-induced DNA break site can result, complicating the phenotypic analysis of the founder mice and their offspring. The extent of mosaicism possible is illustrated by a study in which a cocktail of two distinct *Tyrosinase* (*Tyr*)-specific gRNAs and a Cas9-expressing mRNA were injected into mouse embryos from either of two distinct dark-haired parental lines (i.e. C57BL/6N), and hair color phenotype was assessed in the founders (Yen et al. 2014). The injected embryos developed into albino, mixed pigment, and fully pigmented mice. Deep sequencing showed that a wide variety of *Tyr* alterations occurred in the pups, ranging from single base pair changes, to small (a few bp) and large (hundreds of bp) deletions, insertions, as well as more complex events.

CRISPR-mediated HR in early embryogenesis is a particularly challenging endeavor. Many groups have used CRISPR-Cas9 microinjection to knock-in small, 34 bp *loxP* sequences on either side of a genomic region of interest in an attempt to create conditional alleles (Bishop et al. 2016, Ma et al. 2017, Nakagawa et al. 2016, Yang et al. 2013). Early attempts at this strategy used simultaneous injection of Cas9, two sgRNAs (e.g. one for each *loxP* integration site), and two ssODNs encoding the *loxP* sites in the hope of generating a *loxP*-flanked (e.g. floxed) gene segment. In one example, a knock-in efficiency rate of 16% was reported for the *Mecp2* gene (Yang et al. 2013). However, this rate for the "two-donor floxing method" was not always widely reproducible (Gurumurthy et al. 2019), perhaps because the technique is prone to producing deletions or requires further optimization to make the outcome less variable. Since then, several advances in embryo editing by microinjection have increased the success of *loxP* site knock-in. For example, when attempting to simultaneously insert two *loxP* sites in the genome, the propensity for the sequence between two RNP cut sites to be deleted can be mitigated by sequential microinjection (Horii et al. 2017). In this study, the authors injected Cas9 protein along with the gRNA and ssODN specific for a single *loxP* site into zygotes. Once the embryos reached the 2-cell stage, the reagents corresponding to the second *loxP* site were injected. This allowed the first RNP cut site to be repaired before the second site was cut. The frequency of deletion with sequential injection was less than with simultaneous injection (by 0.55 to 0.63-fold), and the flox frequency was 1.5 to 6.8-fold higher. As might be expected, however, the sequential method was more damaging, producing lower numbers of viable embryos. More recently, long ssDNA donors, comprising both *loxP* sites, have been deployed for conditional mouse production, in a process termed "Easi-CRISPR", which has shown improved efficacy and quality of editing (fewer unwanted mutations at the target site) relative to the dual ssODN strategy (Quadros et al. 2017).

The timing of injection is another critical parameter for successful knock-in. The HR pathway is thought to be most active during the late S-G2 phase of the cell cycle (Hustedt and Durocher 2017). Various groups have taken advantage of this by synchronizing their cells of interest to G2 phase prior to CRISPR delivery, or by fusing Cas9 to Geminin, a protein that accumulates to high levels during phases S and G2 but is selectively degraded during mitosis (Gutschner et al. 2016, Lin et al. 2014, McGarry and Kirschner 1998, Yang et al. 2016). In one example, the authors observed that the G2 phase of 2-cell embryos was exceptionally long (10-12 hrs), and they proceeded to test whether high rates of knock-in could be achieved with injection at this stage (Gu et al. 2018). They obtained a 35% knock-in rate of a *Gata6-Halo* template at the 2-cell stage, in contrast with a 6.5% knock-in frequency at the single cell stage, suggesting a benefit to injecting later in development. A separate group claims it is possible to achieve high knock-in rates by injecting at the single cell stage, if one times it to a specific pronuclear stage (Abe et al. 2020). They obtained their best knock-in efficiencies (54-70%) at the PN3-PN4 stage, which corresponds to S phase. Unexpectedly, injection during PN4-PN5, which aligns with the S-G2 stage thought to be favorable to HDR, produced a low knock-in efficiency in this study.

Overall, a variety of factors make it difficult to conclude one microinjection technique is advantageous over another. The format, source, and concentration of gRNAs, Cas protein/mRNA, and template formats, and sites of integration vary greatly between studies. In one study using ssODNs to create floxed conditional alleles, mutations in *loxP* sites stemmed from errors in oligo synthesis, prompting the authors to suggest alternative providers (Bishop et al. 2016). In experiments that compared linear DNA (PCR product or restriction enzyme-digested plasmid) versus circular plasmid, the linear format was shown to give a higher rate of donor template integration (notably both on- and off-target), particularly if irrelevant vector sequences were cleaved off by restriction digestion beforehand (Yao et al. 2018). Coupling a biotinylated, PCR-derived donor DNA to Cas9 expressed as a fusion with streptavidin (Cas9-mSA) (Lim et al. 2013) has yielded an impressively high (66-95%) KI rate with injection at the 2-cell stage (Gu et al. 2018), though it is unknown whether using this modified Cas9 produces a similarly-high efficiency at the single cell stage. The concentration of the reagents used for gene editing has been observed, in some cases, to negatively impact embryonic development (Horii et al. 2017). Further, HR rates depend on the *locus* being targeted, the homology arm lengths of the template, and the skill of the person performing the microinjections. Microinjection systems have been engineered a piezoelectric actuator, a device that moves the capillary a very short distance at very high speed. This penetrates membranes with minimal cell distortion, yielding higher embryo survival rates (Abe et al. 2020). High-throughput microinjection systems in the form of nanoneedle or nanostraw arrays (Shalek et al. 2010, Wang et al. 2014, Xie et al. 2013) have shown usefulness with immortalized and primary cells, and may show further utility with embryos (Figure 1B). Given the highly technical and laborious aspects of microinjection, numerous labs have turned to electroporation for a higher throughput method requiring less manual skill.

**Electroporation:** Electroporation involves the application of an electrical field to tissues *in vivo*, or to cells suspended in a buffer, causing a transient or permanent increase in membrane permeability. This procedure results in nanometer-sized, water-filled pores that allow polypeptides and highly charged DNA and RNA molecules to pass through (Neumann et al. 1982). For decades, labs have routinely electroporated cells suspended in a cuvette as a supplement to lipid transfection to deliver DNA expression vectors or RNA molecules to primary cells and immortalized cell lines (Figure 1C). It is a high-yield method, allowing millions of cells to be modified at one time. Plasmids and mRNAs encoding Cas protein and gRNAs can be delivered by electroporation, as well as RNP. As described below, the technique can also be applied in a localized fashion to tissues *in vivo*.

The success of electroporation depends on multiple parameters including voltage, pulse space, pulse number and duration, temperature, and buffer composition. Determining the correct combination of parameters can be challenging, especially for primary cells. The downside of the technique is that it can result in substantial cell death caused either by the electrical pulse, or as a result of innate immune sensing of the editing tool, all of which necessitate careful optimization for each combination of cell/tissue type and gene editing modality.

Early studies that attempted to edit primary human immune cells by electroporating them with plasmids expressing Cas9 and associated gRNAs showed modest success (Mandal et al. 2014). Subsequent studies systematically optimized the electroporation parameters for delivering preformed Cas9-RNPs to primary mouse and human T cells and hematopoietic progenitor cells (Gundry et al. 2016, Oh et al. 2019, Seki and Rutz 2018). The power of the method is in its ability to modify millions (potentially billions) of cells simultaneously. As a result of this, electroporation has become critical in the clinic for the recent spate of edited cell therapies. Primary cells are collected from individual patients, electroporated with CRISPR-Cas reagents, and the successfully edited cells are either returned to the subject (autologous transfer) or donated to a recipient (allogeneic transfer). This method has been successfully used as part of the emerging field of edited T cell and hematopoietic stem cell therapies (Esensten et al. 2016). Chimeric antigen receptor (CAR)-expressing T cells, which have become an important tool in treating B cell malignancies, are commonly manufactured by transducing T cells with lentiviral or γ-retroviral vectors encoding the CAR. However, this method of delivery bears the risk of insertional oncogenesis, variable copy number integration, and variegated CAR gene expression, which may lead to high levels of tonic signaling and cell exhaustion (Zhang et al. 2020). These outcomes can be mitigated by directing CAR integration to the T-cell receptor alpha constant (TRAC) *locus* using a TRAC-specific RNP and AAV-encoded CAR transgene (Eyquem et al. 2017, MacLeod et al. 2017). This places the CAR under the control of a native T cell receptor (TCR) promoter, averting the exhausted cell phenotype, and also eliminating endogenous TCR expression. Furthermore, electroporation of RNPs targeting the TCRα and TCRβ constant region *loci*, combined with long dsDNA HR templates, has allowed researchers to replace the endogenous TCR with one capable of recognizing neoantigen-presenting cancer cells in a completely virus-free manner (Roth et al. 2018, Schober et al. 2019).

Electroporation is not limited to cells in a cuvette. Electrical pulses can be applied locally to tissues *in vivo* using electrode-tipped tweezers after injecting gene editing reagents. The concept of applying electric fields in this manner has been in use since at least the 1990s to deliver DNA expression vectors to skin, liver, and testes, among other tissues (Heller et al. 1996, Muramatsu et al. 1996, Nishi et al. 1996). The major advantage of *in vivo* electroporation is that edits can be created quickly, in virtually any genetic background, and at various stages of development or disease. Non-traditional model organisms can also be genetically modified. For example, *in vivo* electroporation has been used to create tissue-specific single and double gene knockouts in the tail of the axolotl, an amphibian known for its ability to regenerate amputated limbs. Injection of Cas9-RNP into the spinal cord of the animal, followed by localized electroporation of the tail, yields efficient (>90%) editing of genes implicated in the regenerative process (Albors and Tanaka 2015, Fei et al. 2016). Chick embryos have been edited in a similar fashion in studies of neural development. The advantage of studying oviparous animals is that their embryos can be surgically exposed and manipulated *in ovo* to achieve spatial and temporal alterations in gene expression during development. *In ovo* electroporation has been used for some time to deliver expression vectors to the embryonic brain and other tissues (Momose et al. 1999, Muramatsu et al. 1996). Plasmids encoding both Cas9 and requisite gRNAs are injected into the lumen of the neural tube of chick embryos, after which electrodes placed on either side of the injected area deliver a pulse, enabling gene edits during early development, which are examined for their phenotypic effects at later stages (Abu-Bonsrah et al. 2016).

Similarly, the rodent retina (Bakondi et al. 2016), brain (Shinmyo et al. 2016), prostate (Leibold et al. 2020), pancreas (Maresch et al. 2016), and oviduct (Takahashi et al. 2015, Ohtsuka et al. 2018) have been edited through *in vivo* electroporation. For example, in a rodent model of retinitis pigmentosa caused by a dominant gain-of-function mutation in the Rho gene (*Rho^{S334}*), subretinal injection of Cas9 and gRNA expression plasmids, followed by localized electroporation of the eye using Tweezertrodes, resulted in successful ablation of the *Rho^{S334}* allele and slowing of retinal degeneration (Bakondi et al. 2016). Plasmids encoding Cas9 and gRNAs, microinjected into the developing mouse brain *in utero,* followed by localized electroporation, have been used to simultaneously ablate multiple genes for rapid analysis during development (Shinmyo et al. 2016). The mouse prostrate has been edited to carry cancer-promoting mutations by injecting plasmid-encoded Cas9 and gRNAs directly into the surgically exposed prostate gland, followed by gland electroporation, generating what the study authors termed electroporation-based genetically engineered mouse models (EPO-GEMMs) (Leibold et al. 2020). Finally, localized injection of Cas9-RNP inside the ovaries of recently mated female mice, followed by local electroporation, is the basis of the Improved-Genome editing via Oviductal Nucleic Acids Delivery method (i-GONAD) (Ohtsuka et al. 2018, Takahashi et al. 2015) (Figure 1D). In this example, the authors compared mRNA-encoded Cas9 with Cas9-RNP, and obtained gene inactivation in 31% of G0 mice with the former, and 97% with the latter. In knock-in experiments using ssODNs, the i-GONAD method was able to achieve 49% editing efficiency (Ohtsuka et al. 2018). Drawbacks of *in vivo* electroporation include cell toxicity and lack of cell type specificity. Akin to the negative effects on cell viability seen with *in vitro* electroporation, *in vivo* electroporation can result in tissue damage and immune cell infiltrates if not optimized properly (Maresch et al. 2016). Electroporation tweezers also do not discriminate between specific cell types within the treatment area. Animals genetically engineered to express Cas9 under the control of cell- and tissue-specific promoters, combined with localized injection and electroporation of the gRNA component, may provide researchers with better control over the editing of desired cell types within an organ.

**Microfluidic membrane deformation.** Microfluidic membrane deformation is similar to electroporation in that it opens transient pores in cell membranes. This promising new method of delivery is less likely than electroporation to cause cellular toxicity, and it has the potential to be high-throughput. Microfluidic membrane deformation, or "cell squeezing", pressurizes cells, pushing them through an opening that is considerably smaller than the cell's diameter. The combination of compression and shear forces results in transient holes in the cell membrane that allow molecules to enter by passive diffusion (Lee et al. 2012, Sharei et al. 2013). The parameters governing its successful use are the diameter of the cells of interest, the dimensions of the constriction through which cells pass (length and width), the number of constrictions, the amount of pressure that is applied, buffer composition (e.g. hypotonic, causing cell swelling), and the properties of the molecule(s) to be delivered. The technique has worked for delivering carbon nanotubes, 15 nm gold nanoparticles, quantum dots, proteins, membrane-impermeable small molecules, Cas9-RNP, and siRNAs to a variety of cell lines, primary cells, and stem cells (Lee et al. 2012, Sharei et al. 2013, Szeto et al. 2015, Li et al. 2017, DiTommaso et al. 2018).

A drawback of this method is that certain molecules may not be able to effectively diffuse into squeezed cells. DNA is one such example. In order to enhance DNA uptake, cell squeezing has been combined with electroporation in a variation termed disruption-and-field-enhanced (DFE) delivery (Ding et al. 2017) (Figure 1E). During electroporation, DNA interacts with the electropermeabilized membrane, forming aggregates that are absorbed by the cytoplasm (Yarmush et al. 2014). These aggregates are thought to be transported to the nucleus by microtubule and actin networks that become activated in response to electroporation (Rosazza et al. 2016). With the DFE technique, cells are first constricted to induce pore formation, and then passed through an electrical field which reversibly ruptures the nuclear envelope and speeds DNA uptake. pDNA reaches the nucleus in as little as 1 hr with this variation on cell squeezing as assessed by green fluorescent protein (GFP) expression from a plasmid in HeLa cells. The speed of GFP expression produced by DFE matched that achieved with microinjection, and was several times faster than that produced by other methods, such as lipofection and conventional electroporation (i.e. the Neon system from ThermoFisher). An important consideration with DFE is identifying buffer conditions optimal for both cell squeezing and electroporation. The electroporation step may also alter cell effector functions and transcriptional profiles (DiTommaso et al. 2018), counteracting the otherwise minimal perturbations induced by cell squeezing. Because of its minor impact on cell function and viability, cell squeezing is of particular relevance for cell-based therapies, since it would preserve cell functionality and greatly reduce manufacturing times. The efficiency with which resting primary T cells can be edited with Cas9-RNP using cell squeezing has been shown to be comparable to electroporation, with few, if any, impacts on cell behavior (DiTommaso et al. 2018), although it remains to be seen if large-scale gene knock-in using HR templates can be conducted with this method.

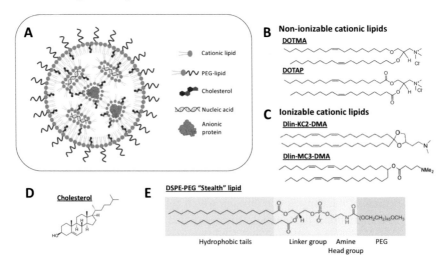

**Figure 2**. A) Example of a lipid nanoparticle, a multi-layered structure composed of permanent or ionizable cationic lipids, with or without conjugated PEG for shielding, cholesterol, and cargo such as nucleic acids or negatively charged proteins. B) Examples of common non-ionizable (permanent) cationic lipids. C) Examples of common ionizable lipids. D) Chemical structure of cholesterol. E) Example of a PEGylated "stealth" lipid.

## Nanoparticle-enabled delivery

### Lipid nanoparticles

Lipids have served as utilitarian delivery vehicles for a host of biomolecules. Cationic liposomes have been used to transfect nucleic acids into cultured cells, providing an effective complement to calcium phosphate and DEAE-dextran transfection techniques (Felgner et al. 1987). Liposome-based formulations enable the delivery of chemotherapeutics, such as Doxil® (Johnson & Johnson) and DaunoXome® (Gilead), for example (Bulbake et al. 2017). Modern lipid nanoparticles (LNPs) are multilayered, liposome-like structures. LNP size, which can approach 1000 nm in diameter, influences uptake and trafficking of the intended cargo (Witzigmann et al. 2020). LNPs are more complex than liposomes in composition, bearing several key components described below (Kanasty et al. 2013) (Figure 2A). The primary ingredient is commonly cationic lipids, which may or may not be ionizable, to which shielding lipids such as poly(ethylene glycol) (PEG), and cholesterol are added. LNPs can be decorated with targeting ligands in order to drive cell type-specific delivery, or formulated with charged molecules that bias LNP uptake by particular organs. While encasing, nucleic acids and other substances in lipids protects them from degradation and enhances their cellular uptake, achieving efficient endosomal escape and tissue specificity remain challenges for this delivery mode. Recently, the first LNP siRNA drug, Onpattro® (Patisiran, Alnylam Pharmaceuticals), was approved for the treatment of the neurodegenerative disease transthyretin (TTR) amyloidosis (Setten et al. 2019). Numerous other LNP formulations suitable for delivering CRISPR editing reagents are undergoing preclinical testing, and Intellia Therapeutics recently announced the first patient has been dosed with their experimental LNP-RNA formulation for CRISPR-Cas9-mediated treatment of TTR (https://clinicaltrials.gov/ct2/show/NCT04601051).

**Cationic lipids:** Non-ionizable cationic lipids have a positively charged head group, which mediates their interaction with highly anionic nucleic acids. Common examples of these types of cationic lipids include DOTMA (1,2-di-O-octadecenyl-3-trimethylammonium propane) and DOTAP (1,2-dioleoyl-3-trimethylammonium propane) (Figure 2B), as well as commercial transfection agents Lipofectamine and RNAiMAX. While they efficiently condense around pDNA, mRNA, and siRNA cargos, so-called "permanent" cationic lipids have several drawbacks. The positively charged head group readily absorbs serum proteins such as immunoglobulins, complement, and serum albumin, which increases the particle's size, decreases its ability to penetrate tissues, decreases endosomal escape and cargo release, and attracts the attention of phagocytes (Liu et al. 2020). Cationic lipids have further been known to cause toxicity by disrupting membranes and inducing reactive oxygen species (Kulkarni et al. 2018). Substituting a portion of the cationic lipids with a neutral lipid such as DOPE (dioleoylphosphatidylethanolamine) helps modulate the amount and types of serum proteins that the particle is able to absorb, and aids in endosomal fusion (Caracciolo 2015).

Ionizable cationic lipids carry a charge that depends on the pH of their environment, which is a function of their $pK_a$. They are able to encapsulate negatively charged nucleic acids when mixed together at a low pH, but at a physiological pH they become neutral in charge. In contrast to permanent cationic lipids, LNPs containing the ionizable variety have a lower affinity for serum proteins, and as a result their half-life in circulation is increased (Kulkarni et al. 2018). Examples of this type of lipid include Dlin-KC2-DMA (2,2-dilinoleyl-4-dimethylaminoethyl-[1,3]-dioxolane) and Dlin-MC3-DMA ((6Z,9Z,28Z,31Z)-heptatriacont-6,9,28,31-tetraene-19-yl 4-(dimethylamino)butanoate)), the latter being a component of the Onpattro® formulation (Figure 2C) (Jayaraman et al. 2012, Semple et al. 2010). Upon arriving at the acidic endosome, the lipids again take on a positive charge.

This causes the anionic phospholipids of the endosomal membrane to "flip-flop" into the LNP complex, forming cationic-anionic lipid pairs. This shape promotes the transition from what were initially apposing lamellar bilayers to an inverted hexagonal phase. The association between the anionic lipids of the endosome with the cationic lipids of the LNP also competes them away from their nucleic acid cargo, further facilitating its release (Hafez et al. 2001, 2000, Xu and Szoka 1996, Zelphati and Szoka 1996).

**Lipid geometry and helper lipids:** Lipid geometry, a term describing lipid shape, governs the way it packs with other lipids to form higher-order assemblies (Cullis and Hope 2017). This, in turn, influences the general topology of the lipid delivery system and the dynamics of membrane fusion. The lipids used in LNPs are comprised by a head group, a linker group, and hydrophobic hydrocarbon tails of varying number (with two being the most common), length, and saturation. The ratio of the volume of the lipid head group to that of its tail determines whether a lipid is more rod-shaped or more cone-shaped. The lamellar phase of lipids, which is a planar structure in which lipid tails face the interior of the plane and heads face the aqueous environment, is generally favored by rod-shaped lipids. The hexagonal phase, which is a set of stacked lipid "tubes" with either heads or tails pointed towards the center, is favored by more conical lipid shapes. Conical-shaped lipids result if a small head group is coupled to flared or kinked tails, or, conversely, if a large head group is coupled to more ordered tails. Screens are common in which researchers swap out various head groups, linkers, and tails of different lengths and saturation to arrive at the most effective combination for delivering their desired cargo into a cell type of interest (Akinc et al. 2008, Mahon et al. 2010, Rosenblum et al. 2020, Wang et al. 2016).

Helper lipids such as 1,2-Dioleoyl-sn-glycero-3-phosphoethanolamine (DOPE), cholesterol, and phosphatidylcholine play a prominent role in dictating the biophysical properties and resulting arrangement of lipids during nanoparticle assembly and membrane fusion (Cheng and Lee 2016). DOPE, for example, has a cone-like shape that facilitates the hexagonal phase occurring during membrane fusion, and as a result is called a "fusogenic" lipid. Cholesterol is a hydrophobic steroid molecule that changes the properties of lipid membranes by increasing phospholipid packing density and thereby enhancing LNP structural integrity (Lee 1977) (Figure 2D). Cholesterol lowers the transition temperature of LNP membranes and aids the transition from the lamellar to the hexagonal phase (Takahashi et al. 1996). By varying the molar ratio of cholesterol in LNP formulations, several-fold increases in DNA delivery have been reported (Pozzi et al. 2012). The cholesterol variant DC-Chol, which is a cationic cholesterol derivative containing a hydrolyzable bond, is used in conjunction with DOPE to reduce LNP toxicity (Li et al. 1996). Naturally-occurring cholesterol analogs, particularly C-24 alkyl phytosterols, may be useful in that they have been shown to enhance transfection rates of LNPs (Patel et al. 2020). Phosphatidylcholines are cylindrical in shape and promote the bilayer phase (Thewalt and Bloom 1992). They also have a high melting temperature and therefore facilitate LNP stability. Thus, helper lipid selection is an important factor influencing lipid arrangements, stability, and cargo release.

**Shielding lipids:** LNPs can aggregate and form non-specific interactions with serum proteins and non-target cells. To reduce the frequency of these non-productive interactions, and prolong the half-life of LNPs in circulation, lipid-anchored PEG may be added during particle formulation (Figure 2E). PEG is a polymer that reduces LNP aggregation and nonspecific interactions through its hydrophilicity and steric repulsion effects (Klibanov et al. 1990). PEGylation has been shown to reduce opsonization (Bazile et al. 1995), shrink LNP size during formulation (Belliveau et al. 2012), and reduce plasma protein adsorption (Gref et al. 2000). The length of the PEG polymer influences its effectiveness as a shield, and can impact cargo encapsulation and LNP stability (Gref et al. 2000, Li et al. 2005, Mao et al. 2006). While PEG is generally thought to be non-immunogenic, there is some controversy surrounding this belief (Schellekens et al. 2013, Verhoef and Anchordoquy 2013). The tradeoff to making LNPs "stealthier" by adding PEG is that PEG can also reduce on-target cellular uptake and release from endosomes. This is because PEG hinders the interaction between the LNP and the endosomal membrane both sterically and electrostatically. This has been termed the "PEG-dilemma" (Hatakeyama et al. 2011). Workarounds include connecting PEG to lipids by cleavable bonds, such as an acid-sensitive bond that causes its release once the endosome is reached (Choi et al. 2003), or a thiolytic bond that can be triggered upon LNP accumulation within a tumor (Kuai et al. 2010). Attaching targeting ligands to PEG, as described in the next section, is another method.

**Targeting ligands:** Targeting ligands enhance the binding or uptake of LNPs by cell types of interest and can be endogenous or exogenous in origin. Endogenous ligands are typically serum proteins that become adsorbed onto the LNP surface while in circulation, such as Apolipoprotein E (ApoE) (Akinc et al. 2010, Bisgaier et al. 1989, Yan et al. 2005), or albumin (Miao et al. 2020). Apolipoproteins are a class of amphipathic proteins that transport fat and cholesterol throughout the body. ApoE enhances the uptake of neutral liposomes by hepatocytes both *in vitro* and *in vivo* (Akinc et al. 2010, Bisgaier et al. 1989, Yan et al. 2005). As evidence of this, RNAi-mediated gene silencing in the liver, when delivered by an ionizable lipid formulation in mice, is nearly abolished in an *apoE*[-/-] genetic background. Silencing is restored upon the addition of ApoE protein to the LNPs prior to injection (Akinc et al. 2010). The receptors that likely mediate ApoE uptake on hepatocytes include low-density lipoprotein (LDL) receptor (LDLR), other LDLR family members, and the scavenger receptor BI. The serum protein albumin is another endogenous protein that may aid the delivery of certain LNPs to hepatocytes (Miao et al. 2020). Proteomic analysis found albumin enriched on LNPs formulated with cKK-E12 and Syn-3, but not A6 lipids. Adding fucoidan, an inhibitor of the albumin receptor, specifically blocked hepatocyte uptake of these albumin-enriched LNPs (Miao et al. 2020).

Exogenous ligands can be added to the liposome during formulation and are commonly attached to PEG. They include a wide range of molecules, including small molecules like anisamide, which is a ligand for the sigma receptors expressed by certain cancer cells (Chono et al. 2008, Li et al. 2008), N-acetylgalactosamine, which is a ligand for the asialoglycoprotein receptor (ASGPR) (Akinc et al. 2010), a peptide from the parasite *Plasmodium* (Longmuir et al. 2006), integrin-attracting RGD peptides (Temming et al. 2005), and antibodies specific to gut leukocyte subsets (Peer et al. 2008). In certain cases, attaching these ligands to long (>45 units) PEG molecules enhances LNP tissue-specificity, as the PEG length is thought to ensure the ligands protrude above the nanoparticle coating. However, these long PEG molecules can also collapse into globular structures that deter ligand recognition. Hence, PEG length should be optimized for the ligand of interest (Stefanick et al. 2013a, 2013b).

Promising results in a peritoneal tumor model were recently seen when LNPs co-encapsulating Cas9 mRNA and gRNA were coated with cell-targeting antibodies (Rosenblum et al. 2020). In a human OV8 peritoneal xenograft model, in which the tumors express high levels of epidermal growth factor receptor (EGFR), anti-EGFR antibodies added to LNPs containing CRISPR-Cas RNAs intended to disrupt the *PLK1* gene, followed by direct intraperitoneal injection, resulted in 82% editing of tumor cells and strong inhibition of tumor growth. A downside of using protein-based targeting ligands is their potential to be immunogenic.

**SORT molecules and barcoded nanoparticles:** A challenge to the clinical use of LNPs as delivery vehicles is their lack of tissue specificity following systemic administration. Rapid clearance of LNPs through the liver favors gene editing in hepatocytes, but targeting other organ systems is, of course, highly desirable. A strategy called Selective ORgan Targeting (SORT) offers a new approach to achieving LNP tropism (Cheng et al. 2020). The SORT method takes a traditional LNP mixture and adds varying quantities of an anionic or cationic "SORT molecule". Adding 18:1 PA (1,2-dioleoyl-*sn*-glycero-3-phosphate), for example, which is an anionic SORT molecule, biases cargo delivery to the spleen, as long as it comprises no more than 40% of the total LNP mixture. DOTAP drives delivery to the lungs, but not spleen or liver, if it comprises 70-90% of the LNP formulation. At DOTAP concentrations lower than 70%, DOTAP SORT LNPs target various combinations of liver, spleen, and lung. Using DOTAP SORT LNPs, the authors were able to skew Cas9-RNP delivery to the liver and the lung, achieving 2.7% and 5.3% editing efficiency of the phosphatase and tensin homolog (*PTEN)* in each tissue, respectively. Higher *PTEN* gene editing efficiencies (up to 15% in the lung) were achieved when Cas9 mRNA/gRNA were used as cargo instead of RNP.

Complementary to the work on SORT molecules is an interesting *in vivo* screening approach aimed at discovering formulations that are retained by specific tissues (Dahlman et al. 2017). Nanoparticle libraries are often screened using cultured cells before promising candidates are tested in animal models. However, *in vitro* cell-based assays can be poor predictors of what nanoparticles will do in a complex, vascularized tissue environment. To more quickly identify cell type-specific LNPs, a method has been developed that pools multiple DNA-barcoded LNPs, each bearing a slightly different formulation. For example, the formulations may differ by the molecular weight of PEG, lipid tail length, or the percentage of a component within the formulation. This mixture is injected intravenously into a mouse. The mouse tissues are harvested several hours later once the LNPs have been cleared from circulation, and deep sequencing is performed. The number of barcode reads in a given tissue, and the ratio of reads between tissues, gives an indication of whether the LNP formulation associated with a particular barcode results in tissue-specific delivery or not. The authors of this study estimate that hundreds of LNP formulations could be simultaneously screened in this fashion.

**LNP formulations of CRISPR-Cas9:** pDNA and mRNA encoding CRISPR-Cas components are readily packaged by permanently cationic lipids or by ionizable cationic lipids if mixed together at a low pH. Finding a way to deliver RNP is desirable because of the liabilities posed by pDNA and mRNA. pDNA can be too stable upon cell entry, causing prolonged Cas9 activity that is associated with a high off-target rate (Fu et al. 2013, Hsu et al. 2013). pDNA also requires transcription and translation, causing a time delay in gene editing. Likewise, Cas9 translation from an mRNA format delays editing, and the mRNA may lack stability or induce innate immune sensors unless chemically modified. RNP, in contrast, acts immediately and is transient enough to pose a much lower off-target risk than pDNA (Zuris et al. 2015). Delivering protein cargoes using lipids can be a complicated endeavor, however. Cationic lipids can be used to package proteins at a neutral pH, but only if those proteins carry a net negative charge. If a protein has a net positive charge, fusing it to supernegatively charged proteins, such as GFP variants to which aspartic and glutamic acid residues have been added, the VP64 transcriptional activation domain, or various epitope tags can facilitate complexation with cationic lipids (Lawrence et al. 2007, Zuris et al. 2015). Adding anionic domains to proteins not only facilitates complexation with cationic lipids, it may also deter protein aggregation (Lawrence et al. 2007).

Cas9 is an example of a protein with a net positive charge at neutral pH. When bound to an gRNA, with its anionic phosphate groups, the complex takes on a net negative charge, making the RNP more amenable to packaging into cationic LNPs than Cas9 alone (Zuris et al. 2015). Cationic lipids such as RNAiMAX, Lipofectamine, or bioreducible lipids such as 8-O14B have been used to deliver Cas9-RNPs to cultured cells, achieving high editing rates (Liang et al. 2015, Wang et al. 2016, Zuris et al. 2015). In one example, researchers used RNAiMAX to deliver Cas9-RNP to cultured human U2OS cells carrying an enhanced GFP (eGFP) reporter and attained 65% reporter gene disruption (Zuris et al. 2015). Here, the measured on-target to off-target ratio was 19-fold higher for RNP delivery over plasmid-encoded RNP components, demonstrating the potential utility of rapid but transient genome exposure to Cas9 (Zuris et al. 2015).

Cationic lipids have also been successfully used to deliver Cas9-RNPs *in vivo* as well. Transgenic Atoh1-GFP mice carry an Atoh1-responsive enhancer upstream of GFP, which is expressed in the hair cells of the cochlea (Lumpkin et al. 2003). Injection of Cas9-RNP-RNAiMAX or Cas9-RNP-Lipofectamine complexes into the cochlear duct of post-natal Atoh1-GFP mice resulted in a 13% and 20% loss of GFP expression, respectively (Zuris et al. 2015). Using ionizable lipids to package RNPs is desirable, as they provide effective endosomal escape. Editing in the liver has been achieved through a biodegradable, ionizable lipid for Cas9-RNP delivery, along with a chemically-stabilized gRNA (Finn et al. 2018, Hendel et al. 2015). Remarkably, a single 3 mg/kg dose given by tail vein injection produced ~70% editing in the liver of the transthyretin (*Ttr*) gene. This resulted in a greater than 97% reduction of serum TTR levels that persisted even 12 months after injection. Packaging of Cas9-RNPs with ionizable lipids is challenging, however, because for efficient incorporation to result, mixing needs to occur at an acidic pH that has been observed to denature the RNP. To address this issue, LNP formulations have been supplemented with cationic lipids to facilitate RNP encapsulation at a neutral pH, preserving RNP structure and function (Wei et al. 2020). RNP-LNPs formulated in this manner outperform complexes made with RNAiMAX and appear to be stable for up to 2 months at 4°C, which has important implications for their clinical use.

**Gold nanoparticles**
Gold nanoparticles (AuNPs) feature a non-toxic, gold particle core (2-250 nm in size) to which a variety of moieties can be attached in a covalent or non-covalent manner

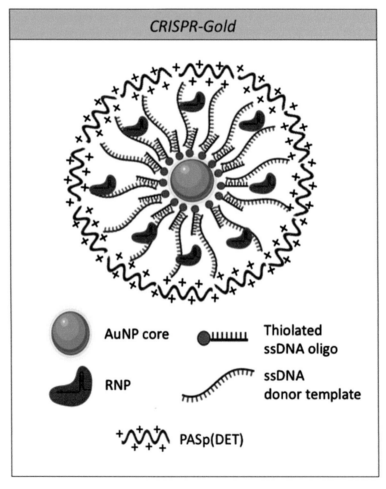

**Figure 3.** Example of a gold nanoparticle, CRISPR-Gold (Lee et al. 2017). A gold core is coated with thiolated single-stranded DNA oligos, which bind homology directed repair templates through complementary base pairing. Cas9-RNP attaches via its natural affinity for DNA. The particle is coated with PASp(DET), a cationic polymer that facilitates endosomal escape.

(Ding et al. 2014, Rana et al. 2012). Covalent attachment of functional moieties to AuNPs takes place via the strong natural interaction gold has with sulfur (S-Au binding), which is ~35% covalent and ~65% electrostatic. RNA and DNA oligos can be directly attached to the particle surface, for example, if they are thiolated (i.e. containing a 5'-SH group) (Giljohann et al. 2009, Lytton-Jean et al. 2011, Seferos et al. 2009). These oligo-AuNPs are taken up by a variety of cell lines despite their strong negative charge, and their uptake appears to be positively correlated with oligo density (Giljohann et al. 2007). Interestingly, nucleic acids attached to gold in this manner appear to be resistant to nuclease digestion, possibly due to interference by charge-balancing $Na^+$ ions that are attracted to the oligos (Seferos et al. 2009). In addition to oligos, many other small, thiolated molecules such as peptides and polyethylene glycols will form self-assembled monolayers when mixed with gold. Each of these molecules can be combined at various ratios to create mixed-monolayer AuNPs with tunable properties (Colangelo et al. 2016).

Non-covalent conjugates to AuNPs are not directly attached to the AuNP core, but rather have some affinity for whatever moiety is already attached. In one example of a non-covalent AuNP, plasmid DNA has been delivered to HEK293T cells using quaternary ammonium-functionalized AuNPs. These are AuNPs for which the gold core is coated with thiol ligands attached via alkyl chains to $NR_4^+$, R being an alkyl or aryl group (Sandhu et al. 2002). The cationic quaternary ammonium radiating out from the AuNP surface creates a net positive charge, which absorbs DNA by associating with its negatively charged backbone. The DNA wraps around the AuNP in what has whimsically been called a "spaghetti and meatballs" formation, and, as an added benefit, is protected from nuclease digestion (Ding et al. 2014, Han et al. 2006). The efficacy of these types of AuNPs for cargo delivery highly depends on several parameters such as what percent of the particle surface is cationic, the AuNP:DNA ratio, and particle hydrophobicity. Another method for creating cationic AuNPs for non-covalent DNA attachment is to attach positively charged amino acids such as lysine to the gold core, to which pDNA molecules have been adhered (Ghosh et al. 2007, 2008). These "lysine-dendronized" AuNPs are intended to resemble nucleosomes in shape, size, and surface functionality, and demonstrate an ability for cell transfection in culture (Ghosh et al. 2008). Arginine has been used in a similar fashion to create a net positive charge on the AuNP surface in order to attract Cas9-RNP fused to a string of negatively charged glutamic acid residues of varying lengths (Mout et al. 2017). Once in the cytosol, the release of nucleic acids from these positively-charged AuNPs is thought to be mediated by glutathione (GSH) (Han et al. 2005). GSH is a natural antioxidant found in the cells of most organisms. It is a tripeptide made of the amino acids cysteine, glutamic acid, and glycine, and is the most abundant thiol species in the cytoplasm. Endogenous GSH becomes incorporated onto the AuNP surface. Owing to its net negative charge, the GSH displaces and releases the DNA or other negatively charged molecules from the cationic ligands surrounding the AuNP.

A downside of AuNPs, like other types of nanoparticles, is their tendency towards "corona" formation (Ahsan et al. 2018, Liu and Peng 2017, Saptarshi et al. 2013). Depending on their composition, AuNPs can interact with lipids, proteins, and nucleic acids within their biological environment. These interactions can be driven by Van der Waals' forces, hydrophobicity, electrostatic interactions, and hydrogen bonding. As observed with lipid nanoparticles, some of these interactions, such as with apolipoproteins, may promote uptake by specific cell types (Ritz et al. 2015). However, the agglomeration of molecules onto the AuNP surface can also dramatically change the particle size, morphology, surface charge, and effectively decrease the rate of cellular uptake. Moreover, proteins attracted to the AuNP may change conformation and become unfolded, exposing new epitopes. This could induce immune recognition and more rapid clearance by macrophages (Saptarshi et al. 2013), ultimately resulting in reduced efficacy, and increased toxicity.

A number of studies (Lee et al. 2018, 2017, Shahbazi et al. 2019, Wang et al. 2017) have shown promise with AuNP-mediated delivery of CRISPR-Cas9 into primary cells and *in vivo*. Two such studies (Lee et al. 2018, 2017) utilize a technology called "CRISPR-Gold", which is a multilayered AuNP capable of delivering both RNP and donor DNA simultaneously (Figure 3). In one formulation, CRISPR-Gold consists of a 15 nm AuNP core covalently attached to a thiolated ssDNA oligo, which contains a region of complementarity to a hybridized ssODN HR donor molecule (Lee et al. 2017). Preformed RNP is then added, binding to the particle through a non-specific electrostatic attraction for gold. The particle is coated with a layer of sodium silicate to increase its negative charge, and, last, is surrounded by the endosomal disruptive polymer poly(N-(N-(2-aminoethyl)-2-aminoethyl) aspartamide) (PAsp(DET)). Once these particles emerge into the cytoplasm from the endosome, intracellular GSH releases the ssDNA-Thiol/RNP/ssODN HR donor complex from the gold core.

In cell-based assays, the CRISPR-Gold nanoparticles were able to facilitate delivery of Cas9-RNP and a homology donor so as to convert BFP to GFP in 11.3% of a HEK293 reporter cell line (Lee et al. 2017). HDR rates fell to 3-4% when the CRISPR-Gold complexes were used on primary human and mouse cells to edit the gene encoding CXCR4, but in comparison to lipofectamine or nucleofection delivery methods, CRISPR-Gold was more efficient and less toxic to cells. A single intramuscular injection of a CRISPR-Gold Cas9-RNP-HR donor complex was also shown to be capable of correcting a mutation in the dystrophin gene in *Dmd^{mdx}* mice, at a rate of 5.4% efficiency 2 weeks-post injection. The CRISPR-Gold injected mice showed significantly increased strength and agility, with minimal off-target toxicity and immunogenicity, as measured by deep sequencing and inflammatory cytokine production.

In a follow-up study (Lee et al. 2018), a similar CRISPR-Gold delivery system, which lacked an HDR template, was deployed for gene targeting in mouse brains. By directly injecting CRISPR-Gold nanoparticles bearing either Cas9 or Cas12a RNPs into the hippocampus of a *Thy1-YFP* mouse model, indels in the *YFP* gene could be observed in 17% and 28% of neurons in the granular layer of the dentate gyrus, respectively. A separate experiment demonstrated STOP codon deletion upstream of a tdTomato reporter in Ai9 mice, thus inducing tdTomato expression, within 10-15% of cells of the hippocampus following intra-hippocampal injection of CRISPR-Gold Cas9 or Cas12a-RNP particles. Interestingly, many "bystander", non-neuronal cell types were modified in this approach, including astrocytes and microglia. More than half the tdTomato⁺ cells were astrocytes, in fact, suggesting there are opportunities for enhancing cell type specific delivery of CRISPR-Gold.

An alternate CRISPR-Cas-AuNP formulation has been used to edit hematopoietic stem and progenitor cells *ex vivo*, with the aim of replacing electroporation and viral transduction as the means for gene editing in the creation of cell-based therapies (Shahbazi et al. 2019). In this example, a 19 nm AuNP core was complexed to gRNA synthesized with an 18-oligo ethylene glycol (OEG) spacer terminating in a thiol group. Recombinant Cas9 or Cas12a nucleases were then added, which bind via their natural affinity for their respective gRNAs. In order to attach an HDR template, these complexes were coated with the polycation polyethylenimine (PEI), providing the electrostatic interaction for DNA, which adheres to the outside of the particle. PEI additionally facilitates lysosomal escape after cellular uptake. In this study, the particles produced editing rates that were comparable to or better than RNP electroporation with relatively lower impacts on cell viability. Formulations bearing Cas12a outperformed Cas9, yielding 17.6% editing at the *CCR5 locus* without a repair template, and 13.4% precision editing with an oligo template. In total, these results suggest gold nanoparticle-based technologies may provide an effective platform for delivering a variety of CRISRP-Cas modalities *in vitro* and *in vivo*.

### Magnetic nanoparticles

Iron oxide nanoparticles, and proteins that naturally store iron, such as ferritin, have been used in a variety of creative contexts to induce biological perturbations with magnetic fields. Membrane-localized ferritin-GFP fusion proteins, for example, associate with co-expressed TRPV1 cation channels fused to a GFP-binding domain. The application of a magnetic field pulls the ferritin which in turn causes the associated TRPV1 channel to open, activating a $Ca^{++}$-responsive reporter

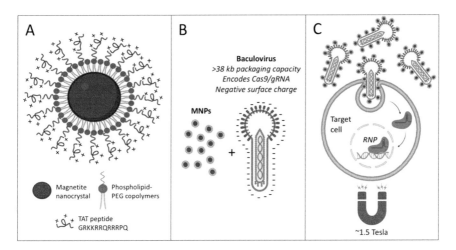

**Figure 4.** Magnetic nanoparticle-baculoviral (MNP-BV) gene delivery system (Zhu et al. 2019). A) Magnetic nanoparticles (MNPs) were created by taking magnetite nanocrystals and coating them with copolymers of phospholipid and PEG, followed by conjugation to the cationic TAT peptide. B) Baculovirus (BV) encoding Cas9 and gRNA is coated with MNPs to create BV-MNPs. C) Applying a magnetic field to MNP-BVs in the presence of cultured cells, or after systemic injection in mice, enhances cell uptake and Cas9 editing rates.

gene system downstream (Stanley et al. 2015). Similar technology fusing ferritin directly to the TRPV1 channel, and delivering this fusion gene to mouse brains using AAV, creates a distinctive "magnetogenetic actuator" in which brain cells can be depolarized *in vivo* upon exposure to a magnetic field (Wheeler et al. 2016). In another application, particles of iron oxide exposed to a magnetic field have been used to disrupt endothelial cell-cell junctions, enabling otherwise excessively-large drug molecules to leave circulation and enter the surrounding tissue (Qiu et al. 2017). A logical extension of studies like these may be to attempt to couple CRISPR-Cas editing reagents to iron-rich particles, and direct them towards cells of interest using magnetic fields.

Magnetic iron oxide nanoparticles (MNPs) (Rohiwal et al. 2020, Zhu et al. 2019), and magneto-electric nanoparticles (MENPs) of cobalt ferrite-barium titanate (Kaushik et al. 2019), have now been used in this regard. For example, iron oxide nanoparticles were coated with a mixture of polyethylenimine (PEI), which is a common transfection reagent (Pandey and Sawant 2016), in addition to pDNAs encoding Cas9/gRNA and an HR template (Rohiwal et al. 2020). The PEI, which is positively charged, was premixed with the pDNA before adding to the MNPs. The resulting PEI-MNPs were given to a HEK293-TLR3 reporter line in the presence or absence of a magnetic field applied under the tissue culture dish for ~1 hour (Rohiwal et al. 2020). These cells carry a non-functional GFP gene containing an ISceI site and stop codon, as well as a BFP gene shifted 2 base pairs in reading frame. In the event of NHEJ, BFP expression results, while HDR leads to GFP expression (Song and Stieger 2017). The percentage of NHEJ resulting from PEI-MNP treatment, while low (4.7%), was not dissimilar to that achieved with lipofectamine transfection of DNA (6%), though there was only a small increase in efficiency with the application of a magnetic field (5.4%) (Rohiwal et al. 2020). The rates of HDR using this system were relatively low (0.8% for Lipofectamine; ~0.5% for PEI-MNPs whether magnetofected or not). Aside from general inefficiency, a potential issue with PEI-MNPs, which is also seen for gold nanoparticles, is the tendency for particle size to increase several-fold in the presence of FBS, as serum proteins nonspecifically bind to the PEI-MNPs.

A separate study made use of a 2-layered culture system mimicking the blood-brain barrier (BBB), comprised of 2 chambers separated by a permeable membrane (Kaushik et al. 2019). The upper side of the membrane was seeded with human brain microvascular endothelial cells (HBMECs), while the lower was populated by immortalized microglial cells carrying a fluorescent reporter under the control of the HIV long terminal repeat (LTR) domain. In order to target the LTR, thus inactivating the reporter, Cas9-RNPs and LTR-specific gRNAs were complexed with cobalt ferrite-barium titanate (e.g. MENPs). The MENPs have a positive surface charge, which simply binds negatively charged RNP via electrostatic interactions. These MENPs can be pulled through the artificial BBB by applying a static magnetic force. The magnetic field is then changed to one with an alternating current. The alternating current is thought to cause the expansion and contraction of ionic bonds on the MENP surface, thereby releasing the RNP from the metallic core. In this system, the type of magnetic field can provide a means for controlled cargo release. LTR expression in cultured cells is reduced from 65% to 40% upon the addition of

MENPs bearing LTR-specific Cas9-RNPs, while the addition of Cas9-RNP alone reduced the percentage of LTR-expressing cells to 48%.

*In vivo* editing has been demonstrated through a combination of MNPs and a custom baculovirus (BV)-based gene delivery device (Figure 4) (Zhu et al. 2019). The MNPs are created by taking ~16 nm magnetite nanocrystals and coating them with copolymers of phospholipid and polyethylene glycol (termed MNP-PEGs) to make them water-dispersible. These MNP-PEGs are then conjugated to the cationic cell-penetrating Trans-Activator of Transcription (TAT) peptide (Torchilin 2008) (Figure 4A). The positively-charged TAT peptide facilitates an interaction between the MNP-PEG and the negatively-charged BV surface via non-specific electrostatic interactions (MNP-BV) (Figure 4B). In the presence of a magnetic field, cell entry by these MNP-BVs was enhanced (Figure 4C). For example, in experiments with Hepa 1-6 cells, MNPs coupled to BV bearing a luciferase reporter gene were able to infect ~60% of the cells in the presence of a magnetic field, versus only ~10% in its absence. Based on experiments using actin depolymerizing agents, it was suggested that changes to the actin network caused by the magnetic field was at least in part responsible for cellular uptake of MNP-BVs.

Since the packaging capacity of BV is very large (>38 kilobases), both Cas9 and the requisite gRNA sequences can be accommodated by this vector. MNP-BV encoding Cas9 and a gRNA targeting the *Vegfr2* gene, when added to Hepa 1-6 cells in a strong magnetic field, could induce ~35% target gene editing by T7E1 assay. When directly injected into tumor tissue, this same delivery system could induce indels within the *Vegfr2* gene in up to 4.7% of mouse tumor cells as assessed by next generation sequencing, although variability was observed, possibly due to heterogeneous distribution of the magnetic field and/or editing reagents. A potential downside of this system is that BV can be inactivated by the host complement system (Chen et al. 2011a, Hofmann and Strauss 1998), although the MNP-BVs are shown to retain activity under conditions of high (50%) adult mouse serum *in vitro*. Given systemically to mice via tail vein injection, the MNP-BVs can introduce CRISPR-Cas9 to seemingly any tissue where a magnet can be overlaid for the hour post-injection, but it does appear challenging to ensure delivery to only one organ system when another is nearby (for example, spleen vs. liver). The key parameters governing MNP-BV efficacy are the MNP to baculoviral vector ratio, the distribution of the MNP-BVs in tissue prior to magnetic field placement, and then the strength, duration, and spatial distribution of the magnetic field.

## Protein transduction

### Cell penetrating peptides

Cell-penetrating peptides (CPPs) are generally small (<30 aa), natural and engineered peptides that translocate across cellular membranes without requiring a receptor. Among the earliest CPPs described is a short, basic region within the TAT protein from the Human Immunodeficiency Virus (HIV) (Frankel and Pabo 1988, Green and Loewenstein 1988), and the third helix of the 60 amino acid-long Drosophila Antennapedia homeodomain, also known as Penetratin (Derossi et al. 1994). A variety of proteins, such as RNase A, β-galactosidase, green-fluorescent protein (GFP), and horseradish peroxidase, have been ferried across cell membranes as fusions to these peptides with retained activity (Fawell et al. 1994, Jones and Sayers 2012, Patel et al. 2019, Taylor and Zahid 2020). These are not exclusively cationic, as certain amphipathic and hydrophobic peptides have this capability as well (Taylor and Zahid 2020). In addition to peptides, full-length proteins engineered to contain patches of cationic residues ("arginine grafting") (Fuchs and Raines 2007), or superpositively charged proteins such as +36 GFP, can also traverse the cell membrane (McNaughton et al. 2009, Thompson et al. 2012). One estimate suggests 2% of the human proteome may have cell-penetrating properties based on each protein's net theoretical charge to molecular weight ratio (Cronican et al. 2011).

The transient nature of CPPs makes them an attractive method for delivering gene editing reagents to cells, and the peptides themselves produce generally few toxic side effects at effective doses. Furthermore, several CRISPR-Cas modalities can be combined with CPPs including pDNA and RNPs (Krishnamurthy et al. 2019, Ramakrishna et al. 2014, Wang et al. 2018, Zhu et al. 2019). For example, positively charged CPPs have long been known to interact with pDNA through electrostatic interactions without compromising CPP membrane permeability (Rudolph et al. 2003). This concept has recently been applied to expression vectors for CRISPR-Cas9 (Wang et al. 2018). For example, a helical CPP with a cationic side-chain terminus was mixed with an expression vector for Cas9 in the presence of PEG. The transfection efficiency of the resulting complexes was comparable or superior to Lipofectamine 3000 and PEI in a panel of cell lines. Further, delivery of a Cas9 and GFP-targeting gRNA expression vector with a linked CPP drove ~30% indel formation when injected into a GFP-expressing, xenograft tumor model (Wang et al. 2018).

Cas9- or Cas12a-RNPs noncovalently coupled to CPPs have shown promise for editing in mouse airway epithelial cells (Krishnamurthy et al. 2019). The peptide in this case, termed S10, was a 34 amino acid peptide based on a fusion between the endoosmolytic peptide CM18, which is derived from insect antimicrobial genes, and a cell penetrating peptide from the HIV TAT protein called PTD4, separated by a flexible glycine-rich linker (Del'Guidice et al. 2018). Nasal administration of either Cas9 or Cas12a-RNPs in combination with S10 produced editing in at least 10% of epithelial cells in both the small and large airways, and produced minimal toxicity in terms of epithelial cell viability, inflammatory cell infiltrates to the lung, and cytokine secretion. Cas-RNP administered without S10 produced no edited airway cells. Other CPP formats, such as helical amphiphilic peptides with reactive hydrazide side chains, when mixed with RNP achieve cleavage rates comparable to Lipofectamine 2000 but with less cellular toxicity (Lostalé-Seijo et al. 2017). Attaching CPPs individually to both Cas9 and the gRNA has shown evidence of activity *in vitro*, but with relatively low efficiency (Ramakrishna et al. 2014).

CPPs have also been coated onto magnetic nanoparticles and then adhered to baculovirus encoding CRISPR-Cas9 (Zhu et al. 2019). Application of a magnetic field facilitates cell entry of the baculovirus via its nanoparticle coating. The downside to using CPPs is that many peptides do not provide cell type specificity, and CPP-fusion proteins can end up sequestered in the endosomal compartment (Vivès et al. 2008). Strategies that have been used to enhance endosomal escape include stearylation (Tai et al. 2015), acylation (Morais et al. 2018), and histidine modification of the peptide N-terminus (Li et al. 2016). CPPs may also lack *in vivo* stability due to proteolytic degradation, and the non-self nature of many CPPs may elicit an immune response.

### Induced transduction by osmocytosis and propanebetaine

CPP fusion proteins work across many different cell types. However, the attachment of a CPP to a heterologous protein can negatively impact its function and alter its subcellular localization (Lundberg and Johansson 2001). Hence, researchers have sought ways to transduce native proteins into primary and immortalized cells. The induced transduction by osmocytosis and propanebetaine (iTOP) method is one way to accomplish this goal (D'Astolfo et al. 2015)[2]. iTOP was discovered serendipitously, when an Oct4-VP16 fusion protein added to Cos-7 cells activated an intracellular reporter, and this activity was found to be dependent on the chemical components of the protein's purification buffer. The relevant buffer components were found to be propanebetaine (also called non-detergent Sulfobetaine-201, or NDSB-201), which is a compound used to reduce aggregation and facilitate protein folding (Vuillard et al. 1995), and sodium chloride (NaCl).

These two components, when added together to cell culture media, resulted in protein uptake by a variety of cell types. Using mouse embryonic fibroblasts (MEFs) preloaded with a beta-lactamase reporter substrate, the authors measured the uptake of recombinant beta-lactamase protein using iTOP, and determined the optimum levels of NaCl and NDSB-201 for this to occur. Importantly, negative impacts on cell proliferation were seen at salt concentrations that produced high levels of protein transduction. This effect of salt concentration on proliferation is known (Kültz et al. 1998), and can be dampened by osmoprotectants such as glycerol, glycine, or taurine. Indeed, the authors found specific combinations of these compounds to be effective at blocking the loss of proliferation caused by iTOP. Certain cell types, such as murine embryonic stem cells, were more sensitive to the hypertonic conditions of iTOP, highlighting the need to optimize the method for each cell type of interest. Interestingly, the NaCl used for iTOP can be substituted with other alkali metal cation-containing salts such as RbCl, KCl, or LiCl, which facilitate protein transduction to different degrees in the presence of NDSB-201. The authors also tested several analogs of NDSB-201, which included the neurotransmitter gamma-amino-butyric acid (GABA), and surprisingly found that GABA produced excellent protein transduction.

iTOP was used to successfully transduce Cas9 protein and gRNA into KBM7 cells targeting an out-of-frame fluorescent dTomato reporter. Cas9-RNP cleavage restored dTomato expression in 56% of the cells. Using human embryonic stem cells, 26% editing was achieved. The mechanism through which iTOP works was explored by treating cells with various inhibitors of cell uptake pathways. The results suggested that iTOP works by triggering micropinocytosis, and requires the $Na^+/H^+$ antiporter NheI (Slc9A1). The iTOP method provides yet another way to introduce CRISPR-Cas9 to cells, but without requiring protein tags, and, given that transcription or translation from a template is not involved, the method gives researchers greater control over the exact amount of Cas9 protein cells receive.

## Closing Remarks

In this chapter, we highlighted several non-viral means for delivering a range of CRISPR-Cas modalities, including the Cas protein, associated gRNAs, and homologous donor DNAs. Depending on the genome modification needs, be it permanent gene disruption, RNA or DNA base editing, or transcription modulation, the tissue-selectivity, durability, and clearance rate of the modalities are critical elements for a successful outcome. As the gene editing toolkit expands, so do the technologies for delivery. While broadly applicable, *in vitro* and *in vivo*, certain cell types may not be amenable to non-viral delivery, in which case virus-based delivery (e.g. lentivirus, AAV, etc.) may be needed. For example, macrophages, which are a non-dividing phagocytic cell type equipped with potent enzymes for degrading proteins and nucleic acids, pose technical difficulties with non-viral methods (Zhang et al. 2009), but can be engineered with high (>80%) efficiency through adenoviral gene delivery (Klichinsky et al. 2020). Non-viral and viral delivery techniques should not be considered mutually exclusive, as it can be advantageous to combine them, such as coupling a Cas9 mRNA in a lipid nanoparticle or Cas9-RNP with an AAV containing an appropriate gRNA expression cassette and HR donor (Yin et al. 2016).

Furthermore, there are lessons to be learned from viruses in the optimization of non-viral methods using biomimicry (Figueroa et al. 2020). For example, the shape (rod-shaped, spherical, filamentous, polyhedral) and surface properties (e.g. glycosylation) of viruses influence their tissue penetration, targeting specificity, and cell internalization (Lico et al. 2016). Understanding how plant viruses such as the tobacco mosaic virus (Pitek et al. 2016) attract relatively few proteins to their surface may inform nanoparticle design and reduce corona formation. Finally, studies of how viruses slip past the mucus barrier may enable oral or inhaled delivery modes for non-viral formulations (Witten and Ribbeck 2017).

We have entered an extended golden age of biology and medicine, driven by ever-improving methods for gene manipulation. Yet, the true potential of these technologies remains untapped, and will only come through future advances for safe, effective, and specific delivery.

## Acknowledgements

Graphics were created using the BioRender application (https://biorender.com/).

## References

Abe, T., Inoue, K., Furuta, Y. and Kiyonari, H. 2020. Pronuclear microinjection during S-Phase increases the efficiency of CRISPR-Cas9-assisted knockin of large DNA donors in mouse zygotes. Cell Reports, 31(7): 107653.

Abu-Bonsrah, K. D., Zhang, D. and Newgreen, D. F. 2016. CRISPR/Cas9 targets chicken embryonic somatic cells in vitro and in vivo and generates phenotypic abnormalities. Scientific Reports, 6(1): 34524.

Ahsan, S. M., Rao, C. M. and Ahmad, Md. F. 2018. Cellular and molecular toxicology of nanoparticles. Advances in Experimental Medicine and Biology, 1048: 175-198.

Akinc, A., Querbes, W., De, S., Qin, J., Frank-Kamenetsky, M., Jayaprakash, K. N., et al. 2010. Targeted delivery of RNAi therapeutics with endogenous and exogenous ligand-based mechanisms. Molecular Therapy, 18(7): 1357-1364.

Akinc, A., Zumbuehl, A., Goldberg, M., Leshchiner, E. S., Busini, V., Hossain, N., et al. 2008. A combinatorial library of lipid-like materials for delivery of RNAi therapeutics. Nature Biotechnology, 26(5): 561-569.

Albors, A. R. and Tanaka, E. M. 2015. Salamanders in regeneration research, methods and protocols. Methods in Molecular Biology, 1290: 115-125.

Anzalone, A. V., Koblan, L. W. and Liu, D. R. 2020. Genome editing with CRISPR-Cas nucleases, base editors, transposases and prime editors. Nature Biotechnology, 38(7): 824-844.

Bakondi, B., Lv, W., Lu, B., Jones, M. K., Tsai, Y., Kim, K. J., et al. 2016. In vivo CRISPR/Cas9 gene editing corrects retinal dystrophy in the S334ter-3 rat model of autosomal dominant retinitis pigmentosa. Molecular Therapy, 24(3): 556-563.

Barnett, F. H., Scharer-Schuksz, M., Wood, M., Yu, X., Wagner, T. E. and Friedlander, M. 2004. Intra-arterial delivery of endostatin gene to brain tumors prolongs survival and alters tumor vessel ultrastructure. Gene Therapy, 11(16): 1283-1289.

Bazile, D., Prud'homme, C., Bassoullet, M., Marlard, M., Spenlehauer, G. and Veillard, M. 1995. Stealth Me.PEG-PLA nanoparticles avoid uptake by the mononuclear phagocytes system. Journal of Pharmaceutical Sciences, 84(4): 493-498.

Belliveau, N. M., Huft, J., Lin, P. J., Chen, S., Leung, A. K., Leaver, T. J., et al. 2012. Microfluidic synthesis of highly potent limit-size lipid nanoparticles for in vivo delivery of siRNA. Molecular Therapy – Nucleic Acids, 1(Nat Biotechnol. 26 2008), e37.

Bisgaier, C. L., Siebenkas, M. V. and Williams, K. J. 1989. Effects of apolipoproteins A-IV and A-I on the uptake of phospholipid liposomes by hepatocytes. The Journal of Biological Chemistry, 264(2): 862-866.

Bishop, K. A., Harrington, A., Kouranova, E., Weinstein, E. J., Rosen, C. J., Cui, X., et al. 2016. CRISPR/Cas9-mediated insertion of loxP sites in the mouse Dock7 gene provides an effective alternative to use of targeted embryonic stem cells. G3: Genes|Genomes|Genetics, 6(7): 2051-2061.

Budker, V. G., Subbotin, V. M., Budker, T., Sebestyén, M. G., Zhang, G. and Wolff, J. A. 2006. Mechanism of plasmid delivery by hydrodynamic tail vein injection. II. Morphological studies. The Journal of Gene Medicine, 8(7): 852-873.

Budker, V., Zhang, G., Danko, I., Williams, P. and Wolff, J. 1998. The efficient expression of intravascularly delivered DNA in rat muscle. Gene Therapy, 5(2): 272-276.

Budker, V., Zhang, G., Knechtle, S. and Wolff, J. A. 1996. Naked DNA delivered intraportally expresses efficiently in hepatocytes. Gene Therapy, 3(7): 593-598.

Bulbake, U., Doppalapudi, S., Kommineni, N. and Khan, W. 2017. Liposomal formulations in clinical use: An updated review. Pharmaceutics, 9(2): 12.

Capecchi, M. R. 2005. Gene targeting in mice: Functional analysis of the mammalian genome for the twenty-first century. Nature Reviews Genetics, 6(6): 507-512.

Caracciolo, G. 2015. Liposome-protein corona in a physiological environment: Challenges and opportunities for targeted delivery of nanomedicines. Nanomedicine: Nanotechnology, Biology and Medicine, 11(3): 543-557.

Charlesworth, C. T., Deshpande, P. S., Dever, D. P., Camarena, J., Lemgart, V. T., Cromer, M. K., et al. 2019. Identification of preexisting adaptive immunity to Cas9 proteins in humans. Nature Medicine, 25(2): 249-254.

Chen, C.-Y., Lin, C.-Y., Chen, G.-Y. and Hu, Y.-C. 2011a. Baculovirus as a gene delivery vector: Recent understandings of molecular alterations in transduced cells and latest applications. Biotechnology Advances, 29(6): 618-631.

Chen, F., Pruett-Miller, S. M., Huang, Y., Gjoka, M., Duda, K., Taunton, J., et al. 2011b. High-frequency genome editing using ssDNA oligonucleotides with zinc-finger nucleases. Nature Methods, 8(9): 753-755.

Cheng, Q., Wei, T., Farbiak, L., Johnson, L. T., Dilliard, S. A. and Siegwart, D. J. 2020. Selective organ targeting (SORT) nanoparticles for tissue-specific mRNA delivery and CRISPR-Cas gene editing. Nature Nanotechnology, 15(4): 313-320.

Cheng, X. and Lee, R. J. 2016. The role of helper lipids in lipid nanoparticles (LNPs) designed for oligonucleotide delivery. Advanced Drug Delivery Reviews, 99(Pt A), 129-137.

Choi, J. S., MacKay, J. A. and Szoka, F. C. 2003. Low-pH-sensitive PEG-stabilized plasmid-lipid nanoparticles: Preparation and characterization. Bioconjugate Chemistry, 14(2): 420-429.

Chono, S., Li, S.-D., Conwell, C. C. and Huang, L. 2008. An efficient and low immunostimulatory nanoparticle formulation for systemic siRNA delivery to the tumor. Journal of Controlled Release, 131(1): 64-69.

Chylinski, K., Makarova, K. S., Charpentier, E. and Koonin, E. V. 2014. Classification and evolution of type II CRISPR-Cas systems. Nucleic Acids Research, 42(10): 6091-6105.

Colangelo, E., Comenge, J., Paramelle, D., Volk, M., Chen, Q. and Lévy, R. 2016. Characterizing self-assembled monolayers on gold nanoparticles. Bioconjugate Chemistry, 28(1): 11-22.

Crespo, A., Peydró, A., Dasí, F., Benet, M., Calvete, J. J., Revert, F., et al. 2005. Hydrodynamic liver gene transfer mechanism involves transient sinusoidal blood stasis and massive hepatocyte endocytic vesicles. Gene Therapy, 12(11): 927-935.

Cronican, J. J., Beier, K. T., Davis, T. N., Tseng, J.-C., Li, W., Thompson, D. B., et al. 2011. A class of human proteins that deliver functional proteins into mammalian cells in vitro and in vivo. Chemistry & Biology, 18(7): 833-838.

Cullis, P. R. and Hope, M. J. 2017. Lipid nanoparticle systems for enabling gene therapies. Molecular Therapy, 25(7): 1467-1475.

Dahlman, J. E., Kauffman, K. J., Xing, Y., Shaw, T. E., Mir, F. F., Dlott, C. C., et al. 2017. Barcoded nanoparticles for high throughput in vivo discovery of targeted therapeutics. Proceedings of the National Academy of Sciences, 114(8): 2060-2065.

D'Astolfo, D. S., Pagliero, R. J., Pras, A., Karthaus, W. R., Clevers, H., Prasad, V., et al. 2015. Efficient intracellular delivery of native proteins. Cell, 161(3): 674-690.

Del'Guidice, T., Lepetit-Stoffaes, J.-P., Bordeleau, L.-J., Roberge, J., Théberge, V., Lauvaux, C., et al. 2018. Membrane permeabilizing amphiphilic peptide delivers recombinant transcription factor and CRISPR-Cas9/Cpf1 ribonucleoproteins in hard-to-modify cells. PLOS ONE, 13(4): e0195558.

Derossi, D., Joliot, A. H., Chassaing, G. and Prochiantz, A. 1994. The third helix of the Antennapedia homeodomain translocates through biological membranes. The Journal of Biological Chemistry, 269(14): 10444-10450.

Ding, X., Stewart, M. P., Sharei, A., Weaver, J. C., Langer, R. S. and Jensen, K. F. 2017. High-throughput nuclear delivery and rapid expression of DNA via mechanical and electrical cell-membrane disruption. Nature Biomedical Engineering, 1(3): 0039.

Ding, Y., Jiang, Z., Saha, K., Kim, C. S., Kim, S. T., Landis, R. F., et al. 2014. Gold nanoparticles for nucleic acid delivery. Molecular Therapy, 22(6): 1075-1083.

DiTommaso, T., Cole, J. M., Cassereau, L., Buggé, J. A., Hanson, J. L. S., Bridgen, D. T., et al. 2018. Cell engineering with microfluidic squeezing preserves functionality of primary immune cells in vivo. Proceedings of the National Academy of Sciences, 115(46): 201809671.

Doudna, J. A. 2020. The promise and challenge of therapeutic genome editing. Nature, 578(7794): 229-236.

Engelholm, L. H., Riaz, A., Serra, D., Dagnæs-Hansen, F., Johansen, J. V., Santoni-Rugiu, E., et al. 2017. CRISPR/Cas9 engineering of adult mouse liver demonstrates that the Dnajb1-Prkaca gene fusion is sufficient to induce tumors resembling fibrolamellar hepatocellular carcinoma. Gastroenterology, 153(6): 1662-1673 e10.

Engreitz, J., Abudayyeh, O., Gootenberg, J. and Zhang, F. 2019. CRISPR tools for systematic studies of RNA regulation. Cold Spring Harbor Perspectives in Biology, 11(8): a035386.

Esensten, J. H., Bluestone, J. A. and Lim, W. A. 2016. Engineering therapeutic T cells: From synthetic biology to clinical trials. Annual Review of Pathology: Mechanisms of Disease, 12(1): 305-330.

Eyquem, J., Mansilla-Soto, J., Giavridis, T., Stegen, S. J. C. van der, Hamieh, M., Cunanan, K. M., et al. 2017. Targeting a CAR to the TRAC locus with CRISPR/Cas9 enhances tumour rejection. Nature, 543(7643): 113-117.

Fawell, S., Seery, J., Daikh, Y., Moore, C., Chen, L. L., Pepinsky, B., et al. 1994. Tat-mediated delivery of heterologous proteins into cells. Proceedings of the National Academy of Sciences, 91(2): 664-668.

Fei, J.-F., Knapp, D., Schuez, M., Murawala, P., Zou, Y., Singh, S. P., et al. 2016. Tissue- and time-directed electroporation of CAS9 protein-gRNA complexes in vivo yields efficient multigene knockout for studying gene function in regeneration. Npj Regenerative Medicine, 1(1): 16002.

Felgner, P. L., Gadek, T. R., Holm, M., Roman, R., Chan, H. W., Wenz, M., et al. 1987. Lipofection: A highly efficient, lipid-mediated DNA-transfection procedure. Proceedings of the National Academy of Sciences, 84(21): 7413-7417.

Figueroa, S. M., Fleischmann, D. and Goepferich, A. 2020. Biomedical nanoparticle design: What we can learn from viruses. Journal of Controlled Release. https://doi.org/10.1016/j.jconrel.2020.09.045

Finn, J. D., Smith, A. R., Patel, M. C., Shaw, L., Youniss, M. R., Heteren, et al. 2018. A single administration of CRISPR/Cas9 lipid nanoparticles achieves robust and persistent in vivo genome editing. Cell Reports, 22(9): 2227-2235.

Frankel, A. D. and Pabo, C. O. 1988. Cellular uptake of the tat protein from human immunodeficiency virus. Cell, 55(6): 1189-1193.

Fu, Y., Foden, J. A., Khayter, C., Maeder, M. L., Reyon, D., Joung, J. K., et al. 2013. High-frequency off-target mutagenesis induced by CRISPR-Cas nucleases in human cells. Nature Biotechnology, 31(9): 822-826.

Fuchs, S. M. and Raines, R. T. 2007. Arginine grafting to endow cell permeability. ACS Chemical Biology, 2(3): 167-170.

Gaj, T., Staahl, B. T., Rodrigues, G. M. C., Limsirichai, P., Ekman, F. K., Doudna, J. A., et al 2017. Targeted gene knock-in by homology-directed genome editing using Cas9 ribonucleoprotein and AAV donor delivery. Nucleic Acids Research, 45(11): gkx154-.

Ghosh, P. S., Han, G., Erdogan, B., Rosado, O., Krovi, S. A. and Rotello, V. M. 2007. Nanoparticles featuring amino acid-functionalized side chains as DNA receptors. Chemical Biology & Drug Design, 70(1): 13-18.

Ghosh, P. S., Kim, C.-K., Han, G., Forbes, N. S. and Rotello, V. M. 2008. Efficient gene delivery vectors by tuning the surface charge density of amino acid-functionalized gold nanoparticles. ACS Nano, 2(11): 2213-2218.

Giljohann, D. A., Seferos, D. S., Patel, P. C., Millstone, J. E., Rosi, N. L. and Mirkin, C. A. 2007. Oligonucleotide loading determines cellular uptake of DNA-modified gold nanoparticles. Nano Letters, 7(12): 3818-3821.

Giljohann, D. A., Seferos, D. S., Prigodich, A. E., Patel, P. C. and Mirkin, C. A. 2009. Gene regulation with polyvalent siRNA-nanoparticle conjugates. Journal of the American Chemical Society, 131(6): 2072-2073.

Green, M. and Loewenstein, P. M. 1988. Autonomous functional domains of chemically synthesized human immunodeficiency virus tat trans-activator protein. Cell, 55(6): 1179-1188.

Gref, R., Lück, M., Quellec, P., Marchand, M., Dellacherie, E., Harnisch, S., et al. 2000. 'Stealth' corona-core nanoparticles surface modified by polyethylene glycol (PEG): influences of the corona (PEG chain length and surface density) and of the core composition on phagocytic uptake and plasma protein adsorption. Colloids and Surfaces B: Biointerfaces, 18(3-4): 301-313.

Gu, B., Posfai, E. and Rossant, J. 2018. Efficient generation of targeted large insertions by microinjection into two-cell-stage mouse embryos. Nature Biotechnology, 36(7): 632-637.

Guen, Y. T. L., Gall, T. L., Midoux, P., Guégan, P., Braun, S. and Montier, T. (2020). Gene transfer to skeletal muscle using hydrodynamic limb vein injection: Current applications, hurdles and possible optimizations. The Journal of Gene Medicine, 22(2): e3150.

Gundry, M. C., Brunetti, L., Lin, A., Mayle, A. E., Kitano, A., Wagner, D., et al. 2016. Highly efficient genome editing of murine and human hematopoietic progenitor cells by CRISPR/Cas9. Cell Reports, 17(5): 1453-1461.

Gurumurthy, C. B., O'Brien, A. R., Quadros, R. M., Adams, J., Alcaide, P., Ayabe, S., et al. 2019. Reproducibility of CRISPR-Cas9 methods for generation of conditional mouse alleles: A multi-center evaluation. Genome Biology, 20(1): 171.

Gutschner, T., Haemmerle, M., Genovese, G., Draetta, G. F. and Chin, L. 2016. Post-translational regulation of Cas9 during G1 enhances homology-directed repair. Cell Reports, 14(6): 1555-1566.

Hafez, I. M., Ansell, S. and Cullis, P. R. 2000. Tunable pH-sensitive liposomes composed of mixtures of cationic and anionic lipids. Biophysical Journal, 79(3): 1438-1446.

Hafez, I., Maurer, N. and Cullis, P. 2001. On the mechanism whereby cationic lipids promote intracellular delivery of polynucleic acids. Gene Therapy, 8(15): 1188-1196.

Hagstrom, J. E., Hegge, J., Zhang, G., Noble, M., Budker, V., Lewis, D. L., et al. 2004. A facile nonviral method for delivering genes and siRNAs to skeletal muscle of mammalian limbs. Molecular Therapy, 10(2): 386-398.

Han, G., Chari, N. S., Verma, A., Hong, R., Martin, C. T. and Rotello, V. M. (2005). Controlled recovery of the transcription of nanoparticle-bound DNA by intracellular concentrations of glutathione. Bioconjugate Chemistry, 16(6): 1356-1359.

Han, G., Martin, C. T. and Rotello, V. M. 2006. Stability of gold nanoparticle-bound DNA toward biological, physical, and chemical agents. Chemical Biology & Drug Design, 67(1): 78-82.

Hatakeyama, H., Akita, H. and Harashima, H. 2011. A multifunctional envelope type nano device (MEND) for gene delivery to tumours based on the EPR effect: A strategy for overcoming the PEG dilemma. Advanced Drug Delivery Reviews, 63(3): 152-160.

Heller, R., Jaroszeski, M., Atkin, A., Moradpour, D., Gilbert, R., Wands, J., et al. 1996. In vivo gene electroinjection and expression in rat liver. FEBS Letters, 389(3): 225-228.

Hendel, A., Bak, R. O., Clark, J. T., Kennedy, A. B., Ryan, D. E., Roy, S., et al. 2015. Chemically modified guide RNAs enhance CRISPR-Cas genome editing in human primary cells. Nature Biotechnology, 33(9): 985-989.

Herweijer, H. and Wolff, J. A. 2007. Gene therapy progress and prospects: Hydrodynamic gene delivery. Gene Therapy, 14(2): 99-107.

Hickman, M. A., Malone, R. W., Lehmann-Bruinsma, K., Sih, T. R., Knoell, D., Szoka, F. C., et al. 1994. Gene expression following direct injection of DNA into liver. Human Gene Therapy, 5(12): 1477-1483.

Hofmann, C. and Strauss, M. 1998. Baculovirus-mediated gene transfer in the presence of human serum or blood facilitated by inhibition of the complement system. Gene Therapy, 5(4): 531-536.

Horii, T., Morita, S., Kimura, M., Terawaki, N., Shibutani, M. and Hatada, I. 2017. Efficient generation of conditional knockout mice via sequential introduction of lox sites. Scientific Reports, 7(1): 7891.

Hsu, P. D., Scott, D. A., Weinstein, J. A., Ran, F. A., Konermann, S., Agarwala, V., et al. 2013. DNA targeting specificity of RNA-guided Cas9 nucleases. Nature Biotechnology, 31(9): 827-832.

Hustedt, N. and Durocher, D. 2017. The control of DNA repair by the cell cycle. Nature Cell Biology, 19(1): 1-9.

Jayaraman, M., Ansell, S. M., Mui, B. L., Tam, Y. K., Chen, J., Du, X., et al. 2012. Maximizing the potency of siRNA lipid nanoparticles for hepatic gene silencing in vivo. Angewandte Chemie International Edition, 51(34): 8529-8533.

Jinek, M., Chylinski, K., Fonfara, I., Hauer, M., Doudna, J. A. and Charpentier, E. 2012. A programmable dual-RNA-guided DNA endonuclease in adaptive bacterial immunity. Science, 337(6096): 816-821.

Jones, A. T. and Sayers, E. J. 2012. Cell entry of cell penetrating peptides: Tales of tails wagging dogs. Journal of Controlled Release, 161(2): 582-591.

Kameda, S., Maruyama, H., Higuchi, N., Iino, N., Nakamura, G., Miyazaki, J., et al. 2004. Kidney-targeted naked DNA transfer by retrograde injection into the renal vein in mice. Biochemical and Biophysical Research Communications, 314(2): 390-395.

Kamimura, K., Yokoo, T., Abe, H., Kobayashi, Y., Ogawa, K., Shinagawa, Y., et al. 2015. Image-guided hydrodynamic gene delivery: Current status and future directions. Pharmaceutics, 7(3): 213-223.

Kanasty, R., Dorkin, J. R., Vegas, A. and Anderson, D. 2013. Delivery materials for siRNA therapeutics. Nature Materials, 12(11): 967-977.

Kaushik, A., Yndart, A., Atluri, V., Tiwari, S., Tomitaka, A., et al. 2019. Magnetically guided non-invasive CRISPR-Cas9/gRNA delivery across blood-brain barrier to eradicate latent HIV-1 infection. Scientific Reports, 9(1): 3928.

Kim, S., Kim, D., Cho, S. W., Kim, J. and Kim, J.-S. 2014. Highly efficient RNA-guided genome editing in human cells via delivery of purified Cas9 ribonucleoproteins. Genome Research, 24(6): 1012-1019.

Kim, S., Koo, T., Jee, H.-G., Cho, H.-Y., Lee, G., Lim, D.-G., et al (2018). CRISPR RNAs trigger innate immune responses in human cells. Genome Research, 28(3): 367-373.

Klibanov, A. L., Maruyama, K., Torchilin, V. P. and Huang, L. (1990). Amphipathic polyethyleneglycols effectively prolong the circulation time of liposomes. FEBS Letters, 268(1): 235-237.

Klichinsky, M., Ruella, M., Shestova, O., Lu, X. M., Best, A., Zeeman, M., et al. 2020. Human chimeric antigen receptor macrophages for cancer immunotherapy. Nature Biotechnology, 38(8): 947-953.

Kobayashi, N, Kuramoto, T., Yamaoka, K., Hashida, M. and Takakura, Y. 2001. Hepatic uptake and gene expression mechanisms following intravenous administration of plasmid DNA by conventional and hydrodynamics-based procedures. The Journal of Pharmacology and Experimental Therapeutics, 297(3): 853-860.

Kobayashi, Naoki, Hirata, K., Chen, S., Kawase, A., Nishikawa, M. and Takakura, Y. 2004. Hepatic delivery of particulates in the submicron range by a hydrodynamics-based procedure: Implications for particulate gene delivery systems. The Journal of Gene Medicine, 6(4): 455-463.

Kowalski, P. S., Rudra, A., Miao, L. and Anderson, D. G. 2019. Delivering the Messenger: Advances in technologies for therapeutic mRNA delivery. Molecular Therapy, 27(4): 710-728.

Krishnamurthy, S., Wohlford-Lenane, C., Kandimalla, S., Sartre, G., Meyerholz, D. K., Théberge, V., et al. 2019. Engineered amphiphilic peptides enable delivery of proteins and CRISPR-associated nucleases to airway epithelia. Nature Communications, 10(1): 4906.

Kuai, R., Yuan, W., Qin, Y., Chen, H., Tang, J., Yuan, M., et al. 2010. Efficient delivery of payload into tumor cells in a controlled manner by TAT and Thiolytic Cleavable PEG co-modified liposomes. Molecular Pharmaceutics, 7(5): 1816-1826.

Kulkarni, J. A., Cullis, P. R. and Meel, R. van der. 2018. Lipid nanoparticles enabling gene therapies: From concepts to clinical utility. Nucleic Acid Therapeutics, 28(3): 146-157.

Kültz, D., Madhany, S. and Burg, M. B. 1998. Hyperosmolality causes growth arrest of murine kidney cells: Induction of GADD45 and GADD153 by osmosensing via stress-activated protein kinase 2. Journal of Biological Chemistry, 273(22): 13645-13651.

Lawrence, M. S., Phillips, K. J. and Liu, D. R. 2007. Supercharging proteins can impart unusual resilience. Journal of the American Chemical Society, 129(33): 10110-10112.

Lee, A. G. 1977. Lipid phase transitions and phase diagrams I. Lipid phase transitions. Biochimica et Biophysica Acta (BBA) – Reviews on Biomembranes, 472(2): 237-281.

Lee, B., Lee, K., Panda, S., Gonzales-Rojas, R., Chong, A., Bugay, V., et al. 2018. Nanoparticle delivery of CRISPR into the brain rescues a mouse model of fragile X syndrome from exaggerated repetitive behaviours. Nature Biomedical Engineering, 2(7): 497-507.

Lee, J., Sharei, A., Sim, W. Y., Adamo, A., Langer, R., Jensen, K. F., et al. 2012. Nonendocytic delivery of functional engineered nanoparticles into the cytoplasm of live cells using a novel, high-throughput microfluidic device. Nano Letters, 12(12): 6322-6327.

Lee, K., Conboy, M., Park, H. M., Jiang, F., Kim, H. J., Dewitt, M. A., et al. 2017. Nanoparticle delivery of Cas9 ribonucleoprotein and donor DNA in vivo induces homology-directed DNA repair. Nature Biomedical Engineering, 1(11): 889-901.

Leibold, J., Ruscetti, M., Cao, Z., Ho, Y.-J., Baslan, T., Zou, M., et al. 2020. Somatic tissue engineering in mouse models reveals an actionable role for WNT pathway alterations in prostate cancer metastasis. Cancer Discovery, 10(7): 1038-1057.

Lewis, D. L. and Wolff, J. A. 2007. Systemic siRNA delivery via hydrodynamic intravascular injection. Advanced Drug Delivery Reviews, 59(2-3): 115-123.

Li, J., Wang, B., Juba, B. M., Vazquez, M., Kortum, S. W., Pierce, B. S., et al. 2017. Microfluidic-enabled intracellular delivery of membrane impermeable inhibitors to study target engagement in human primary cells. ACS Chemical Biology, 12(12): 2970-2974.

Li, S., Gao, X., Son, K., Sorgi, F., Hofland, H. and Huang, L. 1996. DC-Chol lipid system in gene transfer. Journal of Controlled Release, 39(2-3): 373-381.

Li, S.-D., Chen, Y.-C., Hackett, M. J. and Huang, L. 2008. Tumor-targeted delivery of siRNA by self-assembled nanoparticles. Molecular Therapy, 16(1): 163-169.

Li, W., Huang, Z., MacKay, J. A., Grube, S. and Szoka, F. C. 2005. Low-pH-sensitive poly(ethylene glycol) (PEG)-stabilized plasmid nanolipoparticles: Effects of PEG chain length, lipid composition and assembly conditions on gene delivery. The Journal of Gene Medicine, 7(1): 67-79.

Li, X., Liu, Y., Wu, X., Gao, Y., Zhang, J., Zhang, D., et al. 2016. Aptamer-functionalized peptide H3CR5C as a novel nanovehicle for codelivery of fasudil and miRNA-195 targeting hepatocellular carcinoma. International Journal of Nanomedicine, Volume 11: 3891-3905.

Liang, X., Potter, J., Kumar, S., Zou, Y., Quintanilla, R., Sridharan, M., et al. 2015. Rapid and highly efficient mammalian cell engineering via Cas9 protein transfection. Journal of Biotechnology, 208: 44-53.

Lico, C., Giardullo, P., Mancuso, M., Benvenuto, E., Santi, L. and Baschieri, S. 2016. A biodistribution study of two differently shaped plant virus nanoparticles reveals new peculiar traits. Colloids and Surfaces B: Biointerfaces, 148: 431-439.

Lim, K. H., Huang, H., Pralle, A. and Park, S. 2013. Stable, high-affinity streptavidin monomer for protein labeling and monovalent biotin detection. Biotechnology and Bioengineering, 110(1): 57-67.

Lin, S., Staahl, B. T., Alla, R. K. and Doudna, J. A. 2014. Enhanced homology-directed human genome engineering by controlled timing of CRISPR/Cas9 delivery. ELife, 3: e04766.

Liu, C., Zhang, L., Zhu, W., Guo, R., Sun, H., Chen, X., et al (2020). Barriers and strategies of cationic liposomes for cancer gene therapy. Molecular Therapy – Methods & Clinical Development, 18: 751-764.

Liu, F., Song, Y. K. and Liu, D. 1999. Hydrodynamics-based transfection in animals by systemic administration of plasmid DNA. Gene Therapy, 6(7): 1258-1266.

Liu, J. and Peng, Q. 2017. Protein-gold nanoparticle interactions and their possible impact on biomedical applications. Acta Biomaterialia, 55: 13-27.

Longmuir, K. J., Robertson, R. T., Haynes, S. M., Baratta, J. L. and Waring, A. J. 2006. Effective targeting of liposomes to liver and hepatocytes in vivo by incorporation of a plasmodium amino acid sequence. Pharmaceutical Research, 23(4): 759-769.

Lostalé-Seijo, I., Louzao, I., Juanes, M. and Montenegro, J. 2017. Peptide/Cas9 nanostructures for ribonucleoprotein cell membrane transport and gene edition. Chemical Science, 8(12): 7923-7931.

Lumpkin, E. A., Collisson, T., Parab, P., Omer-Abdalla, A., Haeberle, H., Chen, P., et al. 2003. Math1-driven GFP expression in the developing nervous system of transgenic mice. Gene Expression Patterns, 3(4): 389-395.

Lundberg, M. and Johansson, M. 2001. Is VP22 nuclear homing an artifact? Nature Biotechnology, 19(8): 713.

Lytton-Jean, A. K. R., Langer, R. and Anderson, D. G. (2011). Five Years of siRNA Delivery: Spotlight on Gold Nanoparticles. Small, 7(14): 1932-1937.

Ma, X., Chen, C., Veevers, J., Zhou, X., Ross, R. S., Feng, W., et al (2017). CRISPR/Cas9-mediated gene manipulation to create single-amino-acid-substituted and floxed mice with a cloning-free method. Scientific Reports, 7(1): 42244.

MacLeod, D. T., Antony, J., Martin, A. J., Moser, R. J., Hekele, A., Wetzel, K. J., et al. 2017. Integration of a CD19 CAR into the TCR Alpha Chain Locus Streamlines Production of Allogeneic Gene-Edited CAR T Cells. Molecular Therapy, 25(4): 949-961.

Mahon, K. P., Love, K. T., Whitehead, K. A., Qin, J., Akinc, A., Leshchiner, E., et al. 2010. Combinatorial approach to determine functional group effects on lipidoid-mediated siRNA delivery. Bioconjugate Chemistry, 21(8): 1448-1454.

Mandal, P. K., Ferreira, L. M. R., Collins, R., Meissner, T. B., Boutwell, C. L., Friesen, M., et al 2014. Efficient ablation of genes in human hematopoietic stem and effector cells using CRISPR/Cas9. Cell Stem Cell, 15(5): 643-652.

Mao, S., Neu, M., Germershaus, O., Merkel, O., Sitterberg, J., Bakowsky, U., et al. 2006. Influence of polyethylene glycol chain length on the physicochemical and biological properties of poly(ethylene imine)-graft-poly(ethylene glycol) block copolymer/SiRNA polyplexes. Bioconjugate Chemistry, 17(5): 1209-1218.

Maresch, R., Mueller, S., Veltkamp, C., Öllinger, R., Friedrich, M., Heid, I., et al. 2016. Multiplexed pancreatic genome engineering and cancer induction by transfection-based CRISPR/Cas9 delivery in mice. Nature Communications, 7(1): 10770.

McGarry, T. J. and Kirschner, M. W. (1998). Geminin, an inhibitor of DNA replication, is degraded during mitosis. Cell, 93(6): 1043-1053.

McNaughton, B. R., Cronican, J. J., Thompson, D. B. and Liu, D. R. 2009. Mammalian cell penetration, siRNA transfection, and DNA transfection by supercharged proteins. Proceedings of the National Academy of Sciences, 106(15): 6111-6116.

Miao, L., Lin, J., Huang, Y., Li, L., Delcassian, D., Ge, Y., et al. 2020. Synergistic lipid compositions for albumin receptor mediated delivery of mRNA to the liver. Nature Communications, 11(1): 2424.

Miura, H., Gurumurthy, C. B., Sato, T., Sato, M. and Ohtsuka, M. 2015. CRISPR/Cas9-based generation of knockdown mice by intronic insertion of artificial microRNA using longer single-stranded DNA. Scientific Reports, 5(1): 12799.

Momose, T., Tonegawa, † Akane, Takeuchi, J., Ogawa, H., Umesono, K., et al. 1999. Efficient targeting of gene expression in chick embryos by microelectroporation. Development, Growth & Differentiation, 41(3): 335-344.

Moon, S. B., Kim, D. Y., Ko, J.-H. and Kim, Y.-S. 2019. Recent advances in the CRISPR genome editing tool set. Experimental & Molecular Medicine, 51(11): 1-11.

Morais, C. M., Cardoso, A. M., Cunha, P. P., Aguiar, L., Vale, N., Lage, E., et al. 2018. Acylation of the S413-PV cell-penetrating peptide as a means of enhancing its capacity to mediate nucleic acid delivery: Relevance of peptide/lipid interactions. Biochimica et Biophysica Acta (BBA) - Biomembranes, 1860(12): 2619-2634.

Mout, R., Ray, M., Tonga, G. Y., Lee, Y.-W., Tay, T., Sasaki, K., et al. 2017. Direct Cytosolic Delivery of CRISPR/Cas9-ribonucleoprotein for efficient gene editing. ACS Nano, 11(3): 2452-2458.

Muramatsu, T., Mizutani, Y. and Okumura, J. 1996. Live Detection of the firefly luciferase gene expression by bioluminescence in incubating chicken embryos. Nihon Chikusan Gakkaiho, 67(10): 906-909.

Nakagawa, Y., Oikawa, F., Mizuno, S., Ohno, H., Yagishita, Y., Satoh, A., et al. 2016. Hyperlipidemia and hepatitis in liver-specific CREB3L3 knockout mice generated using a one-step CRISPR/Cas9 system. Scientific Reports, 6(1): 27857.

Nelson, C. E. and Gersbach, C. A. 2015. Engineering delivery vehicles for genome editing. Annual Review of Chemical and Biomolecular Engineering, 7(1): 1-26.

Neumann, E., Schaefer-Ridder, M., Wang, Y. and Hofschneider, P. H. (1982). Gene transfer into mouse lyoma cells by electroporation in high electric fields. The EMBO Journal, 1(7): 841-845.

Nishi, T., Yoshizato, K., Yamashiro, S., Takeshima, H., Sato, K., Hamada, K., et al. 1996. High-efficiency in vivo gene transfer using intraarterial plasmid DNA injection following in vivo electroporation. Cancer Research, 56(5): 1050-1055.

Oh, S. A., Seki, A. and Rutz, S. 2019. Ribonucleoprotein Transfection for CRISPR/Cas9-mediated gene knockout in primary T cells. Current Protocols in Immunology, 124(1): e69.

Ohtsuka, M., Sato, M., Miura, H., Takabayashi, S., Matsuyama, M., Koyano, T., et al. 2018. i-GONAD: A robust method for in situ germline genome engineering using CRISPR nucleases. Genome Biology, 19(1): 25.

Pandey, A. P. and Sawant, K. K. 2016. Polyethylenimine: A versatile, multifunctional non-viral vector for nucleic acid delivery. Materials Science and Engineering: C, 68: 904-918.

Pankowicz, F. P., Barzi, M., Kim, K. H., Legras, X., Martins, C. S., Wooton-Kee, C. R., et al. 2018. Rapid disruption of genes specifically in livers of mice using multiplex CRISPR/Cas9 editing. Gastroenterology, 155(6): 1967-1970 e6.

Patel, S., Ashwanikumar, N., Robinson, E., Xia, Y., Mihai, C., Griffith, J. P., et al. 2020. Naturally-occurring cholesterol analogues in lipid nanoparticles induce polymorphic shape and enhance intracellular delivery of mRNA. Nature Communications, 11(1): 983.

Patel, S. G., Sayers, E. J., He, L., Narayan, R., Williams, T. L., Mills, E. M., et al. 2019. Cell-penetrating peptide sequence and modification dependent uptake and subcellular distribution of green florescent protein in different cell lines. Scientific Reports, 9(1): 6298.

Peer, D., Park, E. J., Morishita, Y., Carman, C. V. and Shimaoka, M. 2008. Systemic leukocyte-directed siRNA delivery revealing cyclin D1 as an anti-inflammatory target. Science, 319(5863): 627-630.

Pitek, A. S., Wen, A. M., Shukla, S. and Steinmetz, N. F. 2016. The protein corona of plant virus nanoparticles influences their dispersion properties, cellular interactions, and in vivo fates. Small, 12(13): 1758-1769.

Pozzi, D., Marchini, C., Cardarelli, F., Amenitsch, H., Garulli, C., Bifone, A., et al. 2012. Transfection efficiency boost of cholesterol-containing lipoplexes. Biochimica et Biophysica Acta (BBA) – Biomembranes, 1818(9): 2335-2343.

Qiu, Y., Tong, S., Zhang, L., Sakurai, Y., Myers, D. R., Hong, L., et al. 2017. Magnetic forces enable controlled drug delivery by disrupting endothelial cell-cell junctions. Nature Communications, 8(1): 15594.

Quadros, R. M., Miura, H., Harms, D. W., Akatsuka, H., Sato, T., Aida, T., et al. 2017. Easi-CRISPR: A robust method for one-step generation of mice carrying conditional and insertion alleles using long ssDNA donors and CRISPR ribonucleoproteins. Genome Biology, 18(1): 92.

Ramakrishna, S., Dad, A.-B. K., Beloor, J., Gopalappa, R., Lee, S.-K. and Kim, H. 2014. Gene disruption by cell-penetrating peptide-mediated delivery of Cas9 protein and guide RNA. Genome Research, 24(6): 1020-1027.

Rana, S., Bajaj, A., Mout, R. and Rotello, V. M. 2012. Monolayer coated gold nanoparticles for delivery applications. Advanced Drug Delivery Reviews, 64(2): 200-216.

Renaud, J.-B., Boix, C., Charpentier, M., De Cian, A., Cochennec, J., Duvernois-Berthet, E., et al. 2016. Improved genome editing efficiency and flexibility using modified oligonucleotides with TALEN and CRISPR-Cas9 nucleases. Cell Reports, 14(9): 2263-2272.

Ritz, S., Schöttler, S., Kotman, N., Baier, G., Musyanovych, A., Kuharev, J., et al. 2015. Protein corona of nanoparticles: Distinct proteins regulate the cellular uptake. Biomacromolecules, 16(4): 1311-1321.

Rohiwal, S. S., Dvorakova, N., Klima, J., Vaskovicova, M., Senigl, F., Slouf, M., et al. 2020. Polyethylenimine based magnetic nanoparticles mediated non-viral CRISPR/Cas9 system for genome editing. Scientific Reports, 10(1): 4619.

Rosazza, C., Meglic, S. H., Zumbusch, A., Rols, M.-P. and Miklavcic, D. 2016. Gene electrotransfer: A mechanistic perspective. Current Gene Therapy, 16(2): 98-129.

Rosenblum, D., Gutkin, A., Kedmi, R., Ramishetti, S., Veiga, N., Jacobi, A. M., et al. 2020. CRISPR-Cas9 genome editing using targeted lipid nanoparticles for cancer therapy. Science Advances, 6(47): eabc9450.

Rossmanith, W., Chabicovsky, M., Herkner, K. and Schulte-Hermann, R. 2002. Cellular gene dose and kinetics of gene expression in mouse livers transfected by high-volume tail-vein injection of naked DNA. DNA and Cell Biology, 21(11): 847-853.

Roth, T. L., Puig-Saus, C., Yu, R., Shifrut, E., Carnevale, J., Li, P. J., et al. 2018. Reprogramming human T cell function and specificity with non-viral genome targeting. Nature, 559(7714): 405-409.

Rudolph, C., Plank, C., Lausier, J., Schillinger, U., Müller, R. H. and Rosenecker, J. 2003. Oligomers of the arginine-rich motif of the HIV-1 TAT protein are capable of transferring plasmid DNA into cells. Journal of Biological Chemistry, 278(13): 11411-11418.

Sandhu, K. K., McIntosh, C. M., Simard, J. M., Smith, S. W. and Rotello, V. M. 2002. Gold nanoparticle-mediated transfection of mammalian cells. Bioconjugate Chemistry, 13(1): 3-6.

Saptarshi, S. R., Duschl, A. and Lopata, A. L. 2013. Interaction of nanoparticles with proteins: Relation to bio-reactivity of the nanoparticle. Journal of Nanobiotechnology, 11(1): 26.

Schellekens, H., Hennink, W. E. and Brinks, V. 2013. The immunogenicity of polyethylene glycol: Facts and fiction. Pharmaceutical Research, 30(7): 1729-1734.

Schmid-Burgk, J. L., Gao, L., Li, D., Gardner, Z., Strecker, J., Lash, B., et al. 2020. Highly parallel profiling of Cas9 variant specificity. Molecular Cell, 78(4): 794-800 e8.

Schober, K., Müller, T. R., Gökmen, F., Grassmann, S., Effenberger, M., Poltorak, M., et al. 2019. Orthotopic replacement of T-cell receptor α- and β-chains with preservation of near-physiological T-cell function. Nature Biomedical Engineering, 3(12): 974-984.

Sebestyén, M. G., Budker, V. G., Budker, T., Subbotin, V. M., Zhang, G., Monahan, S. D., et al. 2006. Mechanism of plasmid delivery by hydrodynamic tail vein injection. I. Hepatocyte uptake of various molecules. The Journal of Gene Medicine, 8(7): 852-873.

Seferos, D. S., Prigodich, A. E., Giljohann, D. A., Patel, P. C. and Mirkin, C. A. 2009. Polyvalent DNA Nanoparticle Conjugates Stabilize Nucleic Acids. Nano Letters, 9(1): 308-311.

Seki, A. and Rutz, S. 2018. Optimized RNP transfection for highly efficient CRISPR/Cas9-mediated gene knockout in primary T cells. Journal of Experimental Medicine, 215(3): 985-997.

Semple, S. C., Akinc, A., Chen, J., Sandhu, A. P., Mui, B. L., Cho, C. K., et al. 2010. Rational design of cationic lipids for siRNA delivery. Nature Biotechnology, 28(2): 172-176.

Setten, R. L., Rossi, J. J. and Han, S. 2019. The current state and future directions of RNAi-based therapeutics. Nature Reviews Drug Discovery, 18(6): 421-446.

Shahbazi, R., Sghia-Hughes, G., Reid, J. L., Kubek, S., Haworth, K. G., Humbert, O., et al. 2019. Targeted homology-directed repair in blood stem and progenitor cells with CRISPR nanoformulations. Nature Materials, 18(10): 1124-1132.

Shalek, A. K., Robinson, J. T., Karp, E. S., Lee, J. S., Ahn, D.-R., Yoon, M.-H., et al. 2010. Vertical silicon nanowires as a universal platform for delivering biomolecules into living cells. Proceedings of the National Academy of Sciences, 107(5): 1870-1875.

Sharei, A., Zoldan, J., Adamo, A., Sim, W. Y., Cho, N., Jackson, E., et al. 2013. A vector-free microfluidic platform for intracellular delivery. Proceedings of the National Academy of Sciences, 110(6): 2082-2087.

Shen, B., Zhang, J., Wu, H., Wang, J., Ma, K., Li, Z., et al. 2013. Generation of gene-modified mice via Cas9/RNA-mediated gene targeting. Cell Research, 23(5): 720-723.

Shinmyo, Y., Tanaka, S., Tsunoda, S., Hosomichi, K., Tajima, A. and Kawasaki, H. 2016. CRISPR/Cas9-mediated gene knockout in the mouse brain using in utero electroporation. Scientific Reports, 6(1): 20611.

Song, C.-Q., Jiang, T., Richter, M., Rhym, L. H., Koblan, L. W., Zafra, M. P., et al. 2020. Adenine base editing in an adult mouse model of tyrosinaemia. Nature Biomedical Engineering, 4(1): 125-130.

Song, F. and Stieger, K. 2017. Optimizing the DNA donor template for homology-directed repair of double-strand breaks. Molecular Therapy – Nucleic Acids, 7: 53-60.

Stanley, S. A., Sauer, J., Kane, R. S., Dordick, J. S. and Friedman, J. M. 2015. Remote regulation of glucose homeostasis in mice using genetically encoded nanoparticles. Nature Medicine, 21(1): 92-98.

Stefanick, J. F., Ashley, J. D. and Bilgicer, B. 2013. Enhanced cellular uptake of peptide-targeted nanoparticles through increased peptide hydrophilicity and optimized ethylene glycol peptide-linker length. ACS Nano, 7(9): 8115-8127.

Stefanick, J. F., Ashley, J. D., Kiziltepe, T. and Bilgicer, B. 2013. A systematic analysis of peptide linker length and liposomal polyethylene glycol coating on cellular uptake of peptide-targeted liposomes. ACS Nano, 7(4): 2935-2947.

Szeto, G. L., Egeren, D. V., Worku, H., Sharei, A., Alejandro, B., Park, C., et al. 2015. Microfluidic squeezing for intracellular antigen loading in polyclonal B-cells as cellular vaccines. Scientific Reports, 5(1): 10276.

Tai, Z., Wang, X., Tian, J., Gao, Y., Zhang, L., Yao, C., et al. 2015. Biodegradable stearylated peptide with internal disulfide bonds for efficient delivery of siRNA in vitro and in vivo. Biomacromolecules, 16(4): 1119-1130.

Takahashi, G., Gurumurthy, C. B., Wada, K., Miura, H., Sato, M. and Ohtsuka, M. 2015. GONAD: Genome-editing via oviductal nucleic acids delivery system: A novel microinjection independent genome engineering method in mice. Scientific Reports, 5(1): 11406.

Takahashi, H., Sinoda, K. and Hatta, I. 1996. Effects of cholesterol on the lamellar and the inverted hexagonal phases of dielaidoylphosphatidylethanolamine. Biochimica et Biophysica Acta (BBA) – General Subjects, 1289(2): 209-216.

Taylor, R. E. and Zahid, M. 2020. Cell penetrating peptides, novel vectors for gene therapy. Pharmaceutics, 12(3): 225.

Temming, K., Schiffelers, R. M., Molema, G. and Kok, R. J. 2005. RGD-based strategies for selective delivery of therapeutics and imaging agents to the tumour vasculature. Drug Resistance Updates, 8(6): 381-402.

Thewalt, J. L. and Bloom, M. 1992. Phosphatidylcholine cholesterol phase diagrams. Biophysical Journal, 63(4): 1176-1181.

Thompson, D. B., Cronican, J. J. and Liu, D. R. 2012. Chapter twelve: Engineering and identifying supercharged proteins for macromolecule delivery into mammalian cells. Methods in Enzymology, 503: 293-319.

Torchilin, V. P. 2008. Tat peptide-mediated intracellular delivery of pharmaceutical nanocarriers. Advanced Drug Delivery Reviews, 60(4-5): 548-558.

Verhoef, J. J. F. and Anchordoquy, T. J. 2013. Questioning the use of PEGylation for drug delivery. Drug Delivery and Translational Research, 3(6): 499-503.

Vivès, E., Schmidt, J. and Pèlegrin, A. 2008. Cell-penetrating and cell-targeting peptides in drug delivery. Biochimica et Biophysica Acta (BBA) – Reviews on Cancer, 1786(2): 126-138.

Vuillard, L., Braun-Breton, C. and Rabilloud, T. 1995. Non-detergent sulphobetaines: A new class of mild solubilization agents for protein purification. Biochemical Journal, 305(1): 337-343.

Wang, D., Zhang, F. and Gao, G. 2020. CRISPR-based therapeutic genome editing: Strategies and in vivo delivery by AAV vectors. Cell, 181(1): 136-150.

Wang, H., Yang, H., Shivalila, C. S., Dawlaty, M. M., Cheng, A. W., Zhang, F., et al (2013). One-step generation of mice carrying mutations in multiple genes by CRISPR/Cas-mediated genome engineering. Cell, 153(4): 910-918.

Wang, H.-X., Song, Z., Lao, Y.-H., Xu, X., Gong, J., Cheng, D., et al. 2018. Nonviral gene editing via CRISPR/Cas9 delivery by membrane-disruptive and endosomolytic helical polypeptide. Proceedings of the National Academy of Sciences, 115(19): 201712963.

Wang, M., Zuris, J. A., Meng, F., Rees, H., Sun, S., Deng, P., et al. 2016. Efficient delivery of genome-editing proteins using bioreducible lipid nanoparticles. Proceedings of the National Academy of Sciences, 113(11): 2868-2873.

Wang, P., Zhang, L., Xie, Y., Wang, N., Tang, R., Zheng, W., et al. 2017. Genome editing for cancer therapy: Delivery of Cas9 protein/sgRNA plasmid via a gold nanocluster/lipid core-shell nanocarrier. Advanced Science, 4(11): 1700175.

Wang, Y., Yang, Y., Yan, L., Kwok, S. Y., Li, W., Wang, Z., et al. 2014. Poking cells for efficient vector-free intracellular delivery. Nature Communications, 5(1): 4466.

Warren, L., Manos, P. D., Ahfeldt, T., Loh, Y.-H., Li, H., Lau, F., et al. 2010. Highly efficient reprogramming to pluripotency and directed differentiation of human cells with synthetic modified mRNA. Cell Stem Cell, 7(5): 618-630.

Wei, T., Cheng, Q., Min, Y.-L., Olson, E. N. and Siegwart, D. J. 2020. Systemic nanoparticle delivery of CRISPR-Cas9 ribonucleoproteins for effective tissue specific genome editing. Nature Communications, 11(1): 3232.

Wheeler, M. A., Smith, C. J., Ottolini, M., Barker, B. S., Purohit, A. M., Grippo, et al. 2016. Genetically targeted magnetic control of the nervous system. Nature Neuroscience, 19(5): 756-761.

Wienert, B., Shin, J., Zelin, E., Pestal, K. and Corn, J. E. 2018. In vitro-transcribed guide RNAs trigger an innate immune response via the RIG-I pathway. PLOS Biology, 16(7): e2005840.

Witten, J. and Ribbeck, K. 2017. The particle in the spider's web: Transport through biological hydrogels. Nanoscale, 9(24): 8080-8095.

Witzigmann, D., Kulkarni, J. A., Leung, J., Chen, S., Cullis, P. R. and Meel, R. van der. 2020. Lipid nanoparticle technology for therapeutic gene regulation in the liver. Advanced Drug Delivery Reviews, 159: 344-363.

Wolff, J. A. and Budker, V. 2005. The mechanism of naked DNA uptake and expression. Advances in Genetics, 54(Gene Ther. 37 1996): 1-20.

Xie, X., Xu, A. M., Leal-Ortiz, S., Cao, Y., Garner, C. C. and Melosh, N. A. 2013. Nanostraw-electroporation system for highly efficient intracellular delivery and transfection. ACS Nano, 7(5): 4351-4358.

Xu, Y. and Szoka, F. C. 1996. Mechanism of DNA release from cationic liposome/DNA complexes used in cell transfection †, ‡. Biochemistry, 35(18): 5616-5623.

Xue, W., Chen, S., Yin, H., Tammela, T., Papagiannakopoulos, T., Joshi, N. S., et al. 2014. CRISPR-mediated direct mutation of cancer genes in the mouse liver. Nature, 514(7522): 380-384.

Yan, X., Kuipers, F., Havekes, L. M., Havinga, R., Dontje, B., Poelstra, K., et al. 2005. The role of apolipoprotein E in the elimination of liposomes from blood by hepatocytes in the mouse. Biochemical and Biophysical Research Communications, 328(1): 57-62.

Yang, D., Scavuzzo, M. A., Chmielowiec, J., Sharp, R., Bajic, A. and Borowiak, M. 2016. Enrichment of G2/M cell cycle phase in human pluripotent stem cells enhances HDR-mediated gene repair with customizable endonucleases. Scientific Reports, 6(1): 21264.

Yang, H., Wang, H., Shivalila, C. S., Cheng, A. W., Shi, L. and Jaenisch, R. 2013. One-step generation of mice carrying reporter and conditional alleles by CRISPR/Cas-mediated genome engineering. Cell, 154(6): 1370-1379.

Yao, X., Zhang, M., Wang, X., Ying, W., Hu, X., Dai, P., et al. 2018. Tild-CRISPR allows for efficient and precise gene knockin in mouse and human cells. Developmental Cell, 45(4): 526-536 e5.

Yarmush, M. L., Golberg, A., Serša, G., Kotnik, T. and Miklavčič, D. 2014. Electroporation-based technologies for medicine: Principles, applications, and challenges. Biomedical Engineering, 16(1): 295-320.

Yen, S.-T., Zhang, M., Deng, J. M., Usman, S. J., Smith, C. N., Parker-Thornburg, J., et al. 2014. Somatic mosaicism and allele complexity induced by CRISPR/Cas9 RNA injections in mouse zygotes. Developmental Biology, 393(1): 3-9.

Yin, H., Song, C.-Q., Dorkin, J. R., Zhu, L. J., Li, Y., Wu, Q., et al. 2016. Therapeutic genome editing by combined viral and non-viral delivery of CRISPR system components in vivo. Nature Biotechnology, 34(3): 328-333.

Zelphati, O. and Szoka, F. C. 1996. Mechanism of oligonucleotide release from cationic liposomes. Proceedings of the National Academy of Sciences, 93(21): 11493-11498.

Zhang, G., Gao, X., Song, Y. K., Vollmer, R., Stolz, D. B., Gasiorowski, et al. 2004. Hydroporation as the mechanism of hydrodynamic delivery. Gene Therapy, 11(8): 675-682.

Zhang, Guofeng, Budker, V. and Wolff, J. A. 1999. High levels of foreign gene expression in hepatocytes after tail vein injections of naked plasmid DNA. Human Gene Therapy, 10(10): 1735-1737.

Zhang, X., Edwards, J. P. and Mosser, D. M. 2009. Macrophages and dendritic cells, methods and protocols. Methods in Molecular Biology, 531: 123-143.

Zhang, Y., Li, P., Fang, H., Wang, G. and Zeng, X. 2020. Paving the way towards universal chimeric antigen receptor therapy in cancer treatment: Current landscape and progress. Frontiers in Immunology, 11: 604915.

Zhu, H., Zhang, L., Tong, S., Lee, C. M., Deshmukh, H. and Bao, G. 2019. Spatial control of in vivo CRISPR-Cas9 genome editing via nanomagnets. Nature Biomedical Engineering, 3(2): 126-136.

Zuris, J. A., Thompson, D. B., Shu, Y., Guilinger, J. P., Bessen, J. L., Hu, J. H., et al. 2015. Cationic lipid-mediated delivery of proteins enables efficient protein-based genome editing in vitro and in vivo. Nature Biotechnology, 33(1): 73-80.

# CRISPR Medicine: Advance, Progress and Challenges

Guillermo Ureña-Bailén[1†], Justin S. Antony[1†], Yujuan Hou[1], Janani Raju[1,2,3],
Andres Lamsfus-Calle[1], Alberto Daniel-Moreno[1], Rupert Handgretinger[1]
and Markus Mezger[1]*

[1] University Children's Hospital Department of Paediatrics I, Haematology and Oncology,
University of Tuebingen, Tuebingen 72076, Germany
[2] Department of Neurodegenerative Diseases, Hertie-Institute for Clinical Brain Research
and Center of Neurology, University of Tuebingen, Tuebingen 72076, Germany
[3] German Research Center for Neurodegenerative Diseases (DZNE),
Tuebingen 72076, Germany

## Introduction

The discovery of polymerase chain reaction (PCR) and sequencing technology
are considered as highly valued inventions in molecular biology. The scientific
community keeps expanding the list with new ground-breaking discoveries such as
CRISPR/Cas. In 2015, Science journal announced CRISPR/Cas as *Breakthrough
of the Year,* excitement that gracefully culminated with the primary inventors,
Emmanuelle Charpentier and Jennifer Doudna, receiving the 2020 Nobel Prize
in Chemistry for this discovery. During these five years, CRISPR/Cas technology
has been used by scientists in several fields of biology, including in biomedical
applications to prevent or cure diseases, for diagnosis purposes (e.g. COVID-19), in
the generation of genetically engineered animals and plants, and was even discussed
as an attempt to resurrect extinct species such as the woolly mammoth (Callaway
2015, Yang and Wu 2018, Broughton et al. 2020, Zhu et al. 2020). The steep increase
in the number of CRISPR publications between 2015 and 2020 support the fact that
we are living in the CRISPR era. Although genome editing with programmable
nucleases such as zinc-finger nuclease (ZFN) and TAL effector nuclease (TALEN)
existed several years before the advent of CRISPR technology, the latter offers a
higher level of simplicity with remarkable efficacy and easy access (Antony et al.
2018a, b, Khan 2019). Moreover, the CRISPR technology is entering new avenues

* Corresponding author: Markus.Mezger@med.uni-tuebingen.de
†These co-authors contributed equally to this work

towards ambitious goals such as prime editing, base editing, and epigenetic editing (Liang et al. 2017, Rees and Liu 2018, Anzalone et al. 2019, Gjaltema and Rots 2020).

In 2007, researchers from Danisco (a food bio-products company) first reported CRISPR as a distinctive feature of the bacterial genome involved in resistance to bacteriophages. Moreover, the authors reported that after phage infection, bacteria integrated phage genomic sequences into new spacers into their own genomes, to ensure phage-resistance in a sequence-specific manner mediated by *Cas* genes (Barrangou et al. 2007). Although reported as a bacterial defense system, many scientists (including Emmanuelle Charpentier and Jennifer Doudna) explored the working principle of CRISPR and announced that the technology could be harnessed beyond prokaryotes to any organism with DNA as genetic material (Ishino et al. 2018, Lea and Niakan 2019). Since then, CRISPR has attained massive attention from researchers and the general public. As of now, it is the most discussed scientific method by public media due to its potential controversial applications, such as designer babies and germ-line genome editing (Rose and Brown 2019). Despite these debates, CRISPR/Cas reached clinical trials very rapidly. By December 2020, more than 35 different clinical trials were registered using CRISPR as a method of treatment, and almost 20 different start-up companies were launched using CRISPR-based proprietary platforms (Perez Rojo et al. 2018, Hirakawa et al. 2020).

The use of CRISPR in treating human diseases was massively investigated including blood disorders, cancers, infectious diseases, and eye diseases (Wu et al. 2017, Daniel-Moreno et al. 2019, Xu et al. 2019, He 2020, Lin et al. 2020, Martinez-Lage et al. 2020, Ureña-Bailén et al. 2020). The recent clinical trial data of CTX001 for transfusion-dependent β-thalassemia (TDT; NCT03655678) and sickle cell disease (SCD; NCT03745287) with autologous gene-edited CD34$^+$ hematopoietic stem cells through CRISPR/Cas9 targeting the *BCL11A* enhancer for fetal hemoglobin reactivation demonstrated outstanding efficacy by achieving transfusion independence in the patient with TDT and elimination of vaso-occlusive episodes in the patient with SCD (Frangoul et al. 2020). This chapter will address the current advance and progress of CRISPR medicine along with its existing challenges and possible solutions.

## CRISPR medicine: Advance

Originally discovered in bacteria, the famous Clustered Regularly Interspaced Short Palindromic Repeat (CRISPR) was identified as an adaptive immune system against bacteriophages, i.e. viruses that infect and kill bacteria. It was first described in *Escherichia coli* in 1987 (Ishino et al. 1987) but it was not officially named until 2002 (Jansen et al. 2002) (Figure 1).

The system is composed of CRISPR-associated proteins (Cas proteins) that cut foreign DNA into small pieces, about 20bp long (Mojica et al. 2005, Marraffini and Sontheimer 2008, Hille and Charpentier 2016). These fragments are pasted and concatenated in a CRISPR array. (Haft et al. 2005, Mojica et al. 2005, Makarova et al. 2006, Barrangou et al. 2007). The CRISPR array is transcribed to pre-crRNA (CRISPR RNA), which is further processed to separate the individual crRNAs with the aid of an additional trans-activating crRNA (tracrRNA) (Deltcheva et al. 2011). Once the individual crRNAs are generated, they form a complex with tracrRNA and Cas protein. This complex can recognize DNA sequences that are complementary to the crRNA (Mojica et al. 2005, Makarova et al. 2006, Barrangou et al. 2007). If a specific protospacer adjacent motif (PAM) sequence is present in the target sequence, the Cas protein cuts the invasive DNA molecule and generate a double-strand break (DSB) that promotes its destruction (Mojica et al. 2005, Swarts et al. 2012). After several encounters with the pathogen, the CRISPR system can readily counteract repeat infections.

The CRISPR/Cas system is not only present in bacteria but also in most archaea (Mojica et al. 2000, Ishino et al. 2018). Ironically, recent findings reveal that the system is also incorporated in the genome of certain bacteriophages as a defense mechanism against bacteria (Pausch et al. 2020).

The CRISPR system, widely spread in nature, also finds applications in biotechnology due to the molecular mechanism underlying the CRISPR cutting. The technology derived from the CRISPR/Cas system allows precise gene editing with the help of a guide RNA (gRNA). This synthetic sequence comprises crRNA and tracrRNA in the same molecule, thereby simplifying the targeting process and increasing the editing efficiency (Brouns et al. 2008). When the Cas protein is directed to the target by the gRNA and recognizes the PAM sequence, a DSB is induced in the DNA of the cell (Jinek et al. 2012, Jiang et al. 2013, Nishimasu et al. 2014, Hille and Charpentier 2016). This situation severely compromises cell viability since the cell cycle cannot continue until the cleavage is repaired. To avoid cell death, two different repair mechanisms can be observed: homology-directed repair (HDR) and non-homologous end joining (NHEJ) (Barnes 2001, Dudáš and Chovanec 2004, Jiang et al. 2013, Hoban and Bauer 2016). NHEJ is the most likely to occur, but is imprecise,

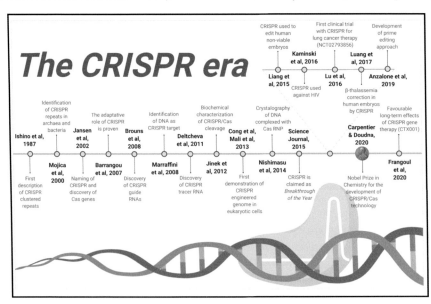

**Figure 1.** Summary of the history of CRISPR/Cas9 technology.

often leading to insertions and deletions of several nucleotides (indel mutations). This mechanism is preferred when a knock-out of the target gene is desired, since indel mutations are very likely to generate a premature stop codon that prevents the correct expression of the protein (Jiang et al. 2013, Hille and Charpentier 2016). On the other hand, HDR can be used for precise editing, which is needed for gene correction or gene addition approaches. It occurs in the presence of a DNA repair template with homologous arms on both sides of the DSB. The break is specifically repaired by homologous recombination (Hille and Charpentier 2016). The use of CRISPR in various diseases in either *in vivo* and *ex vivo* gene editing approaches is illustrated in Figure 2. Likewise, early proof-of-concept studies (2015-2018) concerning the use of CRISPR/Cas technology for treatment and correction of diseases are summarized in Table 1.

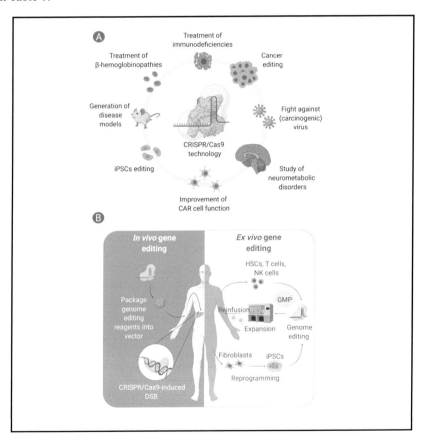

**Figure 2.** A) Medicine and research fields that benefited from the CRISPR/Cas9 technology. B) Gene therapy strategies to treat genetic diseases. The *in vivo* approach is based on the packaging of the CRISPR components inside a suitable vector (nanoparticles, virus, etc.) that will release the cargo inside the target cells. In this way, local genetic modification is expected. The *ex vivo* strategy relies on GMP-grade processing and editing of isolated cells for later reinfusion into the patient. Scheme modified from Hoban et al. 2016 (Copyright by American Society of Hematology).

**Table 1:** Early proof-of-concept studies using CRISPR/Cas technology for the treatment and correction of disease

| Disease | Target | Cells | DOI | Year |
|---|---|---|---|---|
| Duchenne muscular dystrophy | *Dystrophin* | iPSCs | 10.1126/sciadv.1602814 (Zhang et al.) | 2017 |
| Amyotrophic lateral sclerosis | *SOD1 & FUS* | iPSCs | 10.1007/s13238-017-0397-3 (Wang et al.) | 2017 |
| Huntington disease | *HTT* | iPSCs | 10.1016/j.stemcr.2017.01.022 (Xu et al.) | 2017 |
| Recessive dystrophic epidermolysis bullosa | *COL7A1* | iPSCs | 10.1038/npjregenmed.2016.14 (Webber et al.) | 2016 |
| Familial hypercholesterolemia | *LDLR* | iPSCs | 10.1002/hep4.1110 (Omer et al.) | 2017 |
| Hypertrophic cardiomyopathy (HCM) & familial Wolff-Parkinson-White (WPW) syndrome | *PRKAG2* | iPSC-CMs | 10.1016/j.hrthm.2017.09.024 (Ben Jehuda et al.) | 2018 |
| Epidermolysis bullosa | *LAMB3* | Keratinocytes | 10.1016/j.ymthe.2018.07.024 (Benati et al.) | 2018 |
| Fanconi anemia | *FANCD1* | Primary Patient Fibroblasts | 10.3390/ijms18061269 (Skvarova Kramarzova et al.) | 2017 |
| Chronic granulomatous | *NOX2* | HSCs | 10.1126/scitranslmed.aah3480 (De Ravin et al.) | 2017 |
| Juvenile neuronal ceroid lipofuscinosis (Batten disease) | *CLN3* | iPSCs | 10.1089/crispr.2017.0015 (Burnight et al.) | 2018 |
| Leber congenital amaurosis | *CEP290* | Primary fibroblasts | 10.1016/S1525-0016(16)34296-4 (Maeder et al.) | 2015 |

| Disease | Gene | Cell type | Reference | Year |
|---|---|---|---|---|
| Immune-dysregulation polyendocrinopathy-enteropathy-X-linked (IPEX) syndrome | *FOXP3* | Human primary CD4$^+$ T cells | 10.1016/S1525-0016(16)32932-X (Goodwin et al.) | 2016 |
| Fabry disease | *GLA* | Fibroblasts | 10.21037/atm.2017.s043 (Chang et al. 2017b) | 2017 |
| Alzheimer's disease | *PSEN2* | iPSCs | 10.1186/s40478-017-0475-z (Ortiz-Virumbrales et al.) | 2017 |
| Retinitis pigmentosa | *RPGR* | iPSCs | 10.1016/j.stemcr.2018.02.003 (Deng et al.) | 2018 |
| Granular corneal dystrophy | *TGFBI* | Primary corneal keratocytes | 10.1038/s41598-017-16308-2 (Taketani et al) | 2017 |
| Hemophilia B | *F IX* | iPSCs | 10.24920/J1001-9294.2017.032 (He et al.) | 2017 |
| Phenylketonuria | *PAH* | PAH_c.1222C>T COS-7 cells | doi.org/10.1038/srep35794 (Pan et al.) | 2016 |
| Tetrahydrobiopterin (BH4) metabolic disorders | *PTPS* or *DHPR* | iPSCs | 10.1093/hmg/ddw339 (Ishikawa et al.) | 2016 |

## Hematology

### β-hemoglobinopathies

β-hemoglobinopathies are mainly caused by abnormal or absent production of hemoglobin due to mutations in the β-globin gene (*HBB*). A lentiviral gene transfer approach is one of the main strategies to recover the normal phenotype (Negre et al. 2016, Daniel-Moreno et al. 2020, Lamsfus-Calle et al. 2020a). To conduct the treatment, autologous hematopoietic stem cell transplantation (auto-HSCT) is performed. The stem cells are isolated from the blood of the patient and are genetically modified *ex vivo* by a lentiviral vector that incorporates a healthy copy of the *β-globin* gene into the genome. Subsequently, the synthesis of β-globin is activated and the normal production of hemoglobin is restored. It is important to highlight that the lentiviral integration is performed randomly and can drive undesired mutagenesis or oncogenesis. Alternative and safer therapies can be implemented by incorporating the CRISPR/Cas technology (Antony et al. 2018a, b, Lamsfus-Calle et al. 2020b). Some approaches include the correction of the *HBB* gene using a single-stranded oligodeoxynucleotide (ssODN) as a donor template encoding the mutation-free copy of the gene; others are based on the delivery of one healthy copy by adeno-associated virus (AAV) (Dever et al. 2016, Antony et al. 2018b). Another promising gene-editing strategy is the reactivation of normally silenced fetal hemoglobin (HbF) in patients (Akinsheye et al. 2011, Antony et al. 2018a). This approach takes advantage of the higher avidity of HbF for oxygen compared to adult hemoglobin (HbA), resulting in a significant therapeutic effect even with modest efficiency of HbF reactivation. As in the previously mentioned methods, an auto-HSCT is performed; the stem cells are edited *ex vivo* with the CRISPR/Cas9 system and HbF expression is promoted by the disruption of HbF repressors such as BCL11A, a transcription factor involved in the molecular switch of HbF to HbA (Olivieri and Weatherall 1998, Chang et al. 2017a), or by the knock-out of the distal promoter of *HBG1/2*, which can induce a similar outcome since it also regulates the activation of HbF expression (Traxler et al. 2016, Lamsfus-Calle et al. 2020b).

Clinical trial data of CTX001 (autologous transplantation of CRISPR-modified CD34$^+$ HSCs with reactivated expression of HbF) is available for both β-thalassemia and sickle cell disease patients. The patient with β-thalassemia, a 19-year-old female with an IVS-I-110 mutation ($\beta^0/\beta^+$ genotype), had been receiving two-three packed red cell transfusions every month since her birth. Due to the long-term treatment with transfusions, the patient suffered from several side effects including iron overload, inactive hepatitis C, splenomegaly, and osteonecrosis of the skull. One month post-treatment with CTX001, the patient was independent of transfusion for at least 21.5 months. This case report emphasizes the great impact CRISPR Medicine can have on an individual life (Frangoul et al. 2020). Likewise, the patient with SCD, a 33-year-old female with a $\beta^S/\beta^S$ genotype and a single α-globin deletion, suffered an average of seven severe vaso-occlusive episodes every year, resulting in the continuous need for blood transfusions, with liver abnormalities as a consequence. 19 days after the treatment with CTX001, the patient did not exhibit any vaso-occlusive episodes for the last 16.6 months, without the need of packed red cell transfusions (Frangoul et al. 2020).

## Oncology

*Boosting the immunotherapy: CAR-T cells*

The CRISPR technology is an important factor in the improvement of the new generation of chimeric antigen receptor (CAR)-based therapies. The CAR approach takes advantage of the organism's immune cells and boosts their antitumor function. For this purpose, the CAR gene is introduced into the T-cells' genome *ex vivo*, and the effector cells are then reinfused into the patient (Miliotou and Papadopoulou 2018). The CAR protein is designed to specifically recognize the malignancy, as it aims to one protein target that is overexpressed on the membrane of the tumor cells, such as CD19 in B-cell malignancies (Wang et al. 2012, Schneider et al. 2017). However, even though the treatment is targeted against cancer, malignant cells are still able to overcome the antitumor function by suppressing the immune system (Tang et al. 2020). The immunosuppression is based on the inhibitory immune checkpoints, key molecules in immune self-tolerance under physiological conditions. Receptors such as PD1 or CTLA4 restrict the expansion of T-cells when bound to their ligands, which can be upregulated by the tumor cells or their microenvironment in order to evade the immune system (Parry et al. 2005). By targeting and disrupting these molecules with the CRISPR/Cas technology, the antitumor function of T-cells is enhanced by preventing their exhaustion (Liu et al. 2017a, Shi et al. 2017, Ureña-Bailén et al. 2020).

Moreover, CRISPR/Cas editing can improve the safety of the CAR therapy. When autologous treatment is not an option (e.g. due to a low amount of T-cells in the patient's blood), a heterologous approach is necessary. This type of therapy bears the same risks as organ transplantation: graft-versus-host disease (GvHD) and host-versus-graft effect (HvG) are likely to occur (Yang et al. 2015). The universalization of the therapy can ameliorate both iatrogenic effects by disrupting receptors such as TRAC, B2M, and CD52 in the effector cells (Eyquem et al. 2017, Liu et al. 2017a, Jung and Lee 2018, Ureña-Bailén et al. 2020). Moreover, CRISPR/Cas editing can prevent fratricide, a critical issue when treating T-cell malignancies. In this type of leukemia, both effector and cancerous cells share the receptor that is recognized by the CAR; thus, the CAR-T cell becomes a target of the therapy as well. Using CRISPR technology to knock-out CD7 can enhance the anti-tumor activity by removing the CAR target in the effector cells (Cooper et al. 2018, Ureña-Bailén et al. 2020).

*Manipulation of cancer cells*

Therapeutic strategies for solid tumors include the silencing of cancer driving genes and chemo-resistance genes. This has proven to be feasible for treating bladder cancer in mice experiments (Valletta et al. 2015, Cheng et al. 2020). Furthermore, the combination of CRISPR/Cas technology with oncolytic, virus-based intra-tumoral delivery has shown an efficient, even complete, tumor regression in lung cancer mouse models by targeting and disrupting the epidermal growth factor receptor (EGFR) without any detectable off-target effect (Yoon et al. 2020).

Similarly, the use of liposomes has proven to be useful in the delivery of the Cas9 system to disseminated tumors. The disruption of PLK1 mitosis kinase in a murine metastatic ovarian adenocarcinoma model was observed to inhibit tumor growth and increase the survival of affected individuals by 80% (Rosenblum et al. 2020). The targeted release of the CRISPR system in the cancerous cells was promoted with liposomes decorated with antibodies against EGFR, a receptor overexpressed in this type of malignancy. Additionally, extracellular vesicles have been studied as a delivery method for cancer treatment. This approach is biologically safe as they are naturally produced in the cells and constitutes a classical way of cell communication. To direct them to the tumor, a coating with CARs is feasible and can drastically improve the treatment specificity, as observed in the CRISPR-enhanced disruption of MYC oncogene in CD19 leukemia (Xu et al. 2020).

*Fight against carcinogenic viruses*

Viral infections are known to trigger the development of cancer under certain circumstances. In this context, a strong correlation between nasopharyngeal carcinoma and Epstein-Barr virus (EBV) has been observed (Liebowitz 1994, Tsao et al. 2015, Wu et al. 2018). This virus can remain indefinitely in infected cells in the form of episomes; therefore, the risk of new infections and viral-driven oncogenesis is increased. Among all the EBV-derived oncogenic factors, the oncogenic role of latent membrane protein 1 (LMP1) has been extensively studied (Ma et al. 2013, Zhao et al. 2014, He et al. 2016). Its oncogenic potential can last for a life time as it integrates into the genome of the host cell and promotes epithelial-mesenchymal transition (EMT) as well as carcinogenic invasion and proliferation. In a recent investigation, CRISPR-mediated knock-out of LMP1 in cells already infected with EBV was successful to prevent nasopharyngeal carcinoma and recurrent EBV infections (Huo and Hu 2019). Similarly, Hepatitis B virus (HBV) is known to inhibit the tumor suppressor activity of p53, leading to the interruption of cell cycle regulation and the promotion of hepatocellular carcinoma development (Elmore et al. 1997). The CRISPR-mediated disruption of the HBV's covalently closed circular DNA (cccDNA) has proven to be feasible and highly efficient *in vitro*, resulting in full eradication of the viral infection in a stable cell-line (Li et al. 2017).

## Neurology

*Gene editing in human iPSCs*

The underlying causes of complex neurological disorders, such as Alzheimer's and Parkinson's diseases, are not monogenic diseases, but associated with certain genetic variants that usually act in combination with other alleles. As a result, the individual genetic background is crucial for the understanding of the disease in a particular patient. The generation of induced pluripotent stem cells (iPSCs) based models is of great importance for this line of research. With this personalized approach, not only the genetic load is maintained, but the chance to study neurons and brain cells alike without resorting to invasive methods is feasible (Heidenreich and Zhang 2016). The implementation of the CRISPR/Cas system to the experimental workflow can help discern the role of candidate genes in the development of these devastating diseases. Additionally, CRISPR multiplexed gene editing leads the way to study the combinatorial effect of different genes of interest in the correction or progression of the disorder (Yumlu et al. 2019).

### Amyotrophic lateral sclerosis (ALS)

ALS is a severe paralytic disorder caused by motor neuron degeneration in the brain and the spinal cord. The genetic causes underlying the disorder are still unclear but the mutations of certain genes have been associated with the disease (Pasinelli and Brown 2006). The CRISPR-promoted correction of two of them, $SOD1^{+/A272C}$ and $FUS^{+/G1566A}$, has been proved feasible in iPSCs derived from ALS patients (Wang et al. 2017). Moreover, Armstrong et al. designed a correction strategy based on the knock-in of a ssODN assisted by CRISPR/Cas9 technology to correct $tardbp^{A382T}$ and $fus^{R521H}$ in a zebrafish model (Armstrong et al. 2016). Important efforts are being made to identify novel mutations that could shed light on the study of the disease by combining whole genome sequencing and the CRISPR system in a personalized medicine strategy (Yun et al. 2020).

### Alzheimer's disease

Alzheimer's disease is a devastating type of dementia characterized by severe symptoms including disturbances in cognitive areas of visuospatial orientation, memory, and language. In addition, personality changes, decreased judgment ability, wandering, and psychosis are commonly present (Schachter and Davis 2000). A distinguishing feature of this disorder is found in the aberrant cleavage of amyloid precursor protein (APP) by β-secretase-1. This abnormality triggers the production of amyloid β (Aβ) peptide and the subsequent accumulation of amyloid plaques. A significant reduction in amyloid plaques aggregation, as well as the improvement of the symptoms, was observed when *BACE1* (gene encoding for β-secretase-1) was targeted with the CRISPR/Cas9 system in the brain of mouse models (Park et al. 2019). Nevertheless, *BACE1* also presents an important role in synaptic plasticity, and its CRISPR-mediated disruption could provoke undesired side effects (Das and Yan 2017). More research is needed to evaluate the safety of this approach for future application.

## Immunodeficiency

Immunodeficiency refers to a group of clinical syndromes in which the function of the immune system is insufficient or absent due to deficiencies in the immune cells and immune molecules. Immunodeficiency diseases can be classified into primary immunodeficiency (PID), a hereditary disease in which the immune system is damaged because of defects in a certain gene, and secondary immunodeficiency (SID), a condition which occurs after birth and occurs due to environmental factors, such as infection, nutritional disorders, and certain disease states such as acquired immunodeficiency syndrome (AIDS) caused by the human immunodeficiency virus (HIV). This part of the chapter introduces the applications of CRISPR/Cas9 in the treatment of some PIDs and AIDS.

Conventional treatments for PIDs include allogeneic stem cell transplantation (SCT), which still faces important challenges such as the search for a matching donor, immunosuppression, and the risk of graft-versus-host disease. Therefore, it is of great significance to provide an alternative option for eliminating the risks associated with allogeneic SCT (Psatha et al. 2016, Groeschel et al. 2016). In recent years, several studies based on CRISPR gene therapy have achieved encouraging results, which will be described in the following paragraphs.

CRISPR/Cas9 is expected to become the potential treatment method for immunodeficiency disease because of its simplicity in targeting any specific gene of interest. By merely changing the gRNA sequence and the donor sequence, almost any *locus* in the genome can be targeted. This provides researchers with great flexibility and can be used to design gene correction therapies, laying the foundation for the successful treatment of immunodeficiency diseases.

### X-linked severe combined immunodeficiency

Mutations in the gene *IL2RG*, a shared receptor γ chain of IL-2, IL-4, IL-7, IL-9, and IL-15 (γc), cause X-linked severe combined immunodeficiency (X-SCID). In general, X-SCID patients' T- and NK cells are defective; though the numbers of B cells are normal, their functions are impaired. Children suffer from severe bacterial or viral infections soon after birth and in most cases die in infancy (Tan et al. 2015). Research has shown the possibility to correct the mutation for X-SCID via gene therapy in different cell types such as HSCs, iPSCs, and T- cells (Lombardo et al. 2007, Chang et al. 2015, Schiroli et al. 2017, Roth et al. 2018). Pavel-Dinu et al. reported a CRISPR/Cas9-based method to correct the mutations associated with X-SCID (Pavel-Dinu et al. 2019). The codon-optimized *IL2RG* cDNA was integrated into the endogenous *IL2RG* start codon through homologous recombination and obtained gene integration of 45% in CD34[+] HSCs. *In vitro* assays showed the full signal transduction capability of the expressed *IL2Rγ* subunit, and its long-term pluripotency was confirmed by transplanting genome-edited CD34[+] HSCs into mice. This provided a preclinical proof of concept for the specific and effective treatment of X-SCID using CRISPR/Cas9 technology (Pavel-Dinu et al. 2019).

### Wiskott-Aldrich syndrome

Wiskott-Aldrich syndrome (WAS) is a rare X-linked inherited disease which has a disease onset in infancy and presents the clinical manifestations of eczema, repeated infections, and thrombocytopenia. Mutations in the *WAS* gene located on the short arm of the X chromosome result in WAS syndrome (Villa et al. 1995, Galy and Thrasher 2011). Due to this mutation, the immune function of patients declines progressively, where there is a decrease in IgM level and peripheral blood lymphocytes, poor response of polysaccharide antigen-specific antibody, and cellular immune dysfunction. The incidence of lymphoma and autoimmune vasculitis is also high in WAS patients. Rai et al. employed CRISPR/Cas9 to treat Wiskott–Aldrich Syndrome to restore the normal WAS protein expression in primary human HSPCs where they achieved a 61% WAS cDNA knock-in frequency in patient-derived HSPCs. Besides, they found that after successful genetic correction, patient-derived LT-HSC can express WAS protein in the myeloid and lymphatic offspring of NSG mice up to 26 weeks after transplantation which provides an optional strategy to treat WAS by CRISPR-based gene editing with efficacy and potential safety (Rai et al. 2020).

## X-linked hyper-immunoglobulin M syndrome

X-linked hyper-immunoglobulin M (IgM) syndrome (XHIM) is an X-linked, autosomal recessive disease that is characterized by recurrent infections due to the lack of IgG and IgA but increased IgM in the serum (Ramesh et al. 1998). The main clinical manifestations are neutropenia, thrombocytopenia, hemolytic anemia, which may be accompanied by bile duct and liver diseases along with opportunistic infections (Hayward et al. 1997). Kuo et al. used CRISPR/Cas9 to correct mutations in the CD40 ligand found in XHIM. They proceeded with TALEN and CRISPR to correct the mutation in XHIM primary T cells and also to compare the efficiencies of both nucleases (Kuo et al. 2018). Electroporation of peripheral blood CD34+ cells of patients resulted in modification rates of 13.2% ± 3.4% with TALEN mRNA, which was smaller than the rates of targeted integration with CRISPR and the cDNA donor, which was 16.2% ± 4.2% (gRNA, Cas9 mRNA) and 20.8% ± 6.6% (RNP) measured by ddPCR (Kuo et al. 2018). The study demonstrated the enhanced gene targeting effect of the CRISPR system over TALENs as our earlier observation in the *HBB locus* (Antony et al. 2018b).

## Chronic granulomatous disease

Chronic granulomatous disease (CGD) is an X-linked inherited disease that progresses with frequent and recurrent life-threatening bacterial and fungal infections and granuloma formation. Due to the mutations in the *CYBB* gene that encodes gp91[phox] (serves as a catalytic center of NOX2 enzyme), the ability of phagocytes to produce superoxide anions is highly affected. In general, these superoxide anions have several microbicidal and immunoregulatory functions and restoring the function of NOX2 had clinical benefits (Casimir et al. 1991). De Ravin et al. used ssODNs as donor templates and delivered *Cas9* mRNA and sgRNA via electroporation to correct the *CYBB* mutation. With an app. 20% of gene repair, the NOX function and superoxide radical production were restored with successful transplantation in NOD SCID mice for 5 months (De Ravin et al. 2017).

## CRISPR/Cas9 for the treatment of acquired immune deficiency syndrome

The pathogenesis of AIDS is mainly due to the HIV infection in which the function of CD4[+] T lymphocytes is damaged and massively destroyed, leading to cellular immune deficiency. Kaminski et al. demonstrated that it was feasible to remove the incorporated HIV proviral copies in latently infected human cells, opening a hopeful way towards the treatment of AIDS patients (Kaminski et al. 2016).

Additionally, important efforts have been made to prevent HIV infection using CRISPR. The HIV-1 virus attaches to the membrane of target T cells to invade the host immune cells through the CD4 receptor and interacts with CCR5/CXCR4 receptors. Once the virus enters the cell, the viral RNA genome is converted into double-stranded DNA (dsDNA) by HIV reverse transcriptase. The generated DNA enters the host nucleus and integrates into the genome, which produces new viral

RNA that serves as mRNA for protein production and for genomic RNA packaging for new viral particles that are released from the cell. A 32 bp deletion in the *CCR5* allele leads to the inactivation of the gene, which results in high resistance to HIV-1 infection (Bleul et al. 1997). CRISPR-Cas9 can encode the *CCR5/CXCR4* gene or HIV dsDNA formed after reverse transcription of viral RNA and can be integrated with proviral DNA (Wang and Cannon 2016). Some studies are showing that using CRISPR/Cas9 to target *CCR5* promotes the resistance of CD4$^+$ T-lymphocytes against HIV infection (Li et al. 2015, Liu et al. 2017b, Xiao et al. 2019). Interestingly, CCR5-modified autologous CD4$^+$ T-cell infusions with ZFN showed reduced HIV DNA in treated patients (Tebas et al. 2014). The HSC transplantation remains a possible way to cure HIV infection permanently and several studies have investigated the disruption of *CCR5* using ZFN in primates and *ex vivo* models (Holt et al. 2010, Peterson et al. 2016). Sangamo Therapeutics is currently conducting a Phase-1 clinical trial and investigating the CCR5-modified autologous CD34$^+$ HSCs using ZFN for their resistance against HIV entry (NCT02500849).

## CRISPR medicine: Progress

Since the demonstration of CRISPR/Cas in editing mammalian genome in Science Journal by two independent researcher groups (Cong et al. 2013, Mali et al. 2013), the technology created ambitious expectations across basic science, biotechnology, and therapeutic medicine. The CRISPR technology was discussed both positively and negatively across mainstream media which increased hope, fear, and hype (Carroll and Charo 2015). Although several other technologies exhibited similar promising results in preclinical settings including stem cell therapy and gene therapy, they did not shine as revolutionary technology because of their resounding failure in human trials. Clinical studies using retroviral vector gene therapy created blood cancers in treated patients (Hacein-Bey-Abina et al. 2008, Braun et al. 2014). As a consequence, a halo of skepticism was created around CRISPR/Cas9 technology when it was criticized for creating DSB in off-target sites, which could end up in the development of cancer (NM 2018). Moreover, genome editing tools such as ZFN and TALENs, which were previously around for several years, were only used by a handful of researchers and industries, suggesting that the newly-born CRISPR technology was just going to be another gene-editing tool forgotten in the toolbox. On the contrary, CRISPR made a huge impact in science thanks to its versatility. Unlike with ZFN and TALENs, where only a few researchers had access to the technology, the CRISPR reached the hands of scientists across the world through plasmid repository platforms and biotechnological companies that commercialize CRISPR/Cas reagents. The easy access and methodical simplicity of CRISPR made many researchers straightforwardly establish this methodology in their laboratory. As a consequence, this technology platform improved rapidly and was able to address all the limitations systematically, as we will discuss in the next section of this chapter.

Due to encouraging preclinical data ranging from blood disorders to cancers, CRISPR technology moved into clinical trials at lightning speed despite its infancy. As of now, more than 40 clinical trials are ongoing across the globe either as a therapeutic intervention or as a diagnostic tool. We summarized below all the clinical trial data in Table 2 including the details on phase, indication, target, sponsor, and study location. Importantly, a handful of CRISPR technology start-up companies have strong financial support. We listed pioneering CRISPR companies in Table 3 along with details on their leading programs.

## CRISPR medicine: Challenges

CRISPR/Cas technology is not infallible and problems must be overcome for safer use in clinical therapies. One of the main challenges is minimizing off-target effects (Liang et al. 2015). It is known that gRNA's specificity is imperfect and can recognize similar sequences that differ in some nucleotides in comparison with the target sequence. Therefore, unintended editing in other parts of the genome is possible, including in proto-oncogenes, tumor suppressor genes, and essential housekeeping genes (Zhang et al. 2015).

To understand the scope of the off-target effect, several experimental techniques have been developed to detect undesired Cas cleavages (Table 4). Along with them, *in silico* prediction tools offer pre-screening information for each gRNA by comparing its sequence with the whole human genome and finding off-target hotspots. However, the off-targets detected vary from method to method. Therefore, there is a lack of consensus in the scientific community and no standard off-target detection method is available yet. One of the most accepted methods is the GUIDE-seq approach (Tsai et al. 2015). This strategy relies on *in vivo* analysis, which has certain advantages in comparison with the *in vitro* approaches as it accounts for the stability of sgRNA and Cas9 protein. However, the dependence on the integration of a dsODN is a limitation (Lamsfus-Calle et al. 2020a).

To prevent the off-target effect, several efforts have been made to enhance the specificity of the Cas nuclease without a significant loss of efficiency. High fidelity (HiFi) Cas9 was identified from an unbiased bacterial screening of a mutant Cas9 library (Vakulskas et al. 2018). The approach employed two different selective plasmids, one coding a toxic gene susceptible to be targeted by the CRISPR/Cas system (efficiency test) and the other carrying an antibiotic resistance gene with a potential off-target sequence (fidelity test). In this way, the only survivor colonies would exhibit an efficient and specific Cas9 nuclease activity. The research revealed that the introduction of a single point mutation (p.R691A) retained the high on-target activity of Cas9 while reducing the off-target effect (Vakulskas et al. 2018).

Additionally, rationally engineered Cas proteins have been developed for enhanced fidelity. Following structure-guided protein engineering, a modified SpCas9 (eSpCas9) variant was generated with reduced off-target activity and robust on-target cleavage (Slaymaker et al. 2016). By attenuating the helicase activity of the Cas9 protein, DNA mismatches are thermodynamically less favorable and thus, Cas9 specificity is promoted. This was achieved by the neutralization of positive charges of the residues found in the helicase domain, which are responsible for an efficient unwinding of the negative strand of DNA but also the increase of off-target editing. Even though the helicase activity is down-graded, the efficacy of the on-target editing is maintained (Slaymaker et al. 2016).

**Table 2:** List of clinical trials using the CRISPR/Cas system as gene-editing tool

| ID | Disease | Application | Type of nucleases | Target | Phase | Status | Type of study | Sponsor | Country |
|---|---|---|---|---|---|---|---|---|---|
| NCT04535648 | Enterovirus infections | Detection of Enterovirus genotypes | CRISPR/Cas9 | Enterovirus genome | NA | NYR | - | Children's Hospital of Fudan University | China |
| NCT03057912 | Human papillomavirus-related malignant neoplasm | Treatment of HPV-related cervical intraepithelial neoplasia | TALEN, CRISPR/Cas9 | HPV16 and HPV18 E6/E7 genome | I | Unknown | Open-label and triple cohort study | First Affiliated Hospital, Sun Yat-Sen University & Jingchu University of Technology | China |
| NCT04178382 | Pneumonia | PCR diagnosis in patients with open air pneumonia | CRISPR/Cas12a | Genome seq of pathogenic bacteria | NA | Recruiting | Multicenter, randomized controlled trial | Chinese Medical Association | China |
| NCT04426669 | Metastatic gastrointestinal epithelial cancer | Enhancement of antitumor activity in TILs | CRISPR/Cas9 | *CISH* | I-II | Recruiting | Open label | Intima Bioscience, Inc., Masonic Cancer Center, University of Minnesota | US |
| NCT03164135 | Hematological malignancies | Protection against HIV in HIV-infected subjects with hematological malignancies | CRISPR/Cas9 | *CCR5* | NA | Recruiting | Single group assignment | Affiliated Hospital to Academy of Military Medical Sciences, Peking University, Capital Medical University | China |
| NCT04074369 | Pulmonary tuberculosis | Diagnosis for rapid identification of MTB | CRISPR/Cas | MTB genome | NA | Recruiting | Cohort | Huashan Hospital, Wenzhou Central Hospital & Hangzhou Red Cross Hospital | China |

| NCT Number | Condition | Purpose | Technique | Gene/Target | Phase | Status | Study Design | Sponsor | Location |
|---|---|---|---|---|---|---|---|---|---|
| NCT03167450 | Sickle cell disease | Clinical care | CRISPR/Cas | - | NA | Suspended | Family-based | National Human Genome Research Institute (NHGRI), | US |
| NCT04560790 | Refractory herpetic viral keratitis | Disease treatment | CRISPR/Cas9 | BD11 CRISPR/Cas9 mRNA | I-II | Recruiting | Single group assignment | Shanghai BDgene Co., Ltd., Eye & ENT Hospital of Fudan University | China |
| NCT03399448 | Multiple myeloma, synovial sarcoma, myxoid/round cell liposarcoma | Disease treatment | CRISPR/Cas | *TCR & PD1* | I | Terminated | Non-randomized, parallel assignment | University of Pennsylvania, Parker Institute for Cancer Immunotherapy, Tmunity Therapeutics | US |
| NCT03545815 | Multiple solid tumors | Enhancement of CAR-T cell antitumor activity | CRISPR/Cas9 | *TCR & PD1* | I | Recruiting | Single group assignment | Chinese PLA General Hospital | China |
| NCT04037566 | Relapsed or refractory CD19+ leukemia or lymphoma | Enhancement of CAR-T cell antitumor activity | CRISPR/Cas | *HPK1* | I | Recruiting | Single group assignment | Xijing Hospital, Xi'An Yufan Biotechnology Co.,Ltd | China |
| NCT04244656 | Multiple Myeloma | Disease treatment | CRISPR/Cas | *BCMA* | I | Recruiting | Sequential assignment | CRISPR Therapeutics AG | US |
| NCT03655678 | β-thalassemia | Disease correction | CRISPR/Cas | *BCL11A* (CTX001) | I-II | Recruiting | Single group assignment | Vertex Pharma, CRISPR Therapeutics | US & Germany |
| NCT04502446 | T Cell Lymphoma | Disease treatment | CRISPR/Cas | *CD70* | I | Recruiting | Sequential assignment | CRISPR Therapeutics AG | US |
| NCT04035434 | B-cell Malignancy, Non-Hodgkin Lymphoma, B-cell Lymphoma | Disease treatment | CRISPR/Cas | *CD19* | I | Recruiting | Sequential assignment | CRISPR Therapeutics AG | US |

*(Contd.)*

**Table 2:** (*Contd.*)

| ID | Disease | Application | Type of nucleases | Target | Phase | Status | Type of study | Sponsor | Country |
|---|---|---|---|---|---|---|---|---|---|
| NCT03745287 | Sickle Cell Disease | Disease correction | CRISPR/Cas | BCL11A (CTX001) | I-II | Recruiting | Single group assignment | Vertex Pharma, CRISPR Therapeutics | US |
| NCT03728322 | β-thalassemia | Disease correction | CRISPR/Cas9 | HBB | Early phase I | NYR | Single group assignment | Allife Medical Science and Technology Co., Ltd. | China |
| NCT03855631 | Kabuki Syndrome 1 | Disease treatment | CRISPR/Cas9 | Epigenetics of MLL4 | NA | Active, NR | Cohort | University Hospital, Montpellier, Association Française contre les Myopathies Telethon | France |
| NCT03342547 | Gastrointestinal infection | Identification of protein mediators of norovirus infections | CRISPR/Cas9 | Host cellular proteins | NA | Recruiting | Single group assignment | Chinese University of Hong Kong | China |
| NCT04557436 | B cell acute lymphoblastic leukemia | CAR-treatment universalization | CRISPR/Cas9 | CD52 & TRAC | I | NYR | Single group assignment | Great Ormond Street Hospital for Children NHS Foundation Trust, University College of London | UK |
| NCT03747965 | Solid tumors in adults | Enhancement of CAR-T cell antitumor activity | CRISPR/Cas9 | PD1 | I | Recruiting | Single group assignment | Chinese PLA General Hospital | China |
| NCT03166878 | B cell leukemia | CAR-treatment universalization | CRISPR/Cas9 | TCR & B2M | I-II | Recruiting | Single group assignment | Chinese PLA General Hospital | China |

| NCT Number | Condition | Description | System | Gene | Phase | Status | Design | Sponsor | Country |
|---|---|---|---|---|---|---|---|---|---|
| NCT04208529 | β-thalassemia & sickle cell disease | Long-term study of patients treated with CRISPR/Cas9 | CRISPR/Cas9 | *BCL11A* (CTX001) | NA | Enrolled | Cohort | Vertex Pharma, CRISPR Therapeutics AG | US & Germany |
| NCT03606486 | Ovarian cancer | Diagnosis based on CRISPR-Duplex sequencing | CRISPR/Cas | *TP53* | NA | Recruiting | Single group assignment | University of Washington, National Cancer Institute (NCI), Minnesota Ovarian Cancer Alliance | US |
| NCT03081715 | Esophageal cancer | Enhancement of T cell antitumor activity | CRISPR/Cas9 | *PD1* | NA | Completed | Single group assignment | Hangzhou Cancer Hospital, Anhui Kedgene Biotechnology Co., Ltd | China |
| NCT03332030 | Neurofibromatosis type 1, tumors of the CNS | Diagnosis | CRISPR/Cas9 | *NF-1* | NA | Suspended | Cohort | Roger Packer | US |
| NCT03398967 | B Cell Leukemia, B Cell Lymphoma | Disease treatment | CRISPR/Cas9 | CD19-CD20/CD22 CAR integration | I-II | Recruiting | Single group assignment | Chinese PLA General Hospital | China |
| NCT04601051 | Hereditary transthyretin amyloidosis with polyneuropathy (ATTRv-PN) | Disease treatment | CRISPR/Cas9 | *TTR* | I | NYR | Sequential assignment | Intellia Therapeutics | US |
| NCT03681951 | Advanced solid tumors | Drug study for disease treatment | CRISPR/Cas | *RIPK1* | II | Terminated | Sequential assignment | GlaxoSmithKline, Parexel | US |
| NCT04417764 | Advanced hepatocellular carcinoma | Enhancement of CAR-T cell antitumor activity | CRISPR/Cas9 | *PD1* | I | Recruiting | Single group assignment | Central South University | China |

*(Contd.)*

**Table 2:** *(Contd.)*

| ID | Disease | Application | Type of nucleases | Target | Phase | Status | Type of study | Sponsor | Country |
|---|---|---|---|---|---|---|---|---|---|
| NCT03690011 | T-cell acute lymphoblastic leukemia/lymphoma, T-non-Hodgkin lymphoma | Prevention of T cell fratricide | CRISPR/ Cas9 | *CD7* | I | NYR | Single group assignment | Baylor College of Medicine, The Methodist Hospital System | US |
| NCT02863913 | Invasive bladder cancer stage IV | Enhancement of T cell antitumor activity | CRISPR/ Cas9 | *PD1* | I | Withdrawn | Parallel assignment | Peking University | China |
| NCT02867345 | Castration resistant prostate cancer (CRPC) | Enhancement of T cell antitumor activity | CRISPR/ Cas9 | *PD1* | NA | Withdrawn | - | Peking University | China |
| NCT03044743 | EBV associated malignancies | Enhancement of T cell antitumor activity | CRISPR/ Cas9 | *PD1* | I-II | Recruiting | Single group assignment | Yang Yang | China |
| NCT03872479 | Hereditary eye disorders | Disease treatment | CRISPR/ Cas | *CEP290* | I-II | Recruiting | Single group assignment | Allergan, Editas Medicine, Inc. | US |
| NCT04478409 | Familial mediterranean fever | Disease diagnosis | CRISPR/ Cas9 | *MEFV* | NA | NYR | Case-control | Hospices Civils de Lyon | France |
| NCT04122742 | Rubinstein-Taybi syndrome | Diagnosis and detection of CREBBP and EP300 variants | CRISPR/ Cas9 | Acetylation of *CREBBP* and *EP300* | NA | Recruiting | Case-only | University Hospital, Bordeaux | France |

**Table 3:** List of CRISPR products under development by pharmaceutical companies

| Company | Collaboration | Disease | Product name | Cell type | Target | Mechanism | Status | Field of application |
|---------|---------------|---------|--------------|-----------|--------|-----------|--------|---------------------|
| CRISPR Therapeutics | Vertex | Hemoglobinopathies | CTX001 | Autologous HSCs | *BCL11A* | Disruption | Clinical trial | Hemo-globinopathies |
| | Fully owned | CD19⁺ malignancies | CTX110 | Allogenic T cells | *CD19* | Disruption and insertion | Clinical trial | Immuno-Oncology |
| | Fully owned | Multiple myeloma | CTX120 | Allogenic T cells | *BCMA* | Disruption and insertion | Clinical trial | |
| | Fully owned | Solid tumors and hematologic malignancies | CTX130 | Allogenic T cells | *CD70* | Disruption and insertion | Clinical trial | |
| | ViaCyte | Type 1 diabetes mellitus | Allogeneic beta-cell replacement therapy | Immune-evasive stem cell line | Not mentioned | Disruption and insertion | Research | Regenerative Medicine |
| | Fully owned | Glycogen storage disease type Ia (GSD Ia) | Not mentioned | Not mentioned | *G6PC* | Correction/ insertion | Research | *In vivo* approaches |
| | Vertex | Duchenne muscular dystrophy (DMD) | Not mentioned | Not mentioned | *dystrophin* | Various | Research | |
| | Vertex | Myotonic dystrophy type 1 (DM1) | Not mentioned | Not mentioned | *DMPK* | Various | Research | |
| | Vertex | Cystic fibrosis (CF) | Not mentioned | Not mentioned | *CFTR* | Correction/ insertion | Research | |
| Editas | Fully owned | Leber Congenital Amaurosis 10 | EDIT-101 | Not mentioned | *CEP290* | Deletion and insertion | Pre-clinical | *In vivo* approaches |
| | Fully owned | Usher Syndrome 2a | EDIT-102 | Photoreceptor cells | *USH2A* | Deletion and insertion | Research | |

*(Contd.)*

**Table 3:** (*Contd.*)

| Company | Collaboration | Disease | Product name | Cell type | Target | Mechanism | Status | Field of application |
|---|---|---|---|---|---|---|---|---|
| Editas | Fully owned | Autosomal dominant retinitis pigmentosa IV | Not mentioned | Not mentioned | *RHO* | Not mentioned | Early research | *In vivo* approaches |
| | AskBio | Neurological diseases | Not mentioned | Not mentioned | Not mentioned | Not mentioned | Research | |
| | Fully owned | Sickle cell disease | EDIT-301 | Autologous HSCs | *HBG1/2* | Gene disruption | Pre-clinical | Engineered cell medicines |
| | Fully owned | β-thalassemia | Not mentioned | Not mentioned | Not mentioned | Not mentioned | Research | |
| | Collaboration | Cancer | EDIT-201 | Healthy donor NK cells | Not mentioned | Not mentioned | Research | |
| | BlueRock | Cancer | Not mentioned | iPSC NK cells | Not mentioned | Not mentioned | Research | |
| | Fully owned | Cancer | Not mentioned | δ T Cells | Not mentioned | Not mentioned | Research | |
| | Bristol Myers Squibb | Cancer | Not mentioned | αβ T Cells | Not mentioned | Not mentioned | Research | |
| Intellia | Regeneron | Transthyretin amyloidosis | NTLA-2001 | Neurons & cardiomyocytes | *TTFi* | Knockout | Clinical trial | *In vivo* approaches |
| | Fully owned | Hereditary angioedema | NTLA-2002 | Hepatocytes | *KLKB1* | Knockout | Research | |
| | Regeneron | Hemophilia A & B | Not mentioned | Not mentioned | Not mentioned | Insertion | Research | |
| | Fully owned | Not disclosed | Research programs | Not mentioned | Not mentioned | Knockout, insertion, consecutive edits | Research | |

| Company | Partner | Indication | Research programs | Cell type | Target | Various | Research | In vivo / Ex vivo approaches |
|---|---|---|---|---|---|---|---|---|
| Intellia | Regeneron | Not disclosed | | Not mentioned | Not mentioned | | Clinical trial (1/2) | *In vivo approaches* |
| | Novartis | Sickle cell disease | OTQ923 / HIX763 | Hematopoietic stem cells | Not mentioned | | Research | *Ex vivo approaches* |
| | Fully owned | AML | NTLA-5001 | T cells | *TCR* | Not mentioned | Early research | |
| | Fully owned | Solid tumors | Not mentioned | T cells | *TCR* | Not mentioned | | |
| Caribou Biosciences | Fully owned | B cell non-Hodgkin lymphoma (B-NHL) | CB-010 | T cells | *CD19* | Not mentioned | Clinical trial | *Ex vivo approaches* |
| eGenesis | Fully owned | Organ failure | Not mentioned | Not mentioned | Not mentioned | Xenotransplant | Pre-clinical and early research | Engineered xenotransplants |
| Verve Therapeutics | Fully owned | Coronary artery disease | Not mentioned | Not mentioned | *PCSK9* | Not mentioned | Early research | Cardiovascular disease treatment |
| Beam Therapeutics | Fully owned | Sickle cell disease & β-thalassemia | BEAM-101 | Not mentioned | Not mentioned | Fetal hemoglobin activation | Pre-clinical | Hemoglobinopathies |
| | Fully owned | Sickle cell disease | BEAM-102 | Not mentioned | *HBB* | Direct correction of mutation | Pre-clinical | |
| | Fully owned | T-ALL | BEAM-201 | T cells | *CD7* | Multiplex disruption | Pre-clinical | Oncology |
| | Fully owned | AML | Not mentioned | T cells | Not mentioned | Multiplex disruption | Pre-clinical | Oncology |
| Graphite Bio | Fully owned | Sickle cell disease | Not mentioned | HSCs | Not mentioned | Homologous directed repair | Pre-clinical | Hemoglobinopathies |
| Excision Bio Therapeutics | Fully owned | Anti-viral | Not mentioned | Not mentioned | *HIV/SARS CoV2/HBV/JCV* | Viral RNA cleavage | Pre-clinical | Infectious Diseases |

**Table 4:** List of unbiased available methods for off-target detection

| Abbreviation | Acronym | Methodology | Sequencing method | Sensitivity | Fidelity | Advantages | Limitations |
|---|---|---|---|---|---|---|---|
| LAM-HTGTS | Linear amplification–mediated high-throughput genome-wide sequencing | Prey and bait method: Translocation of DSB due to nuclease to a fixed 'bait' DSB generated by the yeast I-SceI nuclease | NGS | High | Unbiased | • Scalable<br>• Inexpensive<br>• Reproducible<br>• Less time consuming (1 week)<br>• Nucleotide-level resolution | • Misrepresentation of highly repetitive sequences<br>• Translocations are rare events<br>• Does not provide information about all the DSB (only prey DSB)<br>• Detects only translocable DSBs<br>• Cannot provide absolute cutting rates |
| CHIP-seq | Chromatin Immunoprecipitation followed by high-throughput sequencing | Based on DNA:protein interaction (Target DNA: gRNA/dCas9) | High-throughput sequencing or Sanger sequencing | Low | Unbiased | • Provides information about where Cas9 binds in the genome | • Not reliable predictors of off-target cleavage by CRISPR/Cas9<br>• Amplification of off target sites based on cleavage assay<br>• Does not provide precise off target sites |
| CIRCLE-seq | Circularization for *in vitro* reporting of cleavage effects by sequencing | Circularization of DNA fragments by intermolecular ligation of sheared genome followed linearization for adaptor ligation, PCR amplification and sequencing | NGS | High | Unbiased | • Does not require reference genome sequences<br>• Rapid, and comprehensive<br>• Less background | • Requires an *in vivo* method to verify the off-targets screened. Serves as an initial screen method<br>• Higher false-positive rate |

| Method | Full Name | Description | Sequencing | | Bias | Advantages | Disadvantages |
|---|---|---|---|---|---|---|---|
| IDLV capture | Integrative-Deficient Lentiviral Vector Capture | Integration of IDLV with an antibiotic resistance gene at the DSB site generated by nucleases | Deep sequencing | High | Unbiased | • Off target detection frequency (< 1%) | • Not all off targets are captured<br>• Relies on IDLV delivery |
| SITE-seq | Selective enrichment and Identification of tagged genomic DNA ends by sequencing | A biochemical procedure which involves selective tagging, enrichment and sequencing of cleaved genomic DNA followed by mapping with a reference genome | Amplicon sequencing | High | Unbiased | • Biochemical mapping of Cas9 cleavage sites<br>• Does not rely on cellular events (DNA repair, translocations) | • Requires other cellular off target detecting methods to have a complete information of an edited genome |
| BLISS | Breaks Labeling *In Situ* and sequencing | Labelling directly the DSBs in a fixed sample and linear amplification by *in vitro* transcription followed by quantifying with unique molecular identifiers | NGS | High | Unbiased | • Direct labelling of DSBs and identification of their genomic location by NGS<br>• Requires less sample input<br>• Scalability | • Determines breaks at a specific moment in time |
| GUIDE-seq | Genome-wide, unbiased identification of DSBs enabled by sequencing | Integration of dsODN at break points, amplification of integrated dsODN followed by sequencing | NGS | High | Unbiased | • Genome wide off target detection<br>• Identifies Cas9 independent break sites | • dsODN are toxic to cells<br>• Transfection efficiency is low<br>• Poor chromatin accessibility |

*(Contd.)*

**Table 4:** (*Contd.*)

| Abbreviation | Acronym | Methodology | Sequencing method | Sensitivity | Fidelity | Advantages | Limitations |
|---|---|---|---|---|---|---|---|
| Digenome-seq | Digenome-sequencing | *In vitro* digestion of genome and computational identification and arrangement of 5'ends at the cleavage ends | Whole genome sequencing | High | Unbiased | • Detection frequency 0.1%<br>• Comparable results to other cell based off target detection methods<br>• Reproducible | • Sequencing depth<br>• Not usually a common method of choice |
| Eva-CRISPR | Quantitative Evaluation of CRISPR/Cas9-mediated editing | Ligation-based dosage-sensitive method which uses multiplex ligation-based probe amplification (MLPA) assay design | Sanger sequencing | High | Unbiased | • Detection of all mutations<br>• Locates mutations in difficult genomic regions<br>• Unbiased sgRNA screening method | • Not for whole genome analysis<br>• Not reliable for mutations with less frequency (5%) |
| DISCOVER-seq | Discovery of *in situ* Cas off-targets and verification by sequencing | Based on CHIP seq (especially the binding of MRE11 at off target site) with a custom software called BLENDER | Amplicon-NGS | High | Unbiased | • Provides information about the molecular action and dynamics of Cas-nucleases<br>• Detects translocations and large deletions<br>• Single base substitutions | • Question of false positive results<br>• Very-low-frequency off-targets are not detected |

| CHANGE-seq | Circularization for high-throughput analysis of nuclease genome-wide effects by sequencing | Tagmentation-based method | NGS | High | Unbiased | • DNA repair machinery independent detection<br>• Simultaneous reading of both sides of cleavage site | • No *in vivo* off-target detection is provided |

Similarly, attempts were made to increase HDR levels in human cells. This pathway is notably inefficient as very specific conditions must be met, including the availability of a homologous repair template and inducing the CRISPR modification during S- and G2-phase of the cell cycle. Therefore, NHEJ is usually handling the repair of CRISPR-induced DSBs (Frit et al. 2014, Liu et al. 2019). By using small molecules, timing the delivery of the CRISPR/Cas system, and especially by modulating proteins involved in the DNA repair pathways to down-regulate NHEJ while promoting HDR, the efficiency of this pathway can be optimized (Antony et al. 2018a, Liu et al. 2019). This enhancement can greatly increase the grade of gene correction or addition, and subsequently, the efficacy of CRISPR-improved cellular therapies.

Another challenge for CRISPR editing is found in its transport to the target cells. Big plasmids containing the Cas protein and gRNAs cannot be easily delivered due to size limitations. Very recently, a minimal functional CRISPR-Cas system was discovered in the genomes of huge bacteriophages. The size of the Cas protein (known as CasΦ) is about 70kDa, which is much smaller than conventional Cas nucleases (Pausch et al. 2020). The implementation of this protein in CRISPR-based strategies can potentially increase the efficiency of the approach.

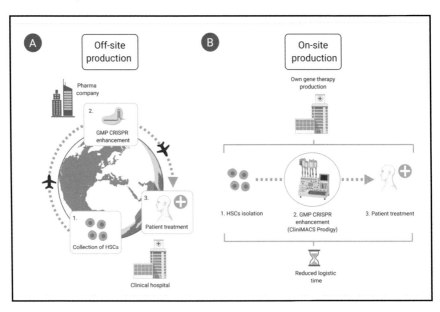

**Figure 3.** A) Off-site gene therapy production. Due to the lack of GMP-processing installations, an external company is needed for the CRISPR improvement of the patient cells and international cooperation is frequently required. B) On-site gene therapy production. As an alternative for the classical off-site production, this choice drastically reduces the logistic time since the cellular product is generated in the same clinic where the patient is receiving treatment. GMP-standard technologies such as CliniMACS Prodigy (Miltenyi Biotec) offer automated cell processing and are compatible with gene editing protocols. This kind of biomedical equipment is already available in the market and can be feasibly integrated into the hospital routine.

The efficiency of CRISPR-based therapies can be enhanced not only by improving the editing technology but also the fitness of the edited cells. The reported efficiency in two CRISPR-improved CAR-T cell-based clinical trials for the treatment of lung cancer (Stadtmauer et al. 2020, Lu et al. 2020) was much lower by comparison with 45-90% indels in cells such as primary T cells from healthy donor, and with pre-clinical models presented by gene-edited companies. Apart from the technology employed in these trials, the fitness of the base-line T cells may have played an important role in the low editing efficacy, since the cell phenotype could have been adversely affected by prior cytotoxic therapy (Das et al. 2019, Nirali et al. 2019). Future studies should consider the adoption of the more advanced technology and selection of patients with high-quality cells.

An important issue to address is based on the logistic problems to deliver safe gene therapies to the patients. A good manufacturing practice (GMP) standard is required in the production of cellular products to ensure maximum safety and minimize therapy-related risks. Therefore, an external collaboration between the hospital and properly equipped pharmacological companies is typically required (Figure 3A). This practice is highly inconvenient as it importantly enlarges the logistic times, so an on-site production could be explored (Coopman 2014). Innovative biomedical technologies such as CliniMACS Prodigy (Miltenyi Biotec) are GMP-compatible and already available in the market. The CRISPR protocols can be feasibly integrated into the automated routine of the device, offering an ideal solution for clinical hospitals with qualified personnel to produce the cellular therapy (Figure 3B).

## Acknowledgements

This work was supported by Clinician Scientist Program Tübingen (N°. 440-0-0), Stiftung Förderverein für krebskranke Kinder Tübingen e.V., Stefan Morsch Stiftung and the University Children's Hospital of Tübingen. Figures were created with BioRender.com.

## References

Akinsheye, I., Alsultan, A., Solovieff, N., Ngo, D., Baldwin, C. T., Sebastiani, P., et al. 2011. Fetal hemoglobin in sickle cell anemia. Blood, 118(1): 19-27.

Antony, J. S., Haque, A. K. M. A., Lamsfus-Calle, A., Daniel-Moreno, A., Mezger, M. and Kormann, M. S. D. 2018a. CRISPR/Cas9 system: A promising technology for the treatment of inherited and neoplastic hematological diseases. Advances in Cell and Gene Therapy. 1(1): e10.

Antony, J. S., Latifi, N., Haque, A. K. M. A., Lamsfus-Calle, A., Daniel-Moreno, A., Graeter, S., et al. 2018b. Gene correction of HBB mutations in CD34(+) hematopoietic stem cells using Cas9 mRNA and ssODN donors. Molecular and Cellular Pediatrics, 5(1): 9.

Anzalone, A. V., Randolph, P. B., Davis, J. R., Sousa, A. A., Koblan, L. W., Levy, J. M. et al. 2019. Search-and-replace genome editing without double-strand breaks or donor DNA. Nature, 576(7785): 149-157.

Armstrong, G. A. B., Liao, M., You, Z., Lissouba, A., Chen, B. E. and Drapeau, P. 2016. Homology directed knockin of point mutations in the zebrafish tardbp and fus genes in ALS using the CRISPR/Cas9 system. PloS One, 11(3): e0150188-e0150188.

Barnes, D. E. 2001. Non-homologous end joining as a mechanism of DNA repair. Current Biology, 11(12): R455-R457.

Barrangou, R., Fremaux, C., Deveau, H., Richards, M., Boyaval, P., Moineau, S., et al. 2007. CRISPR provides acquired resistance against viruses in prokaryotes. Science, 315(5819): 1709.

Benati, D., Miselli, F., Cocchiarella, F., Patrizi, C., Carretero, M., Baldassarri, S., et al. 2018. CRISPR/Cas9-mediated in situ correction of LAMB3 gene in keratinocytes derived from a junctional epidermolysis bullosa patient. Molecular Therapy, 26(11): 2592-2603.

Ben Jehuda, R., Eisen, B., Shemer, Y., Mekies, L. N., Szantai, A., Reiter, I., et al. 2018. CRISPR correction of the PRKAG2 gene mutation in the patient's iPSC-derived cardiomyocytes eliminates the electrophysiological and structural abnormalities. Heart Rhythm, 15(2): 267-276.

Bleul, C. C., Wu, L., Hoxie, J. A., Springer, T. A. and Mackay, C. R. 1997. The HIV coreceptors CXCR4 and CCR5 are differentially expressed and regulated on human T-lymphocytes. Proceedings of the National Academy of Sciences, 94(5): 1925.

Braun, C. J., Witzel, M., Paruzynski, A., Boztug, K., von Kalle, C., Schmidt, M., et al. 2014. Gene therapy for Wiskott-Aldrich Syndrome-Long-term reconstitution and clinical benefits, but increased risk for leukemogenesis. Rare Diseases (Austin, Tex.), 2(1): e947749-e947749.

Broughton, J. P., Deng, X., Yu, G., Fasching, C. L., Servellita, V., Singh, J., et al. 2020. CRISPR–Cas12-based detection of SARS-CoV-2. Nature Biotechnology, 38(7): 870-874.

Brouns, S. J. J., Jore, M. M., Lundgren, M., Westra, E. R., Slijkhuis, R. J. H., Snijders, A. P. L., et al. 2008. Small CRISPR RNAs guide antiviral defense in prokaryotes. Science, (New York, N.Y.) 321(5891): 960-964.

Burnight, E. R., Bohrer, L. R., Giacalone, J. C., Klaahsen, D. L., Daggett, H. T., East, J. S., et al. 2018. CRISPR-Cas9-mediated correction of the 1.02 kb common deletion in CLN3 in induced pluripotent stem cells from patients with batten disease. The CRISPR Journal, 1(1): 75-87.

Callaway, E. 2015. Mammoth genomes hold recipe for Arctic elephants. Nature, 521(7550) (1476-4687 (Electronic)): 18-19.

Carroll, D. and Charo, R. A. 2015. The societal opportunities and challenges of genome editing. Genome Biology, 16: 242.

Casimir, C. M., Bu-Ghanim, H. N., Rodaway, A. R., Bentley, D. L., Rowe, P. and Segal, A. W. 1991. Autosomal recessive chronic granulomatous disease caused by deletion at a dinucleotide repeat. Proceedings of the National Academy of Sciences of the United States of America, 88(7): 2753-2757.

Chang, C.W., Lai, Y.S., Westin, E., Khodadadi-Jamayran, A., Pawlik, K. M., Lamb, L. S., et al. 2015. Modeling human severe combined immunodeficiency and correction by CRISPR/ Cas9-enhanced gene targeting. Cell Reports, 12(10): 1668-1677.

Chang, K.H., Smith, S. E., Sullivan, T., Chen, K., Zhou, Q., West, J. A., et al. 2017a. Long-term engraftment and fetal globin induction upon BCL11A gene editing in bone-marrow-derived CD34(+) hematopoietic stem and progenitor cells. Molecular Therapy: Methods & Clinical Development, 4: 137-148.

Chang, S. K., Lu, Y. H., Chen, Y. R., Hsieh, Y. P., Lin, W. J., Hsu, T. R., et al. 2017b. AB043. Correction of the GLA IVS4+ 919 G> A mutation with CRISPR/Cas9 deletion strategy in fibroblasts of Fabry disease. Annals of Translational Medicine, 5(Suppl 2).

Cheng, X., Fan, S., Wen, C. and Du, X. 2020. CRISPR/Cas9 for cancer treatment: Technology, clinical applications and challenges. Briefings in Functional Genomics, 19(3): 209-214.

Cong, L., Ran, F. A., Cox, D., Lin, S., Barretto, R., Habib, N., et al. 2013. Multiplex genome engineering using CRISPR/Cas systems. Science, 339(6121)(1095-9203 (Electronic)): 819-823.

Cooper, M. L., Choi, J., Staser, K., Ritchey, J. K., Devenport, J. M., Eckardt, K. et al. 2018. An "off-the-shelf" fratricide-resistant CAR-T for the treatment of T cell hematologic malignancies. Leukemia, 32(9): 1970-1983.

Coopman, K. and Medcalf, N. 2014. From production to patient: Challenges and approaches for delivering cell therapies. StemBook [Internet].

Daniel-Moreno, A., Lamsfus-Calle, A., Raju, J., Antony, J. S., Handgretinger, R. and Mezger, M. 2019. CRISPR/Cas9-modified hematopoietic stem cells—present and future perspectives for stem cell transplantation. Bone Marrow Transplantation, 54(12): 1940-1950.

Daniel-Moreno, A., Lamsfus-Calle, A., Wilber, A., Chambers, C. B., Johnston, I., Antony, J. S., et al. 2020. Comparative analysis of lentiviral gene transfer approaches designed to promote fetal hemoglobin production for the treatment of β-hemoglobinopathies. Blood Cells, Molecules, and Diseases, 84(1096-0961 (Electronic)).

Das, B. and Yan, R. 2017. Role of BACE1 in Alzheimer's synaptic function. Translational Neurodegeneration, 6: 23.

Das, R.K., Vernau, L., Grupp, S. A. and Barret, D. M. 2019. Naïve T-cell deficits at diagnosis and after chemotherapy impair cell therapy potential in pediatric cancers. Cancer Discovery, 9, 492-499.

Deng, W. L., Gao, M. L., Lei, X. L., Lv, J. N., Zhao, H., He, K. W., et al. 2018. Gene correction reverses ciliopathy and photoreceptor loss in iPSC-derived retinal organoids from retinitis pigmentosa patients. Stem Cell Reports, 10(4): 1267-1281.

De Ravin, S. S., Li, L., Wu, X., Choi, U., Allen, C., Koontz, S., et al. 2017. CRISPR-Cas9 gene repair of hematopoietic stem cells from patients with X-linked chronic granulomatous disease. Science Translational Medicine, 9(372)(1946-6242 (Electronic)).

Deltcheva, E., Chylinski, K., Sharma, C. M., Gonzales, K., Chao, Y., Pirzada, Z. A., et al. 2011. CRISPR RNA maturation by trans-encoded small RNA and host factor RNase III. Nature, 471(7340): 602-607.

Dever, D. P., Bak, R. O., Reinisch, A., Camarena, J., Washington, G., Nicolas, C. E., et al. 2016. CRISPR/Cas9 β-globin gene targeting in human haematopoietic stem cells. Nature, 539(7629): 384-389.

Dudáš, A. and Chovanec, M. 2004. DNA double-strand break repair by homologous recombination. Mutation Research/Reviews in Mutation Research, 566(2): 131-167.

Elmore, L. W., Hancock, A. R., Chang, S. F., Wang, X. W., Chang, S., Callahan, C. P., et al. 1997. Hepatitis B virus X protein and p53 tumor suppressor interactions in the modulation of apoptosis. Proceedings of the National Academy of Sciences of the United States of America, 94(26): 14707-14712.

Eyquem, J., Mansilla-Soto, J., Giavridis, T., van der Stegen, S. J. C., Hamieh, M., Cunanan, K. M., et al. 2017. Targeting a CAR to the TRAC *locus* with CRISPR/Cas9 enhances tumour rejection. Nature, 543(7643): 113-117.

Frangoul, H., Altshuler, D., Cappellini, M. D., Chen, Y.-S., Domm, J., Eustace, B. K., et al. 2020. CRISPR-Cas9 gene editing for sickle cell disease and β-thalassemia. New England Journal of Medicine, 384(3): 252-260.

Frit, P., Barboule, N., Yuan, Y., Gomez, D. and Calsou, P. 2014. Alternative end-joining pathway(s): Bricolage at DNA breaks. DNA Repair (Amst). 17: 81-97.

Galy, A. and Thrasher, A. J., 2011. Gene therapy for the Wiskott-Aldrich syndrome. Current Opinion in Allergy and Clinical Immunology, 11(6)(1473-6322 (Electronic)): 545-550.

Gjaltema, R. A. F. and Rots, M. G. 2020. Advances of epigenetic editing. Current Opinion in Chemical Biology, 57: 75-81.

Goodwin, M., de Sio, F. S., Dever, D., Porteus, M., Roncarolo, M. G. and Bacchetta, R. 2016. Gene editing as a therapeutic approach to treat IPEX syndrome. Molecular Therapy, 24: S51.

Groeschel, S., Kühl, J. S., Bley, A. E., Kehrer, C., Weschke, B., Döring, M., et al. 2016. Long-term outcome of allogeneic hematopoietic stem cell transplantation in patients with juvenile metachromatic leukodystrophy compared with nontransplanted control patients. JAMA Neurology, 73(9)(2168-6157 (Electronic)): 1133-1140.

Hacein-Bey-Abina, S., Garrigue, A., Wang, G. P., Soulier, J., Lim, A., Morillon, E., et al. 2008. Insertional oncogenesis in 4 patients after retrovirus-mediated gene therapy of SCID-X1. The Journal of Clinical Investigation, 118(9): 3132-3142.

Haft, D. H., Selengut, J., Mongodin, E. F. and Nelson, K. E. 2005. A guild of 45 CRISPR-associated (Cas) protein families and multiple CRISPR/Cas subtypes exist in prokaryotic genomes. PLOS Computational Biology, 1(6): e60.

Hayward, A. R., Levy, J., Facchetti, F., Notarangelo, L., Ochs, H. D., Etzioni, A., et al. 1997. Cholangiopathy and tumors of the pancreas, liver, and biliary tree in boys with X-linked immunodeficiency with hyper-IgM. The Journal of Immunology, 158(2)(0022-1767 (Print)): 977-983.

He, J., Tang, F., Liu, L., Chen, L., Li, J., Ou, D., et al. 2016. Positive regulation of TAZ expression by EBV-LMP1 contributes to cell proliferation and epithelial-mesenchymal transition in nasopharyngeal carcinoma. Oncotarget, 8(32): 52333-52344.

He, Q., Wang, H. H., Cheng, T., Yuan, W. P., Ma, Y. P., Jiang, Y. P., et al. 2017. Genetic correction and hepatic differentiation of hemophilia B-specific human induced pluripotent stem cells. Chinese Medical Sciences Journal, 27; 32(3): 135-144.

He, S. 2020. The first human trial of CRISPR-based cell therapy clears safety concerns as new treatment for late-stage lung cancer. Signal Transduction and Targeted Therapy, 5(1): 168.

Heidenreich, M. and Zhang, F. 2016. Applications of CRISPR-Cas systems in neuroscience. Nature Reviews. Neuroscience, 17(1): 36-44.

Hille, F. and Charpentier, E. 2016. CRISPR-Cas: Biology, mechanisms and relevance. Philosophical Transactions of the Royal Society B: Biological Sciences, 371(1707): 20150496.

Hirakawa, M. P., Krishnakumar, R., Timlin, J. A., Carney, J. P. and Butler, K. S. 2020. Gene editing and CRISPR in the clinic: Current and future perspectives. Bioscience Reports, 40(4): BSR20200127.

Hoban, M. D. and Bauer, D. E. 2016. A genome editing primer for the hematologist. Blood, 127(21): 2525-2535.

Holt, N., Wang, J., Kim, K., Friedman, G., Wang, X., Taupin, V., et al. 2010. Human hematopoietic stem/progenitor cells modified by zinc-finger nucleases targeted to CCR5 control HIV-1 in vivo. Nature Biotechnology, 28(8)(1546-1696 (Electronic)): 839-847.

Huo, H. and Hu, G. 2019. CRISPR/Cas9-mediated LMP1 knockout inhibits Epstein-Barr virus infection and nasopharyngeal carcinoma cell growth. Infect Agent Cancer, 14(30).

Ishikawa, T., Imamura, K., Kondo, T., Koshiba, Y., Hara, S., Ichinose, H., et al. 2016. Genetic and pharmacological correction of aberrant dopamine synthesis using patient iPSCs with BH4 metabolism disorders. Human Molecular Genetics, 25(23): 5188-5197.

Ishino, Y., Krupovic, M. and Forterre, P. 2018. History of CRISPR-Cas from encounter with a mysterious repeated sequence to genome editing technology. Journal of Bacteriology, 200(7): e00580-00517.

Ishino, Y., Shinagawa, H., Makino, K., Amemura, M. and Nakata, A. 1987. Nucleotide sequence of the iap gene, responsible for alkaline phosphatase isozyme conversion in Escherichia coli, and identification of the gene product. Journal of Bacteriology, 169(12): 5429-5433.

Jansen, R., Embden, J. D. A. V., Gaastra, W. and Schouls, L. M. 2002. Identification of genes that are associated with DNA repeats in prokaryotes. Molecular Microbiology, 43(6): 1565-1575.

Jiang, W., Bikard, D., Cox, D., Zhang, F. and Marraffini, L. A. 2013. RNA-guided editing of bacterial genomes using CRISPR-Cas systems. Nature Biotechnology, 31(3): 233-239.

Jinek, M., Chylinski, K., Fonfara, I., Hauer, M., Doudna, J. A. and Charpentier, E. 2012. A programmable dual-RNA-guided DNA endonuclease in adaptive bacterial immunity. Science, 337(6096): 816.

Jung, I.Y. and Lee, J. 2018. Unleashing the therapeutic potential of CAR-T cell therapy using gene-editing technologies. Molecules and Cells, 41(8): 717-723.

Kaminski, R., Chen, Y., Fischer, T., Tedaldi, E., Napoli, A., Zhang, Y., et al. 2016. Elimination of HIV-1 genomes from human T-lymphoid cells by CRISPR/Cas9 gene editing. Scientific Reports, 6(1): 22555.

Khan, S. H. 2019. Genome-editing technologies: Concept, pros, and cons of various genome-editing techniques and bioethical concerns for clinical application. Molecular Therapy – Nucleic Acids, 16: 326-334.

Kuo, C. Y., Long, J. D., Campo-Fernandez, B., de Oliveira, S., Cooper, A. R., Romero, Z., et al. 2018. Site-specific gene editing of human hematopoietic stem cells for X-linked hyper-IgM syndrome. Cell Reports, 23(9)(2211-1247 (Electronic)): 2606-2616.

Lamsfus-Calle, A., Daniel-Moreno, A., Ureña-Bailén, G., Raju, J., Antony, J. S., Handgretinger, R., et al. 2020a. Hematopoietic stem cell gene therapy: The optimal use of lentivirus and gene editing approaches. Blood Reviews, 40(1532-1681 (Electronic)).

Lamsfus-Calle, A., Daniel-Moreno, A., Antony, J. S., Epting, T., Heumos, L., Baskaran, P., et al. 2020b. Comparative targeting analysis of KLF1, BCL11A, and HBG1/2 in CD34(+) HSPCs by CRISPR/Cas9 for the induction of fetal hemoglobin. Scientific Reports, 10(2045-2322 (Electronic)).

Lea, R. and Niakan, K.K. 2019. Human germline genome editing. Nature Cell Biology, 21(12): 1479-1489.

Li, C., Guan, X., Du, T., Jin, W., Wu, B., Liu, Y., et al. 2015. Inhibition of HIV-1 infection of primary CD4+ T-cells by gene editing of CCR5 using adenovirus-delivered CRISPR/Cas9. Journal of General Virology, 96(8)(1465-2099 (Electronic)): 2381-2393.

Li, H., Sheng, C., Wang, S., Yang, L., Liang, Y., Huang Y., et al. 2017. Removal of integrated Hepatitis B virus DNA using CRISPR-Cas9. Frontiers in Cellular and Infection Microbiology, (2235-2988 (Electronic)).

Liang, P., Ding, C., Sun, H., Xie, X., Xu, Y., Zhang, X., et al. 2017. Correction of β-thalassemia mutant by base editor in human embryos. Protein & Cell, 8(11): 811-822.

Liang, P., Xu, Y., Zhang, X., Ding, C., Huang, R., Zhang, Z., et al. 2015. CRISPR/Cas9-mediated gene editing in human tripronuclear zygotes. Protein & Cell, 6(5): 363-372.

Liebowitz, D. 1994. Nasopharyngeal carcinoma: The Epstein-Barr virus association. Seminars in Oncology, 21(3)(0093-7754 (Print)): 376-381.

Lin, F.L., Wang, P.Y., Chuang, Y.F., Wang, J.H., Wong, V.H.Y., Bui, B.V., et al. 2020. Gene therapy intervention in neovascular eye disease: A recent update. Molecular Therapy, 28(10): 2120-2138.

Liu, X., Zhang, Y., Cheng, C., Cheng, A. W., Zhang, X., Li, N., et al. 2017a. CRISPR-Cas9-mediated multiplex gene editing in CAR-T cells. Cell Research, 27(1): 154-157.

Liu, Z., Chen, S., Jin, X., Wang, Q., Yang, K., Li, C., et al. 2017b. Genome editing of the HIV co-receptors CCR5 and CXCR4 by CRISPR-Cas9 protects CD4(+) T cells from HIV-1 infection. Cell & Bioscience, 7: 47.

Liu, M., Rehman, S., Tang, X., Gu, K., Fan, Q., Chen, D., et al. 2019. Methodologies for improving HDR efficiency. Front Genet, 9: 691.

Lombardo, A., Genovese, P., Beausejour, C. M., Colleoni, S., Lee, Y., Kim, K. A., et al. 2007. Gene editing in human stem cells using zinc finger nucleases and integrase-defective lentiviral vector delivery. Nature Biotechnology, 25(11)(1087-0156 (Print)): 1298-1306.

Lu, Y., Xue, J., Deng, T., Zhou, X., Yu, K., Deng, L., et al. 2020. Safety and feasibility of CRISPR-edited T cells in patients with refractory non-small-cell lung cancer. Nature Medicine, 26, 732-740.

Ma, X., Xu, Z., Yang, L., Xiao, L., Tang, M., Lu, J., et al. 2013. EBV-LMP1-targeted DNAzyme induces DNA damage and causes cell cycle arrest in LMP1-positive nasopharyngeal carcinoma cells. International Journal of Oncology, 43(5): 1541-1548.

Maeder, M. L., Shen, S., Burnight, E. R., Gloskowski, S., Mepani, R., Friedland, A. E., et al. 2015. Therapeutic correction of an LCA-causing splice defect in the CEP290 gene by CRISPR/Cas-mediated genome editing. Molecular Therapy, 23: S273-S274.

Makarova, K.S., Grishin, N.V., Shabalina, S.A., Wolf, Y.I. and Koonin, E.V. 2006. A putative RNA-interference-based immune system in prokaryotes: Computational analysis of the predicted enzymatic machinery, functional analogies with eukaryotic RNAi, and hypothetical mechanisms of action. Biology Direct, 1(1): 7.

Mali, P., Yang, L., Esvelt, K.M., Aach, J., Guell, M., DiCarlo, J.E., et al. 2013. RNA-guided human genome engineering via Cas9. Science, 339(1095-9203 (Electronic)): 823-826.

Marraffini, L. A. and Sontheimer, E. J. 2008. CRISPR interference limits horizontal gene transfer in staphylococci by targeting DNA. Science, 322(1095-9203 (Electronic)): 1843-1845.

Martinez-Lage, M., Torres-Ruiz, R., Puig-Serra, P., Moreno-Gaona, P., Martin, M.C., Moya, F.J., et al. 2020. In vivo CRISPR/Cas9 targeting of fusion oncogenes for selective elimination of cancer cells. Nature Communications, 11(1): 5060.

Miliotou, A. N. and Papadopoulou, L. C. 2018. CAR T-cell therapy: A new era in cancer immunotherapy. Current Pharmaceutical Biotechnology, 19(1)(1873-4316 (Electronic)): 5-18.

Mojica, F. J. M., Díez-Villaseñor, C., Soria, E. and Juez, G. 2000. Biological significance of a family of regularly spaced repeats in the genomes of archaea, bacteria and mitochondria. Molecular Microbiology, 36(1)(0950-382X (Print)): 244-246.

Mojica, F. J. M., Díez-Villaseñor, C., García-Martínez, J. and Soria, E. 2005. Intervening sequences of regularly spaced prokaryotic repeats derive from foreign genetic elements. Journal of Molecular Evolution, 60(2): 174-182.

Negre, O., Eggimann, A. V., Beuzard, Y., Ribeil, J. A., Bourget, P., Borwornpinyo, S., et al. 2016. Gene therapy of the β-hemoglobinopathies by lentiviral transfer of the β(A(T87Q))-globin gene. Human Gene Therapy, 27(2)(1557-7422 (Electronic)): 148-165.

Nirali, N. S. and Terry, J. F. 2019. Mechanisms of resistance to CAR T cell therapy. Nature Reviews Clinical Oncology, 16, 372-385.

Nishimasu, H., Ran, F. A., Hsu, P. D., Konermann, S., Shehata, S. I., Dohmae, N., et al. 2014. Crystal structure of Cas9 in complex with guide RNA and target DNA. Cell, 156(5): 935-949.

NM. 2018. Keep off-target effects in focus. Nature Medicine, 24(8): 1081-1081.

Olivieri, N. F. and Weatherall, D. J. 1998. The therapeutic reactivation of fetal haemoglobin. Human Molecular Genetics, 7(0964-6906 (Print)): 1655-1658.

Omer, L., Hudson, E. A., Zheng, S., Hoying, J. B., Shan, Y. and Boyd, N. L. 2017. CRISPR correction of a homozygous low-density lipoprotein receptor mutation in familial hypercholesterolemia induced pluripotent stem cells. Hepatology Communications, 1(9): 886-898.

Ortiz-Virumbrales, M., Moreno, C. L., Kruglikov, I., Marazuela, P., Sproul, A., Jacob, S., et al. 2017. CRISPR/Cas9-correctable mutation-related molecular and physiological phenotypes in ipSC-derived Alzheimer's PSEN2 N141I neurons. Acta Neuropathologica Communications, 5(1): 1-20.

Pan, Y., Shen, N., Jung-Klawitter, S., Betzen, C., Hoffmann, G. F., Hoheisel, J. D., et al. 2016. CRISPR RNA-guided Fok I nucleases repair a PAH variant in a phenylketonuria model. Scientific Reports, 6(1): 1-7.

Park, H., Oh, J., Shim, G., Cho, B., Chang, Y., Kim, S., et al. 2019. In vivo neuronal gene editing via CRISPR-Cas9 amphiphilic nanocomplexes alleviates deficits in mouse models of Alzheimer's disease. Nature Neurosciences, 22(4)(1546-1726 (Electronic)): 524-528.

Parry, R. V., Chemnitz, J. M., Frauwirth, K. A., Lanfranco, A. R., Braunstein, I., Kobayashi, S. V., et al. 2005. CTLA-4 and PD-1 receptors inhibit T-cell activation by distinct mechanisms. Molecular and Cellular Biology, 25(21): 9543-9553.

Pasinelli, P. and Brown, R. H. 2006. Molecular biology of amyotrophic lateral sclerosis: Insights from genetics. Nature Reviews Neuroscience, 7(9)(1471-003X (Print)): 710-723.

Pausch, P., Al-Shayeb, B., Bisom-Rapp, E., Tsuchida, C. A., Li, Z., Cress, B. F., et al. 2020. CRISPR-CasΦ from huge phages is a hypercompact genome editor. Science, 369(6501): 333.

Pavel-Dinu, M., Wiebking, V., Dejene, B. T., Srifa, W., Mantri, S., Nicolas, C. E., et al. 2019. Gene correction for SCID-X1 in long-term hematopoietic stem cells. Nature Communications, 10(1): 1634.

Perez Rojo, F., Nyman, R. K. M., Johnson, A. A. T., Navarro, M. P., Ryan, M. H., Erskinem W., et al. 2018. CRISPR-Cas systems: Ushering in the new genome editing era. Bioengineered, 9(1): 214-221.

Peterson, C. W., Wang, J., Norman, K. K., Norgaard, Z. K., Humbert, O., Tse, C. K., et al. 2016. Long-term multilineage engraftment of autologous genome-edited hematopoietic stem cells in nonhuman primates. Blood, 127(20): 2416-2426.

Psatha, N., Karponi, G. and Yannaki, E. 2016. Optimizing autologous cell grafts to improve stem cell gene therapy. Experimental Hematology, 44(7): 528-539.

Rai, R., Romito, M., Rivers, E., Turchiano, G., Blattner, G., Vetharoy, W., et al. 2020. Targeted gene correction of human hematopoietic stem cells for the treatment of Wiskott-Aldrich Syndrome. Nature Communications, 11(1): 4034.

Ramesh, N., Seki, M., Notarangelo, L. D. and Geha, R. S. 1998. The hyper-IgM (HIM) syndrome. Springer Seminars in Immunopathology, 19: 383-399.

Rees, H. A. and Liu, D. R. 2018. Base editing: Precision chemistry on the genome and transcriptome of living cells. Nature Reviews Genetics, 19(12): 770-788.

Rose, B. I. and Brown, S. 2019. Genetically modified babies and a first application of Clustered Regularly Interspaced Short Palindromic Repeats (CRISPR-Cas9). Obstetrics & Gynecology, 134(1).

Rosenblum, D., Gutkin, A., Kedmi, R., Ramishetti, S., Veiga, N., Jacobi, A. M., et al. 2020. CRISPR-Cas9 genome editing using targeted lipid nanoparticles for cancer therapy. Science Advances, 6(47): eabc9450.

Roth, T. L., Puig-Saus, C., Yu, R., Shifrut, E., Carnevale, J., Li, P. J., et al. 2018. Reprogramming human T cell function and specificity with non-viral genome targeting. Nature, 559(1476-4687 (Electronic)): 405-409.

Schachter, A. S. and Davis, K. L. 2000. Alzheimer's disease. Dialogues in Clinical Neuroscience, 2(2): 91-100.

Schiroli, G. A., Ferrari, S. A., Conway, A., Jacob, A. A., Capo, V. A., Albano, L., et al. 2017. Preclinical modeling highlights the therapeutic potential of hematopoietic stem cell gene editing for correction of SCID-X1. Science Translational Medicine, 9(411)(1946-6242 (Electronic)).

Schneider, D., Xiong, Y., Wu, D., Nölle, V., Schmitz, S., Haso, W., et al. 2017. A tandem

CD19/CD20 CAR lentiviral vector drives on-target and off-target antigen modulation in leukemia cell lines. Journal for Immunotherapy of Cancer, 5: 42.

Shi, L., Meng, T., Zhao, Z., Han, J., Zhang, W., Gao, F. et al. 2017. CRISPR knock out CTLA-4 enhances the anti-tumor activity of cytotoxic T lymphocytes. Gene, 636(1879-0038 (Electronic)): 36-41.

Skvarova Kramarzova, K., Osborn, M. J., Webber, B. R., DeFeo, A. P., McElroy, A. N., Kim, C. J., et al. 2017. CRISPR/Cas9-mediated correction of the FANCD1 gene in primary patient cells. International Journal of Molecular Sciences, 18(6): 1269.

Slaymaker, I. M., Gao, L., Zetsche, B., Scott, D. A., Yan, W. X. and Zhang, F. 2016. Rationally engineered Cas9 nucleases with improved specificity. Science (New York, N.Y.) 351(6268): 84-88.

Stadtmauer, E. A., Fraietta, J. A., Davis, M. M., Cohen, A. D., Weber, K. L., Lancaster, E. et al. 2020. CRISPR-engineered T cells in patients with refractory cancer. Science, 367(6481), eaba7365.

Swarts, D. C., Mosterd, C., van Passel, M. W. J. and Brouns, S. J. J. 2012. CRISPR interference directs strand specific spacer acquisition. PLOS ONE, 7(4): e35888.

Taketani, Y., Kitamoto, K., Sakisaka, T., Kimakura, M., Toyono, T., Yamagami, S., et al. 2017. Repair of the TGFBI gene in human corneal keratocytes derived from a granular corneal dystrophy patient via CRISPR/Cas9-induced homology-directed repair. Scientific Reports, 7(1): 1-7.

Tan, W., Yu, S., Lei, J., Wu, B. and Wu, C. 2015. A novel common gamma chain mutation in a Chinese family with X-linked severe combined immunodeficiency (X-SCID; T(-)NK(-)B(+)). Immunogenetics, 67(11-12)(1432-1211 (Electronic)): 629-639.

Tang, S., Ning, Q., Yang, L., Mo, Z. and Tang, S. 2020. Mechanisms of immune escape in the cancer immune cycle. International Immunopharmacology, 86: 106700.

Tebas, P., Stein, D., Tang, W. W., Frank, I., Wang, S. Q., Lee, G. et al. 2014. Gene editing of CCR5 in autologous CD4 T cells of persons infected with HIV. New England Journal of Medicine, 370(10): 901-910.

Traxler, E. A., Yao, Y., Wang, Y.-D., Woodard, K. J., Kurita, R., Nakamura, Y., et al. 2016. A genome-editing strategy to treat β-hemoglobinopathies that recapitulates a mutation associated with a benign genetic condition. Nature Medicine, 22(9): 987-990.

Tsai, S. Q., Zheng, Z., Nguyen, N. T., Liebers, M., Topkar, V. V., Thapar, V., et al. 2015. GUIDE-seq enables genome-wide profiling of off-target cleavage by CRISPR-Cas nucleases. Nature Biotechnology, 33(2): 187-197.

Tsao, S. W., Tsang, C. M., To, K. F. and Lo. K. W. 2015. The role of Epstein-Barr virus in epithelial malignancies. Journal of Pathology, 235(2)(1096-9896 (Electronic)).

Ureña-Bailén, G., Lamsfus-Calle, A., Daniel-Moreno, A., Raju, J., Schlegel, P., Seitz, C. et al. 2020. CRISPR/Cas9 technology: Towards a new generation of improved CAR-T cells for anticancer therapies. Briefings in Functional Genomics, 19(3): 191-200.

Vakulskas, C. A., Dever, D. P., Rettig, G. R., Turk, R., Jacobi, A. M., Collingwood, M. A., et al. 2018. A high-fidelity Cas9 mutant delivered as a ribonucleoprotein complex enables efficient gene editing in human hematopoietic stem and progenitor cells. Nature Medicine, 24(8): 1216-1224.

Valletta, S., Dolatshad, H., Bartenstein, M., Yip, B. H., Bello, E., Gordon, S., et al. 2015. ASXL1 mutation correction by CRISPR/Cas9 restores gene function in leukemia cells and increases survival in mouse xenografts. Oncotarget, 6(42)(1949-2553 (Electronic)).

Villa, A., Notarangelo, L., Macchi, P., Mantuano, E., Cavagni, G., Brugnoni, D., et al. 1995. X-linked thrombocytopenia and Wiskott-Aldrich syndrome are allelic diseases with mutations in the WASP gene. Nature Genetics, 9(4): 414-417.

Wang, C. A.-O. and Cannon, P. M. 2016. The clinical applications of genome editing in HIV. Blood, 127(21)(1528-0020 (Electronic)): 2546-2552.

Wang, K., Wei, G. and Liu, D. 2012. CD19: A biomarker for B cell development, lymphoma diagnosis and therapy. Experimental Hematology & Oncology, 1(1): 36.

Wang, L., Yi, F., Fu, L., Yang, J., Wang, S., Wang, Z., et al. 2017. CRISPR/Cas9-mediated targeted gene correction in amyotrophic lateral sclerosis patient iPSCs. Protein & Cell, 8(5): 365-378.

Webber, B., Osborn, M., McElroy, A., Twaroski, K., Lonetree, C., DeFeo, A. P., et al. 2016. CRISPR/Cas9-based genetic correction for recessive dystrophic epidermolysis bullosa. NPJ Regenerative Medicine, 1: 16014 .

Wu, L., Li, C. and Pan, L. 2018. Nasopharyngeal carcinoma: A review of current updates. Experimental and Therapeutic Medicine, 15(4): 3687-3692.

Wu, W., Tang, L., D'Amore, P. A. and Lei, H. 2017. Application of CRISPR-Cas9 in eye disease. Experimental Eye Research, 161(1096-0007 (Electronic)): 116-123.

Xiao, Q., Chen, S., Wang, Q., Liu, Z., Liu, S., Deng, H., et al. 2019. CCR5 editing by Staphylococcus aureus Cas9 in human primary CD4(+) T cells and hematopoietic stem/progenitor cells promotes HIV-1 resistance and CD4(+) T cell enrichment in humanized mice. Retrovirology, 16(1)(1742-4690 (Electronic)).

Xu, X., Tay, Y., Sim, B., Yoon, S. I., Huang, Y., Ooi, J., et al. 2017. Reversal of phenotypic abnormalities by CRISPR/Cas9-mediated gene correction in Huntington disease patient-derived induced pluripotent stem cells. Stem Cell Reports, 8(3): 619-633.

Xu, L., Wang, J., Liu, Y., Xie, L., Su, B., Mou, D., et al. 2019. CRISPR-edited stem cells in a patient with HIV and acute lymphocytic leukemia. New England Journal of Medicine, 381(13): 1240-1247.

Xu, Q., Zhang, Z., Zhao, L., Qin, Y., Cai, H., Geng, Z., et al. 2020. Tropism-facilitated delivery of CRISPR/Cas9 system with chimeric antigen receptor-extracellular vesicles against B-cell malignancies. Journal of Controlled Release, 326: 455-467.

Yang, H. and Wu, Z. 2018. Genome editing of pigs for agriculture and biomedicine. Frontiers in Genetics, 9: 360.

Yang, Y., Jacoby, E. and Fry, T. J. 2015. Challenges and opportunities of allogeneic donor-derived CAR T cells. Current Opinion in Hematology, 22(6)(1531-7048 (Electronic)): 509-515.

Yoon, A. R., Jung, B.-K., Choi, E., Chung, E., Hong, J., Kim, J.-S., et al. 2020. CRISPR-Cas12a with an oAd induces precise and cancer-specific genomic reprogramming of EGFR and efficient tumor regression. Molecular Therapy, 28(10): 2286-2296.

Yumlu, S., Bashir, S., Stumm, J. and Kühn, R. 2019. Efficient gene editing of human induced pluripotent stem cells using CRISPR/Cas9. Methods in Molecular Biology, 1961(1940-6029 (Electronic)): 137-151.

Yun, Y., Hong, S.-A., Kim, K.-K., Baek, D., Lee, D., Londhe, A. M. et al. 2020. CRISPR-mediated gene correction links the ATP7A M1311V mutations with amyotrophic lateral sclerosis pathogenesis in one individual. Communications Biology, 3(1): 33.

Zhang, X.-H., Tee, L. Y., Wang, X.-G., Huang, Q.-S. and Yang, S.-H. 2015. Off-target effects in CRISPR/Cas9-mediated genome engineering. Molecular Therapy – Nucleic Acids, 4: e264.

Zhang, Y., Long, C., Li, H., McAnally, J. R., Baskin, K. K., Shelton, J. M., et al. 2017. CRISPR-Cpf1 correction of muscular dystrophy mutations in human cardiomyocytes and mice. Science Advances, 3(4): e1602814.

Zhao, Y., Pang, Y. Ty, Wang, S., Kang, H., Ding, W.-B., Yong, W.-W., et al. 2014. LMP1 stimulates the transcription of eIF4E to promote the proliferation, migration and invasion of human nasopharyngeal carcinoma. The FEBS Journal, 281(13)(1742-4658 (Electronic)): 3004-3018.

Zhu, H., Li, C. and Gao, C. 2020. Applications of CRISPR-Cas in agriculture and plant biotechnology. Nature Reviews Molecular Cell Biology, 21(11): 661-677.

# Subject Index